Invitation to Linear Algebra

TEXTBOOKS in MATHEMATICS

PUBLISHED TITLES

ABSTRACT ALGEBRA: A GENTLE INTRODUCTION
Gary L. Mullen and James A. Sellers

ABSTRACT ALGEBRA: AN INTERACTIVE APPROACH, SECOND EDITION
William Paulsen

ABSTRACT ALGEBRA: AN INQUIRY-BASED APPROACH
Jonathan K. Hodge, Steven Schlicker, and Ted Sundstrom

ADVANCED LINEAR ALGEBRA
Hugo Woerdeman

ADVANCED LINEAR ALGEBRA
Nicholas Loehr

ADVANCED LINEAR ALGEBRA, SECOND EDITION
Bruce Cooperstein

APPLIED ABSTRACT ALGEBRA WITH MAPLE™ AND MATLAB®, THIRD EDITION
Richard Klima, Neil Sigmon, and Ernest Stitzinger

APPLIED DIFFERENTIAL EQUATIONS: THE PRIMARY COURSE
Vladimir Dobrushkin

A BRIDGE TO HIGHER MATHEMATICS
Valentin Deaconu and Donald C. Pfaff

COMPUTATIONAL MATHEMATICS: MODELS, METHODS, AND ANALYSIS WITH MATLAB® AND MPI, SECOND EDITION
Robert E. White

A COURSE IN DIFFERENTIAL EQUATIONS WITH BOUNDARY VALUE PROBLEMS, SECOND EDITION
Stephen A. Wirkus, Randall J. Swift, and Ryan Szypowski

A COURSE IN ORDINARY DIFFERENTIAL EQUATIONS, SECOND EDITION
Stephen A. Wirkus and Randall J. Swift

DIFFERENTIAL EQUATIONS: THEORY, TECHNIQUE, AND PRACTICE, SECOND EDITION
Steven G. Krantz

PUBLISHED TITLES CONTINUED

DIFFERENTIAL EQUATIONS: THEORY, TECHNIQUE, AND PRACTICE WITH BOUNDARY VALUE PROBLEMS
Steven G. Krantz

DIFFERENTIAL EQUATIONS WITH APPLICATIONS AND HISTORICAL NOTES, THIRD EDITION
George F. Simmons

DIFFERENTIAL EQUATIONS WITH MATLAB®: EXPLORATION, APPLICATIONS, AND THEORY
Mark A. McKibben and Micah D. Webster

DISCOVERING GROUP THEORY: A TRANSITION TO ADVANCED MATHEMATICS
Tony Barnard and Hugh Neill

DISCRETE MATHEMATICS, SECOND EDITION
Kevin Ferland

ELEMENTARY NUMBER THEORY
James S. Kraft and Lawrence C. Washington

EXPLORING CALCULUS: LABS AND PROJECTS WITH MATHEMATICA®
Crista Arangala and Karen A. Yokley

EXPLORING GEOMETRY, SECOND EDITION
Michael Hvidsten

EXPLORING LINEAR ALGEBRA: LABS AND PROJECTS WITH MATHEMATICA®
Crista Arangala

EXPLORING THE INFINITE: AN INTRODUCTION TO PROOF AND ANALYSIS
Jennifer Brooks

GRAPHS & DIGRAPHS, SIXTH EDITION
Gary Chartrand, Linda Lesniak, and Ping Zhang

INTRODUCTION TO ABSTRACT ALGEBRA, SECOND EDITION
Jonathan D. H. Smith

INTRODUCTION TO ANALYSIS
Corey M. Dunn

INTRODUCTION TO MATHEMATICAL PROOFS: A TRANSITION TO ADVANCED MATHEMATICS, SECOND EDITION
Charles E. Roberts, Jr.

INTRODUCTION TO NUMBER THEORY, SECOND EDITION
Marty Erickson, Anthony Vazzana, and David Garth

LINEAR ALGEBRA, GEOMETRY AND TRANSFORMATION
Bruce Solomon

PUBLISHED TITLES CONTINUED

MATHEMATICAL MODELLING WITH CASE STUDIES: USING MAPLE™ AND MATLAB®, THIRD EDITION
B. Barnes and G. R. Fulford

MATHEMATICS IN GAMES, SPORTS, AND GAMBLING–THE GAMES PEOPLE PLAY, SECOND EDITION
Ronald J. Gould

THE MATHEMATICS OF GAMES: AN INTRODUCTION TO PROBABILITY
David G. Taylor

A MATLAB® COMPANION TO COMPLEX VARIABLES
A. David Wunsch

MEASURE AND INTEGRAL: AN INTRODUCTION TO REAL ANALYSIS, SECOND EDITION
Richard L. Wheeden

MEASURE THEORY AND FINE PROPERTIES OF FUNCTIONS, REVISED EDITION
Lawrence C. Evans and Ronald F. Gariepy

NUMERICAL ANALYSIS FOR ENGINEERS: METHODS AND APPLICATIONS, SECOND EDITION
Bilal Ayyub and Richard H. McCuen

ORDINARY DIFFERENTIAL EQUATIONS: AN INTRODUCTION TO THE FUNDAMENTALS
Kenneth B. Howell

PRINCIPLES OF FOURIER ANALYSIS, SECOND EDITION
Kenneth B. Howell

REAL ANALYSIS AND FOUNDATIONS, FOURTH EDITION
Steven G. Krantz

RISK ANALYSIS IN ENGINEERING AND ECONOMICS, SECOND EDITION
Bilal M. Ayyub

SPORTS MATH: AN INTRODUCTORY COURSE IN THE MATHEMATICS OF SPORTS SCIENCE AND SPORTS ANALYTICS
Roland B. Minton

TRANSFORMATIONAL PLANE GEOMETRY
Ronald N. Umble and Zhigang Han

TEXTBOOKS in MATHEMATICS

Invitation to Linear Algebra

David C. Mello

Johnson & Wales University

Providence, Rhode Island, USA

CRC Press is an imprint of the
Taylor & Francis Group an informa business
A CHAPMAN & HALL BOOK

CRC Press
Taylor & Francis Group
6000 Broken Sound Parkway NW, Suite 300
Boca Raton, FL 33487-2742

First issued in paperback 2022

ISBN 13: 978-1-03-247687-2 (pbk)
ISBN 13: 978-1-4987-7956-2 (hbk)

DOI: 10.1201/9781315153193

Publisher's Note
The publisher has gone to great lengths to ensure the quality of this reprint but points out that some imperfections in the original copies may be apparent.

Library of Congress Cataloging-in-Publication Data

Names: Mello, David C. (David Cabral), 1950-.
Title: Invitation to linear algebra / David C. Mello.
Description: Boca Raton : CRC Press, 2017. |
Includes bibliographical references and index.
Identifiers: LCCN 2016053247 | ISBN 9781498779562 (978-1-4987-7956-2
Subjects: LCSH: Algebras, Linear—Textbooks.
Classification: LCC QA184.2 .M45 2017 | DDC 512/.5--dc23
LC record available at https://lccn.loc.gov/2016053247

Visit the Taylor & Francis Web site at
http://www.taylorandfrancis.com

and the CRC Press Web site at
http://www.crcpress.com

Contents

Contents

Unit IV: More About Vector Spaces

Unit V: Linear Transformations

Unit VI: Matrix Diagonalization

Contents

Unit VII: Complex Vector Spaces

Unit VIII: Advanced Topics

Unit IX: Applications

Contents

Appendix

To the Instructor

This book is intended as an introductory course in linear algebra for sophomore or junior students majoring in mathematics, computer science, economics, and the physical sciences. Lofty assumptions have not been made about the level of mathematical preparation of the typical student; in fact, it is only assumed that the typical student has completed a standard calculus course, and is familiar with basic integration techniques.

As a fellow instructor, I know that this course is probably the typical student's first encounter with the requirement of formulating mathematical proofs, and dealing with mathematical formalism. For this reason, each definition has been carefully stated, and detailed proofs of the key theorems have been provided. More importantly, in each proof, the motivation for each step has been given, along with the "intermediate steps" that are normally omitted in most texts.

Unlike most books of this type, the book has been organized into "lessons" rather than chapters. This has been done to limit the size of the mathematical morsels that must be digested by your students during each class, and to make it easier for you, the instructor, to budget class time. Most lessons can be covered in a standard class period.

Considerably more material has been provided than is normally covered in a first course. For example, several advanced topics such as Jordan canonical form, and matrix power series have been included. This is to make the book more flexible, and allow you to choose enrichment material which may reflect your interest and that of your students.

In addition, numerous applications of the course material to mathematics and to science appear in the exercises. The special applications, consisting of the application of linear algebra to both linear and nonlinear dynamical systems appear in Lessons 36-38 at the end of the text.

I would like to thank my colleague Dr. Adam Hartman, for carefully reviewing the manuscript, and for his many helpful comments. If you should have any ideas or suggestions for improving the book, please feel free to email me directly at **dmello@jwu.edu**.

David C. Mello
Providence, Rhode Island
October, 2016

To the Student

This book has been written with you, the student, in mind. It has been designed to help you learn the key elements of linear algebra in an enjoyable fashion. Hopefully, it will give you a glimpse of the intrinsic beauty of this subject, and how it can be used in your chosen field of study.

For most students such as yourself, linear algebra is probably your first encounter with formal mathematics and in constructing proofs of mathematical propositions. In using the book, you should pay close attention to the definitions of new concepts, and to the various theorems; these items have been clearly designated throughout the entire text.

Whenever possible, you should try to work through the proof of each theorem. Pay attention to each step, and make sure it makes sense to you, before proceeding to the next step. If you do this, you will be rewarded with a deeper understanding of the course material, and it will make the exercises involving proofs much easier.

I often tell my students that learning mathematics is like learning to play a musical instrument. Listening to someone else play a beautiful melody might be enjoyable, but it doesn't help you master the instrument yourself. So, you have to take the time to actually "do mathematics" in order to play its music.

But exactly how do you "do mathematics?" Here are some helpful hints for learning and doing mathematics:

1. Always read the text with a pencil in hand, and take the time to work through each example provided in the text. Numerous examples have been provided to help you master the course material.

2. Before attempting to solve a problem that involves constructing your own proof, look at the relevant *definitions* of the concepts involved so you can clearly understand exactly what has to be demonstrated.

3. Be sure to do the homework problems that are assigned by your instructor. If you are unable to do some of the assigned work, or if you are confused about some point, then don't get discouraged, but be sure to ask your instructor for help.

4. Prepare for each class by reading the assigned material in the text *before* class. If you take the time to do this, you'll find that you will spend less time taking notes, and more time really understanding and enjoying each class.

I sincerely hope that you will find the textbook to be extremely helpful and that you will enjoy "doing mathematics." Don't be afraid of new concepts if you don't fully understand them at your first reading; a great physicist once said that "if you give mathematics a chance, and if you learn to love mathematics, then it will love you back!"

I: Matrices and Linear Systems

Arthur Cayley (1821-1895)
(Portrait in London by Barraud and Jerrard)

It is interesting to note that the general rules of matrix algebra were first elucidated by the English mathematician Arthur Cayley, Sadlerian professor of mathematics at Cambridge University. In an important paper, published in 1855, Cayley formally defined the concept of a matrix, introduced the rules of matrix algebra, and examined many of the important properties of matrices.

Although the basic properties of matrices were known prior to Cayley's work, he is generally acknowledged as the creator of the theory of matrices because he was the first mathematician to treat matrices as distinct mathematical objects in their own right, and to elucidate the formal rules of matrix algebra. Cayley received many honors for his mathematical work and made major contributions to the theory of determinants, linear transformations, the analytic geometry of n dimensional spaces, and the theory of invariants.

1. Introduction to Matrices

In this lesson, we'll learn about a brand new type of mathematical object called a *matrix*. As we will see in the lessons that follow, matrices are very useful in many different applications, and they are particularly useful in helping us to solve systems of linear equations.

A **matrix** is a rectangular array of objects that are called **elements** of the given matrix. We usually use a *capital letter* to denote any given matrix, and we enclose its elements with two brackets. Take a look at the first example.

Example-1: The following rectangular arrays are matrices:

$$A = \begin{bmatrix} -1 & 3 \\ 2 & 6 \end{bmatrix}, \quad B = \begin{bmatrix} -3 \\ 2 \end{bmatrix}, \quad C = \begin{bmatrix} -3 & 4 & 7 \\ 2 & 0 & 5 \end{bmatrix}$$

$$D = \begin{bmatrix} -1 \\ 2 \\ 3 \end{bmatrix}, \quad E = \begin{bmatrix} 0 & 1 & 4 \\ 1 & 0 & 3 \\ 9 & 6 & 7 \end{bmatrix}, \quad F = \begin{bmatrix} -1 & 0 & 3 \end{bmatrix}$$

We say that a given matrix is an $r \times c$ matrix [read "r by c"], or has the **size** or **dimension** $r \times c$, if that matrix has exactly r rows and c columns. In the example above, we see that A is a 2×2 matrix, B is a 2×1 matrix, C is a 2×3 matrix, D is a 3×1 matrix, E is a 3×3 matrix, and F is a 1×3 matrix. *Observe that when we specify the size or dimensions of a matrix, we always specify the number of rows first, and then we specify the number of columns.*

A matrix is said to be a **square matrix** if it has the same number of rows and columns. The adjective "square" comes from the fact that a square matrix has a square shape. In the example above, it is easy to see that both A and E are square matrices.

A matrix is sometimes called a **column vector** if it just has one column, and a **row vector** if it has only one row. In the above example, we see that the matrices B and D are column vectors, while the matrix F is a row vector.

In the discussion that follows, we will see that matrices have an algebra of their own, and in many ways this algebra resembles the algebra of real numbers. We shall first define what it means for two matrices to be equal.

Definition: *(Matrix Equality)*

Let A and B both be $m \times n$ matrices, then we say that "A equals B" and agree to write $A = B$ if and only if each element of A is equal to each corresponding element of B.

Thus, for two matrices to be equal two things must happen: (1) the matrices must have the *same size*, and (2) the corresponding elements of the two matrices must be equal to each other. Take a look at the next example.

Example-2: Given that:

$$\begin{bmatrix} 2x & 3y \\ 4z & 5w \end{bmatrix} = \begin{bmatrix} 4 & 9 \\ 4 & 25 \end{bmatrix},$$

solve for x, y, z and w.

Solution: Since the two matrices are equal, then their corresponding elements must be equal. Consequently, we have:

$$2x = 4, \ \ 3y = 9, \ \ 4z = 4, \ \ 5w = 25$$

We conclude that $x = 2$, $y = 3$, $z = 1$ and $w = 5$.

We now turn our attention to the addition and subtraction of matrices, and the operation of scalar multiplication, i.e., multiplying a matrix by a number (or scalar).

Definition: *(Matrix Addition)*

Let A and B *both* be $m \times n$ matrices. Then their **sum**, denoted by $A + B$, is the $m \times n$ matrix obtained by adding the corresponding elements of A and B.

Observe that we can add any two given matrices only when they have the *same size*; otherwise, we say that their sum is *not defined*.

Example-3: Perform the following matrix addition:

$$\begin{bmatrix} 1 & 2 \\ 3 & 4 \end{bmatrix} + \begin{bmatrix} -1 & -2 \\ -3 & -4 \end{bmatrix}$$

Solution: Since both matrices have the same size, their sum is defined. Now, we just add their corresponding elements:

$$\begin{bmatrix} 1 & 2 \\ 3 & 4 \end{bmatrix} + \begin{bmatrix} -1 & -2 \\ -3 & -4 \end{bmatrix} = \begin{bmatrix} 1+(-1) & 2+(-2) \\ 3+(-3) & 4+(-4) \end{bmatrix} = \begin{bmatrix} 0 & 0 \\ 0 & 0 \end{bmatrix}$$

Since each element of our answer is zero, we call this a zero matrix. In general, a **zero matrix** is any matrix whose elements are all zero.

Definition: *(Matrix Subtraction)*

Let A and B *both* be $m \times n$ matrices. Then their **difference**, denoted by $A - B$, is the $m \times n$ matrix which is obtained by subtracting the corresponding elements of A and B.

In the above definition, we see that the subtraction of any two matrices only makes sense when both matrices have the *same dimensions*; otherwise, we say that their difference is not defined. In this regard, matrix subtraction is similar to the addition of matrices.

Example-4: Perform the following matrix subtraction:

$$\begin{bmatrix} 1 & 2 \\ 3 & 4 \end{bmatrix} - \begin{bmatrix} -1 & -2 \\ -3 & -4 \end{bmatrix}$$

Solution: Since both matrices have the same dimensions, their difference is defined. We now subtract their corresponding elements:

$$\begin{bmatrix} 1 & 2 \\ 3 & 4 \end{bmatrix} - \begin{bmatrix} -1 & -2 \\ -3 & -4 \end{bmatrix} = \begin{bmatrix} 1-(-1) & 2-(-2) \\ 3-(-3) & 4-(-4) \end{bmatrix} = \begin{bmatrix} 2 & 4 \\ 6 & 8 \end{bmatrix}$$

Finally, we will examine the operation which consists of multiplying any given matrix by a number. Since both real and complex numbers are commonly called **scalars** (to distinguish them from matrices), this new operation is called **scalar multiplication.** Let's take a look at the last definition of this lesson.

Definition: *(Scalar Multiplication)*

Let A be an $m \times n$ matrix, and let c be any number - either real or complex. Then the matrix cA is the $m \times n$ matrix obtained by multiplying each element of A by the number c.

Example-5: Given that:

$$A = \begin{bmatrix} 1 & 2 \\ 0 & 3 \end{bmatrix} \text{ and } B = \begin{bmatrix} -1 & -2 \\ -3 & 5 \end{bmatrix}$$

Evaluate each of the following expressions:
(a) $A + 2B$
(b) $2A - 4B$
(c) $3A + B$

Solution:
(a) Using the definitions of scalar multiplication and matrix addition, we get:

$$A + 2B = \begin{bmatrix} 1 & 2 \\ 0 & 3 \end{bmatrix} + 2\begin{bmatrix} -1 & -2 \\ -3 & 5 \end{bmatrix}$$

$$= \begin{bmatrix} 1 & 2 \\ 0 & 3 \end{bmatrix} + \begin{bmatrix} -2 & -4 \\ -6 & 10 \end{bmatrix}$$

$$= \begin{bmatrix} -1 & -2 \\ -6 & 13 \end{bmatrix}$$

(b) Similarly, we obtain:

$$2A - 4B = 2\begin{bmatrix} 1 & 2 \\ 0 & 3 \end{bmatrix} - 4\begin{bmatrix} -1 & -2 \\ -3 & 5 \end{bmatrix}$$

$$= \begin{bmatrix} 2 & 4 \\ 0 & 6 \end{bmatrix} - \begin{bmatrix} -4 & -8 \\ -12 & 20 \end{bmatrix}$$

$$= \begin{bmatrix} 6 & 12 \\ 12 & -14 \end{bmatrix}$$

(c) Proceeding in a similar fashion, we obtain:

$$3A + B = 3\begin{bmatrix} 1 & 2 \\ 0 & 3 \end{bmatrix} + \begin{bmatrix} -1 & -2 \\ -3 & 5 \end{bmatrix}$$

$$= \begin{bmatrix} 3 & 6 \\ 0 & 9 \end{bmatrix} + \begin{bmatrix} -1 & -2 \\ -3 & 5 \end{bmatrix}$$

$$= \begin{bmatrix} 2 & 4 \\ -3 & 14 \end{bmatrix}$$

Exercise Set 1

In Exercises 1-20, use the matrices:

$$A = \begin{bmatrix} -1 & 2 \\ 3 & 1 \end{bmatrix}, \quad B = \begin{bmatrix} 4 & -3 \\ -1 & 3 \end{bmatrix}, \quad C = \begin{bmatrix} 0 & 1 \\ 0 & 2 \end{bmatrix}$$

$$D = \begin{bmatrix} 1 \\ 0 \end{bmatrix}, \quad E = \begin{bmatrix} 0 \\ 1 \end{bmatrix}$$

to perform the indicated operations. If an operation is not defined, then explain why.

1. $A + B$

2. $A - B$

3. $A + B + C$

4. $2(A + B)$

5. $2(B + C)$

6. $A + (B + C)$

7. $(A + B) + C$

8. $(A - B) + (B - A)$

9. $3(A - A)$

10. $A + E$

11. $3E + 7D$

12. $4(D - E)$

13. $4D - 4E$

14. $B + D$

15. $2(3A)$

16. $(2 \cdot 3)A$

17. $A + 2B + 3C$

18. $2(D - E) + 2E - 2D$

19. $2(A + B + C) - 2A - 2B - 2C$

20. $3A + 2B + E$

21. Given that:

$$\begin{bmatrix} 2x+1 & 2y \\ 3z-1 & 4w \end{bmatrix} = \begin{bmatrix} 3 & 4 \\ 8 & 12 \end{bmatrix}$$

find x, y, z and w.

22. Given that:

$$\begin{bmatrix} 2x+y & 0 \\ 2 & 2x-y \end{bmatrix} = \begin{bmatrix} 3 & 0 \\ 2 & 1 \end{bmatrix}$$

find x and y.

23. It is known that a certain 2×2 matrix X satisfies the equation:

$$2X + \begin{bmatrix} 0 & 1 \\ 2 & 3 \end{bmatrix} = \begin{bmatrix} 2 & 3 \\ 4 & 7 \end{bmatrix}$$

Find the matrix X.

In Exercises 24-26, use the 2×2 matrices:

$$A = \begin{bmatrix} 1 & 4 \\ 2 & 3 \end{bmatrix} \text{ and } B = \begin{bmatrix} 3 & 6 \\ 1 & 9 \end{bmatrix}$$

to verify each of the following statements:

24. Matrix addition is **commutative**, i.e.,

$$A + B = B + A$$

25. Scalar multiplication is **distributive**:

$$c(A + B) = cA + cB \quad \text{for any scalar } c$$

26. If the **additive inverse** of A is the matrix defined by:

$$-A = -1 \cdot A$$

then the matrix $-A$ has the property that:

$$A + (-A) = (-A) + A = 0$$

where 0 is denotes the 2×2 zero matrix.

27. An unknown 2×2 matrix X satisfies the equation:

$$4X + \begin{bmatrix} 1 & 0 \\ 2 & -3 \end{bmatrix} = 2X - \begin{bmatrix} 1 & 4 \\ 2 & 3 \end{bmatrix}$$

Find the matrix X.

2. Matrix Multiplication

In this lesson, we shall discuss the basis aspects of matrix multiplication. It will be convenient to first introduce a useful notational convention, which is called **double subscript notation.** We can use this shorthand notation to represent any element of a given matrix.

Definition: *(Double Subscript Notation)*

For any $m \times n$ matrix A, the shorthand notation:

$$A = [a_{ij}] \quad \text{for} \begin{cases} i = 1,2,3,\ldots,m \\ j = 1,2,3,\ldots,n \end{cases} \tag{2.1}$$

shall mean that the matrix A consists of the elements a_{ij}, where the *first* subscript i denotes the *row* of any given element, and the *second* subscript j denotes its *column*. In general, we can write:

$$A = [a_{ij}] = \begin{bmatrix} a_{11} & a_{12} & \ldots & a_{1n} \\ a_{21} & a_{22} & \ldots & a_{2n} \\ \ldots & \ldots & \ldots & \ldots \\ a_{m1} & a_{m2} & \ldots & a_{mn} \end{bmatrix} \tag{2.2}$$

Example-1: Given

$$A = [a_{ij}] = \begin{bmatrix} 0 & -1 \\ 3 & 2 \end{bmatrix},$$

we have $a_{11} = 0$, $a_{12} = -1$, $a_{21} = 3$, and $a_{22} = 2$.

Example-2: Given the matrix

$$B = [b_{ij}] = \begin{bmatrix} 1 & 0 & -1 \\ 2 & 3 & -2 \end{bmatrix}$$

we have $b_{11} = 1, b_{12} = 0, b_{13} = -1, b_{21} = 2$, etc.

Before we define matrix multiplication in a general way, we shall first define the **dot product** of a row vector and a column vector. The result of this new operation is always a scalar; that is, a real or complex number.

Definition: *(Dot Product)*

Let **r** be a $1 \times p$ row vector:

$$\mathbf{r} = \begin{bmatrix} r_1 & r_2 & r_3 & \dots & r_p \end{bmatrix}$$

and let **c** be a $p \times 1$ column vector:

$$\mathbf{c} = \begin{bmatrix} c_1 \\ c_2 \\ c_3 \\ \dots \\ c_p \end{bmatrix}$$

then the **dot product** of **r** and **c**, denoted by $\mathbf{r} \cdot \mathbf{c}$, is the *scalar (number)* given by:

$$\mathbf{r} \cdot \mathbf{c} = r_1 c_1 + r_2 c_2 + r_3 c_3 + \ldots + r_p c_p \tag{2.3}$$

In other words, to obtain the dot product $\mathbf{r} \cdot \mathbf{c}$, we multiply each element of the row vector **r** by the corresponding element of the column vector **c** and add all of the resulting products.

Example-3: Calculate the dot product:

$$\begin{bmatrix} 2 & 5 & 4 \end{bmatrix} \cdot \begin{bmatrix} -1 \\ 2 \\ 3 \end{bmatrix}$$

Solution: Using the above definition, we multiply each element of the row vector by the corresponding element of the column vector, and then add the resulting products:

$$\begin{bmatrix} 2 & 5 & 4 \end{bmatrix} \cdot \begin{bmatrix} -1 \\ 2 \\ 3 \end{bmatrix} = (2)(-1) + (5)(2) + (4)(3) = 20$$

It turns out that two general matrices A and B can be multiplied together under certain conditions. If the number of *columns* in A is equal to the number of *rows* in B, then we say that the product AB is *defined*.

Definition: *(Matrix Multiplication)*

Let A be an $m \times p$ matrix, and B be a $p \times n$ matrix. Furthermore, let $\text{row}_i(A)$ denote the i-th row of A, and $\text{col}_j(B)$ denote the j-th column of B. Then the product of A and B, denoted by AB, is the $m \times n$ matrix $C = [c_{ij}]$ whose typical element c_{ij} is obtained by taking the *dot product* of the i-th row of A, and the j-th column of B. Symbolically, we can write:

$$c_{ij} = \text{row}_i(A) \cdot \text{col}_j(B) \quad \text{where} \begin{cases} i = 1,2,3,\ldots,m \\ j = 1,2,3,\ldots,n \end{cases}, \tag{2.4}$$

or equivalently, for each *fixed* value of i and j,

$$c_{ij} = \sum_{k=1}^{p} a_{ik} b_{kj} \quad \text{where} \begin{cases} i = 1,2,3,\ldots,m \\ j = 1,2,3,\ldots,n \end{cases} \tag{2.5}$$

Example-4: Perform the indicated multiplications:

(a)

$$\begin{bmatrix} 1 & 2 \\ 3 & 4 \end{bmatrix} \begin{bmatrix} -1 & 0 \\ -2 & 3 \end{bmatrix}$$

(b)

$$\begin{bmatrix} -1 & 0 \\ -2 & 3 \end{bmatrix} \begin{bmatrix} 1 & 2 \\ 3 & 4 \end{bmatrix}$$

(c)

$$\begin{bmatrix} 2 & 3 \\ 1 & 4 \end{bmatrix} \begin{bmatrix} 0 \\ 1 \end{bmatrix}$$

Solution:

(**a**) Here we have a 2×2 matrix times a 2×2 matrix. According to the previous definition, the size of the product will be a 2×2 matrix as well. Also, according to this definition, we must multiply each row of the first matrix by each column of the second matrix:

$$\begin{bmatrix} 1 & 2 \\ 3 & 4 \end{bmatrix} \begin{bmatrix} -1 & 0 \\ -2 & 3 \end{bmatrix} = \begin{bmatrix} 1(-1)+2(-2) & 1(0)+2(3) \\ 3(-1)+4(-2) & 3(0)+4(3) \end{bmatrix} = \begin{bmatrix} -5 & 6 \\ -11 & 12 \end{bmatrix}$$

(**b**) Once again, we have a 2×2 matrix times a 2×2 matrix so the size of the product will be a 2×2 matrix as well. Once again, we multiply each row of the first matrix by each column of the second matrix to get:

$$\begin{bmatrix} -1 & 0 \\ -2 & 3 \end{bmatrix}\begin{bmatrix} 1 & 2 \\ 3 & 4 \end{bmatrix} = \begin{bmatrix} -1(1)+0(3) & -1(2)+0(4) \\ -2(1)+3(3) & -2(2)+3(4) \end{bmatrix} = \begin{bmatrix} -1 & -2 \\ 7 & 8 \end{bmatrix}$$

Observe that this answer was different from the one we got in (a). *This means that matrix multiplication is not commutative; in general, the order in which we multiply any two matrices matters.*

(**c**) Here we have a 2×2 matrix times a 2×1 matrix. Since the number of columns in the first matrix equals the number of rows in the second matrix, then their product is defined. According to the above definition, the size of their product will be a 2×1 matrix. We obtain:

$$\begin{bmatrix} 2 & 3 \\ 1 & 4 \end{bmatrix}\begin{bmatrix} 0 \\ 1 \end{bmatrix} = \begin{bmatrix} 2(0)+3(1) \\ 1(0)+4(1) \end{bmatrix} = \begin{bmatrix} 3 \\ 4 \end{bmatrix}$$

Definition: *(Kronecker Delta and Identity Matrix)*

The **Kronecker delta** δ_{ij} of order n, is defined by

$$\delta_{ij} = \left\{ \begin{array}{l} 1, \text{ if } i = j \\ 0, \text{ if } i \neq j \end{array} \right\} \text{ for } i,j = 1,2,3,\ldots,n \tag{2.6}$$

An **identity matrix** I of order n is a *square* $n \times n$ matrix such that $I = [\delta_{ij}]$.

In other words, a square matrix is said to be an identity matrix if all of its elements on the **main diagonal** (going from the top left corner to the bottom right corner) are all equal to one and all of its remaining elements are equal to zero.

Example-5: The following matrices are identity matrices:

$$I = \begin{bmatrix} 1 & 0 \\ 0 & 1 \end{bmatrix} \text{ and } I = \begin{bmatrix} 1 & 0 & 0 \\ 0 & 1 & 0 \\ 0 & 0 & 1 \end{bmatrix}.$$

Identity matrices possess a very important property: when any square matrix is multiplied by the identity matrix of the same size, the product is always the matrix you started with. This idea is expressed in the following theorem:

Theorem 2.1: *(Multiplication by an Identity Matrix)*

Let A be an $n \times n$ matrix, and let I be the $n \times n$ identity matrix. Then,

$$AI = IA = A \tag{2.7}$$

That is, *I is the identity element for matrix multiplication, and multiplying any given square matrix by I is commutative.*

Proof: Let $A = [a_{ij}]$, and $I = [\delta_{ij}]$. Furthermore, let $[AI]_{ij}$ denote a typical element of the product AI. Clearly,

$$[AI]_{ij} = \sum_{k=1}^{n} a_{ik}\delta_{kj} = a_{ij}$$
$$\Rightarrow AI = A$$

since for each *fixed* value of i and j, the Kronecker delta δ_{kj} is *non-zero* only when $k = j$. Similarly,

$$[IA]_{ij} = \sum_{k=1}^{n} \delta_{ik}A_{kj} = a_{ij}$$
$$\Rightarrow IA = A$$

Consequently, we conclude that $AI = IA = A$ as claimed. □.

Before we conclude this lesson, we summarize the basic rules of matrix algebra for easy reference.

Theorem 2.2: *(Rules of Matrix Algebra)*

If the sizes of the matrices $A, B,$ and C are such that the indicated operations are defined, then the following basic rules of matrix algebra are valid:

$A + B = B + A$	(Commutative Law for Addition)
$(A + B) + C = A + (B + C)$	(Associative Law for Addition)
$A(B + C) = AB + AC$	(Left Distributive Law)
$(B + C)A = BA + CA$	(Right Distributive Law)
$(AB)C = A(BC)$	(Associative Law for Multiplication)
$c(A + B) = cA + cB$	for any scalar c.

Proof: We prove only the left distributive law. Let A be an $m \times p$ matrix, and assume that both B and C are $p \times n$ matrices. Using the definitions of matrix addition and multiplication, we can write

$$[A(B+C)]_{ij} = \sum_{k=1}^{p} a_{ip}(b_{pj} + c_{pj}) = \sum_{k=1}^{p} a_{ip}b_{pj} + \sum_{k=1}^{p} a_{ip}c_{pj}$$

so that

$$[A(B+C)]_{ij} = [AB]_{ij} + [AC]_{ij}$$

and consequently, $A(B+C) = AB + AC$ as claimed. The proofs of the remaining properties are easy, and are left as exercises. □.

Roughly speaking, the previous theorem says that matrices almost obey the same laws as real numbers; however, there are two important exceptions:

(1) Matrix multiplication is *not commutative* in general,
(2) The **zero factor property** of real numbers is not satisfied; that is, if we have matrices such that $AB = 0$, we cannot safely conclude that either $A = 0$ or $B = 0$.

Exercise Set 2

In Exercises 1-20, use the square matrices:

$$A = \begin{bmatrix} 1 & 2 \\ -2 & 3 \end{bmatrix} \quad B = \begin{bmatrix} 0 & -2 \\ 10 & 4 \end{bmatrix}$$

$$C = \begin{bmatrix} -5 & 2 \\ -3 & 1 \end{bmatrix} \quad D = \begin{bmatrix} 3 \\ -1 \end{bmatrix} \quad I = \begin{bmatrix} 1 & 0 \\ 0 & 1 \end{bmatrix}$$

to perform the indicated operations.

1. AB

2. BA

3. AC

4. CA

5. $A(B+C)$

6. $AB + AC$

7. ABC

8. CAB

9. BCA

10. A^2 [For any square matrix A, we define $A^2 = AA$.]

11. $(A+B)(A-B)$

12. $(A+B)^2$ [See Exercise 10]

13. $(A-I)^2$

14. $(A+I)(A-I)$

15. $(A+C)(A-I)$

16. $(A+B)D$

17. CD

18. ID

19. IA

20. AI

21. Use the matrices $A, B,$ and C (above) to verify that matrix multiplication is *associative*; that is, show that:

$$A(BC) = (AB)C$$

22. Use the matrices $A, B,$ and C (above) to verify that matrix multiplication *distributes* over matrix addition; that is,

$$A(B+C) = AB + AC$$

23. If A is any square matrix, then use the rules of matrix algebra to show that:

$$(A+I)(A-I) = A^2 - I$$

24. If A and B are any $n \times n$ matrices, then use the rules of matrix algebra to show that:

$$(A+B)^2 = A^2 + AB + BA + B^2$$

Why can't we just combine the terms AB and BA?

25. The **transpose** of an $m \times n$ matrix A, denoted by A^T, is the $n \times m$ matrix whose *columns* are obtained from the *rows* of A. For example, if

$$A = \begin{bmatrix} 1 & 2 \\ -2 & 3 \end{bmatrix} \text{ then } A^T = \begin{bmatrix} 1 & -2 \\ 2 & 3 \end{bmatrix}$$

If B is any 2×2 matrix, show that:

$$(B^T)^T = B$$

26. Let A and B both be 2×2 matrices. Show that:

$$(AB)^T = B^T A^T$$

27. *(Pauli Spin Matrices)* An important collection of square matrices that occur in the non-relativistic theory of electron spin is called the **Pauli spin matrices.** These matrices are defined as:

$$\sigma_0 = \begin{bmatrix} 1 & 0 \\ 0 & 1 \end{bmatrix}, \quad \sigma_1 = \begin{bmatrix} 0 & 1 \\ 1 & 0 \end{bmatrix}$$

$$\sigma_2 = \begin{bmatrix} 0 & -i \\ i & 0 \end{bmatrix}, \quad \sigma_3 = \begin{bmatrix} 1 & 0 \\ 0 & -1 \end{bmatrix}$$

By direct calculation, show that (for $i,j = 0,1,2,3$):

(a)

$$\sigma_i^2 = I$$

(b)

$$\sigma_i \sigma_j + \sigma_j \sigma_i = 2\delta_{ij} I$$

3. Additional Topics in Matrix Algebra

In this lession, we shall explore some additional topics in matrix algebra which prove useful in science and engineering. Also, we shall examine some new types of special matrices, and additional matrix operations.

Definition: *(Transpose)*

Let A be an $m \times n$ matrix; then the **transpose** of A, denoted by A^T, is the $n \times m$ matrix whose rows are formed from the successive columns of A. In other words, if $A = [a_{ij}]$ and $A^T = [b_{ij}]$ is the transpose of A, then $b_{ij} = a_{ji}$.

Note that if $B = A^T$, then we form the *rows* of B by using the *columns* of A. A simple example should clarify this idea.

Example-1: Given the matrices $A, B,$ and C, where

$$A = \begin{bmatrix} 1 & 0 & 3 \end{bmatrix}, \quad B = \begin{bmatrix} -1 & 0 & 2 \\ 3 & 2 & 4 \end{bmatrix}, \quad \text{and } C = \begin{bmatrix} 4 \\ -3 \\ 2 \end{bmatrix},$$

we find that their transposes are:

$$A^T = \begin{bmatrix} 1 & 0 & 3 \end{bmatrix}^T = \begin{bmatrix} 1 \\ 0 \\ 3 \end{bmatrix},$$

$$B^T = \begin{bmatrix} -1 & 0 & 2 \\ 3 & 2 & 4 \end{bmatrix}^T = \begin{bmatrix} -1 & 3 \\ 0 & 2 \\ 2 & 4 \end{bmatrix}$$

$$C^T = \begin{bmatrix} 4 \\ -3 \\ 2 \end{bmatrix}^T = \begin{bmatrix} 4 & -3 & 2 \end{bmatrix}$$

From the above example, we see that the transpose of a row vector is a column vector, and conversely, the transpose of a column vector is always a row vector. The transpose obeys several important properties that we summarize in the following theorem.

Theorem 3.1: *(Properties of Matrix Transposition)*

If the sizes of the matrices A and B are such that the indicated operations are defined, then the following basic rules of matrix transposition are valid:

$$(A^T)^T = A \tag{3.1a}$$

$$(A+B)^T = A^T + B^T \tag{3.1b}$$

$$(cA)^T = cA^T \text{ for any scalar } c \tag{3.1c}$$

$$(AB)^T = B^T A^T \tag{3.1d}$$

Proof: We only prove (3.1d). Assume that A is an $m \times p$ matrix, B is a $p \times n$ matrix, and let $C = AB$. Furthermore, let $A^T = (a'_{ij})$, $B^T = (b'_{ij})$, and $C^T = (c'_{ij})$. Finally, let $[B^T A^T]_{ij}$ denote a typical element of $B^T A^T$.
We can write

$$[B^T A^T]_{ij} = \sum_{k=1}^{p} b'_{ik} a'_{kj} = \sum_{k=1}^{p} a_{jk} b_{ki}$$
$$= c_{ji} = c'_{ij} = [AB]^T_{ij}$$

so that $B^T A^T = (AB)^T$ as claimed. The proofs of the remaining properties are left as exercises for the reader. □.

There are special types of square matrices, called *symmetric matrices* and *anti-symmetric matrices* that often appear in the application of matrices to important problems in physics and engineering. We define them as follows:

Definition: *(Symmetric and Anti-Symmetric Matrices)*

A *square* matrix A is **symmetric** if

$$A^T = A,$$

while it is said to be **anti-symmetric** if

$$A^T = -A$$

We note, in passing, that if a square matrix $A = [a_{ij}]$ is anti-symmetric, then in particular, we must have $a_{ii} = -a_{ii}$ so that all of its diagonal elements must be zero. On the other hand, a square matrix is symmetric if the elements of the given matrix are symmetric with respect to the main diagonal.

Example-2: Consider the matrices

$$A = \begin{bmatrix} 2 & 1 \\ 1 & 3 \end{bmatrix} \quad \text{and} \quad B = \begin{bmatrix} 0 & 3 \\ -3 & 0 \end{bmatrix}$$

Here, A is symmetric, since

$$A^T = \begin{bmatrix} 2 & 1 \\ 1 & 3 \end{bmatrix} = A$$

while the matrix B is anti-symmetric, since

$$B^T = \begin{bmatrix} 0 & -3 \\ 3 & 0 \end{bmatrix} = -B$$

It turns out that every square matrix has a split personality in the sense that it can be written as the sum of a symmetric matrix and an anti-symmetric matrix. This is the content of the next theorem.

Theorem 3.2: *(Decomposition of a Square Matrix)*

Every square matrix A can be written as the sum of a symmetric matrix A_{sym} and an anti-symmetric matrix A_{anti}; that is,

$$A = A_{sym} + A_{anti} \tag{3.2a}$$

where

$$A_{sym} = \frac{1}{2}(A + A^T) \text{ and } A_{anti} = \frac{1}{2}(A - A^T). \tag{3.2b}$$

Proof: Observe that

$$A_{sym}^T = \frac{1}{2}(A + A^T)^T = \frac{1}{2}(A^T + A) = \frac{1}{2}(A + A^T) = A_{sym}$$
$$A_{anti}^T = \frac{1}{2}(A - A^T)^T = \frac{1}{2}(A^T - A) = -\frac{1}{2}(A - A^T) = -A_{anti}$$

so A_{sym} and A_{anti} are symmetric and anti-symmetric matrices, respectively. Finally, if we add these two matrices together, we obtain

$$\begin{aligned} A_{sym} + A_{anti} &= \frac{1}{2}(A + A^T) + \frac{1}{2}(A - A^T) \\ &= \frac{1}{2}A + \frac{1}{2}A \\ &= A \end{aligned}$$

as claimed. □.

Upon completing this lesson, we introduce one last important matrix operation, called computing the *trace* of a *square* matrix. The trace of any square matrix is always a scalar (number), and often appears in the application of matrices to physics and in the application of group theory to physical problems.

Definition: *(Trace of a Matrix)*

Given a square matrix A of order n, the **trace** of A, denoted by $Tr(A)$, is defined as the sum of the *diagonal* elements of A; that is,

$$Tr(A) = \sum_{i=1}^{n} a_{ii} = a_{11} + a_{22} + \ldots + a_{nn} \qquad (3.3)$$

So, according to (3.3), if we want to find the trace of any square matrix, then all we have to do is add all of its elements along the main diagonal. The trace of a square matrix has several useful properties that are given in the last theorem of this lesson.

Theorem 3.3: *(Properties of the Trace)*

Let A and B be square matrices of the same size, then

$$Tr(A + B) = Tr(A) + Tr(B) \qquad (3.4a)$$

$$Tr(cA) = cTr(A) \text{ for any scalar } c \qquad (3.4b)$$

$$Tr(AB) = Tr(BA) \qquad (3.4c)$$

$$Tr(A) = Tr(A^T) \qquad (3.4d)$$

Proof: The proofs of (3.4a), (3.4b) and (3.4d) are easy, and are left as exercises for the reader. Let's prove (3.4c). To this end, let $[AB]_{ij}$ denote a typical element of AB, and observe that

$$Tr(AB) = \sum_{i=1}^{n}[AB]_{ii} = \sum_{i=1}^{n}\left(\sum_{j=1}^{n} a_{ij}b_{ji}\right)$$

On the other hand, we can write

$$Tr(BA) = \sum_{j=1}^{n}[BA]_{jj} = \sum_{j=1}^{n}\left(\sum_{i=1}^{n} b_{ji}a_{ij}\right) = \sum_{i=1}^{n}\left(\sum_{j=1}^{n} a_{ij}b_{ji}\right)$$

Note that in the last step, we interchanged the order of summation. The result follows upon comparing the right hand sides of the above equations. □.

Exercise Set 3

In Exercises 1-11, use the square matrices:

$$A = \begin{bmatrix} 1 & 2 \\ -2 & 3 \end{bmatrix}, \; B = \begin{bmatrix} 0 & -1 \\ 3 & 4 \end{bmatrix},$$

$$C = \begin{bmatrix} 3 & 5 \end{bmatrix}, \quad D = \begin{bmatrix} 1 \\ 3 \end{bmatrix}$$

to perform the indicated operations:

1. $A^T + B^T$

2. $(A + B)^T$

3. $(AB)^T$

4. $B^T A^T$

5. $Tr(AB)$

6. $Tr(BA)$

7. $Tr(A^T)$

8. $Tr(A)$

9. $Tr(A^2)$

10. CC^T

11. $D^T D$

12. Show that the Pauli spin matrix σ_2 given by

$$\sigma_2 = \begin{bmatrix} 0 & -i \\ i & 0 \end{bmatrix}$$

is anti-symmetric, while the remaining Pauli spin matrices

$$\sigma_0 = \begin{bmatrix} 1 & 0 \\ 0 & 1 \end{bmatrix}, \quad \sigma_1 = \begin{bmatrix} 0 & 1 \\ 1 & 0 \end{bmatrix}, \text{ and } \sigma_3 = \begin{bmatrix} 1 & 0 \\ 0 & -1 \end{bmatrix}$$

are all symmetric.

13. Given that

$$A = \begin{bmatrix} 0 & 1 & -1 \\ 2 & 5 & 3 \\ 4 & 0 & 1 \end{bmatrix},$$

write A as the sum of a symmetric and anti-symmetric matrix.

14. Provide proofs of properties (3.1a), (3.1b), and (3.1c).

15. A square matrix $D = [d_{ij}]$ is said to be a **diagonal matrix** if $d_{ij} = 0$ for all $i \neq j$. If D is a diagonal matrix of size n, show that for any natural number k,

$$Tr(D^k) = \sum_{i=1}^{n} d_{ii}^k$$

16. Prove properties (3.4a) and (3.4b).

4. Introduction to Linear Systems

One of the most important applications of matrices is their use in solving systems of linear equations. In this lesson, we'll discuss a method called the **Gauss-Jordan method** which is useful in solving linear systems.

An $m \times n$ **linear system**, or a **system of** m **linear equations in** n **unknowns**, is a collection of linear equations of the form

$$\begin{aligned} a_{11}x_1 + a_{12}x_2 + \ldots + a_{1n}x_n &= b_1 \\ a_{21}x_1 + a_{22}x_2 + \ldots + a_{2n}x_n &= b_2 \\ \vdots \quad\quad \vdots \quad\quad\quad\quad \vdots \quad &\quad \vdots \\ a_{m1}x_1 + a_{m2}x_2 + \ldots + a_{mn}x_n &= b_m \end{aligned} \tag{4.1}$$

Here, the mn-quantities a_{ij}, along with the m-quantities $b_1, \ldots, b_m$, are given constants, and our goal is to find values of the unknowns $x_1, \ldots, x_n$ that will satisfy each equation in (4.1).

A **solution** to the linear system (4.1) is a set of n numbers, say $c_1, c_2, \ldots, c_n$ such that when we substitute the values

$$x_1 = c_1,\ x_2 = c_2, \ldots,\ x_n = c_n$$

into the system, each equation in the system is satisfied. If the linear system (4.1) has *at least one* solution, then it is called a **consistent system**; if it has no solutions, then it is said to be an **inconsistent system**.

Example-1: Consider the 3×3 linear system

$$\begin{aligned} x_1 + x_2 + x_3 &= 2 \\ 2x_1 + x_2 + x_3 &= 1 \\ x_1 - 2x_2 - x_3 &= 0 \end{aligned}$$

It turns out that the set of values $x_1 = -1, x_2 = -4$, and $x_3 = 7$ is a solution of this system since when we substitute these values into the system we obtain the true statements:

$$\begin{aligned} (-1) + (-4) + (7) &= 2 \\ 2(-1) + (-4) + (7) &= 1 \\ (-1) - 2(-4) - (7) &= 0 \end{aligned}$$

Thus, the system is a *consistent system.*

If it so happens that the quantities $b_1 = b_2 = \ldots = b_n = 0$ in (4.1), then we say that (4.1) is a **homogeneous system**; otherwise, we refer to it as **non-homogeneous system.** It is clear that every $m \times n$ *homogeneous* system always has at least one solution; namely,

$$x_1 = 0, x_2 = 0, \ldots, x_n = 0$$

which is called the **trivial solution.** As we shall see, such a system may or may not have additional **non-trivial solutions.**

Now, we can use the definition of matrix multiplication to rewrite the system (4.1) in matrix form as

$$\begin{bmatrix} a_{11} & a_{12} & \cdots & a_{1n} \\ a_{21} & a_{22} & \cdots & a_{2n} \\ \vdots & \vdots & & \vdots \\ a_{m1} & a_{m2} & \cdots & a_{mn} \end{bmatrix} \begin{bmatrix} x_1 \\ x_2 \\ \vdots \\ x_n \end{bmatrix} = \begin{bmatrix} b_1 \\ b_2 \\ \vdots \\ b_n \end{bmatrix} \tag{4.2}$$

or even more neatly as

$$A\mathbf{x} = \mathbf{b} \tag{4.3}$$

Here, we have made the identifications:

$$A = \begin{bmatrix} a_{11} & a_{12} & \cdots & a_{1n} \\ a_{21} & a_{22} & \cdots & a_{2n} \\ \vdots & \vdots & & \vdots \\ a_{m1} & a_{m2} & \cdots & a_{mn} \end{bmatrix}, \quad \mathbf{x} = \begin{bmatrix} x_1 \\ x_2 \\ \vdots \\ x_n \end{bmatrix}, \quad \mathbf{b} = \begin{bmatrix} b_1 \\ b_2 \\ \vdots \\ b_n \end{bmatrix} \tag{4.4}$$

Observe that the elements of the matrix A are simply the *coefficients* of the unknowns, the column vector $\mathbf{x}$ contains the unknowns themselves, and the column vector $\mathbf{b}$ contains the constants on the right-hand side of the system.

Example-2: We can write the linear system of the previous example in matrix form as

$$\begin{bmatrix} 1 & 1 & 1 \\ 2 & 1 & 1 \\ 1 & -2 & -1 \end{bmatrix} \begin{bmatrix} x_1 \\ x_2 \\ x_3 \end{bmatrix} = \begin{bmatrix} 2 \\ 1 \\ 0 \end{bmatrix}$$

While there are many ways of solving the system (4.2), one very efficient method of solution is **Gauss-Jordan elimination.** Before we can illustrate how this method works, however, we need some definitions.

Definition: *(Reduced Row Echelon Form)*

An $m \times n$ matrix is said to be in **reduced row echelon form** if it satisfies each of the following properties:

(a) Any **zero row**, i.e., any row of the matrix which consists of all zeros, appears at the *bottom* of the matrix.

(b) If a row is not a zero-row, then the first *non-zero* entry of that row is 1. We call this the **leading 1**.

(c) Given any two successive non-zero rows, the leading 1 of the lower row appears to the right of the leading 1 in the upper row.

(d) Each column of the matrix that contains a leading 1 has all zeros elsewhere.

If a matrix only satisfies (a), (b), and (c) but not property (d), then we shall say that it is in **row-echelon form**.

Example-3: The matrices

$$\begin{bmatrix} 1 & 0 & 0 \\ 0 & 1 & 0 \\ 0 & 0 & 0 \end{bmatrix}, \begin{bmatrix} 1 & 0 & 0 & 2 \\ 0 & 1 & 0 & 1 \\ 0 & 0 & 1 & 2 \\ 0 & 0 & 0 & 0 \end{bmatrix}, \begin{bmatrix} 1 & 0 & 0 \\ 0 & 1 & 0 \\ 0 & 0 & 1 \end{bmatrix}$$

are all in reduced row echelon form. On the other hand, the matrices

$$\begin{bmatrix} 1 & 3 & 0 \\ 0 & 1 & 0 \\ 0 & 0 & 0 \end{bmatrix}, \begin{bmatrix} 1 & 2 & 2 & 2 \\ 0 & 1 & 1 & 1 \\ 0 & 0 & 1 & 2 \\ 0 & 0 & 0 & 0 \end{bmatrix}, \begin{bmatrix} 1 & 5 & 0 \\ 0 & 1 & 0 \\ 0 & 0 & 0 \end{bmatrix}$$

are all in row echelon form.

From the previous example, it's easy to see that a matrix in reduced row echelon form must have zeros both above and below each leading 1 (with the exception of its first row), but a matrix in row echelon form only needs to have zeros *below* each of its leading 1's.

Now, the basic idea behind Gauss-Jordan elimination is to perform an appropriate sequence of algebraic operations on the system in (4.2) so that the matrix A is transformed to reduced row echelon form, since when it is in this simplified form, we can easily find the solution to the system. These permissible operations are known as *elementary row operations*.

Definition: *(Elementary Row Operations)*

Let A be an $m \times n$ matrix whose rows are denoted by $R_1, R_2, \ldots, R_m$. An **elementary row operation** on A is any one of the following permissible operations:

(a) We can interchange any two rows of A. If, for example, we interchange the i-th and j-th rows of A we shall denote this operation by $R_i \leftrightarrow R_j$.

(b) We can multiply any row of A by a *non-zero* constant. For example, if we multiply the i-th row of A by a non-zero constant c, then we shall denote this operation by cR_i.

(c) We can multiply any row of A by a constant and then add the result to another row of A. For example, if we multiply the i-th row of A by a constant λ and add the result to the j-th row of A, we shall denote this operation by $\lambda R_i + R_j$.

Let's see how we can use a sequence of elementary row operations to solve an actual linear system.

Example-4: Solve the linear system

$$\begin{aligned} 2x_1 + 3x_2 &= 1 \\ 3x_1 + x_2 &= -2 \end{aligned}$$

Solution:

Step 1: First, we simplify our work by rewriting the system in the shorthand form:

$$\left[\begin{array}{cc|c} 2 & 3 & 1 \\ 3 & 1 & -2 \end{array}\right]$$

This matrix - without the unknowns explicitly listed - is called an **augmented matrix**. Note that the coefficients of the unknowns appear to the left of the vertical bar, and the column of constants appears to right of it. Our goal is to use some sequence of elementary row operations to reduce the matrix on the left of the vertical bar to reduced row echelon form.

Step 2: We begin in the first column. We want the leading 2 to be a 1, so we multiply row 1 by 1/2:

$$\left[\begin{array}{cc|c} 2 & 3 & 1 \\ 3 & 1 & -2 \end{array}\right] \overset{\frac{1}{2}R_1}{\Longrightarrow} \left[\begin{array}{cc|c} 1 & 3/2 & 1/2 \\ 3 & 1 & -2 \end{array}\right]$$

Next, we need to make the 3 a zero. We can do this as follows:

$$\left[\begin{array}{cc|c} 1 & 3/2 & 1/2 \\ 3 & 1 & -2 \end{array}\right] \overset{-3R_1 + R_2}{\Rightarrow} \left[\begin{array}{cc|c} 1 & 3/2 & 1/2 \\ 0 & -7/2 & -7/2 \end{array}\right]$$

Step 3: We now work on the second column. We want the –7/2 to be a leading 1, so we multiply row 2 by –2/7:

$$\left[\begin{array}{cc|c} 1 & 3/2 & 1/2 \\ 0 & -7/2 & -7/2 \end{array}\right] \overset{-\frac{2}{7}R_2}{\Rightarrow} \left[\begin{array}{cc|c} 1 & 3/2 & 1/2 \\ 0 & 1 & 1 \end{array}\right]$$

Step 4: We now want to make the 3/2 in the second column a 0. We can do this as follows:

$$\left[\begin{array}{cc|c} 1 & 3/2 & 1/2 \\ 0 & 1 & 1 \end{array}\right] \overset{-\frac{3}{2}R_2 + R_1}{\Rightarrow} \left[\begin{array}{cc|c} 1 & 0 & -1 \\ 0 & 1 & 1 \end{array}\right]$$

Step 5: Notice that the last augmented matrix (in the previous step) corresponds to the linear system:

$$\begin{aligned} 1x_1 + 0x_2 &= -1 \\ 0x_1 + 1x_2 &= 1 \end{aligned}$$

We can now read off the solution as $x_1 = -1$, and $x_2 = 1$.

Observe that, in the previous example, when we performed elementary row operations on the augmented matrix, we were just performing the same operations on the corresponding equations, to obtain an **equivalent system** of equations at each step along the way.

Example-5: *(A linear system with a unique solution.)* Use the Gauss-Jordan method to solve the system:

$$\begin{aligned} x + y + z &= 2 \\ 2x + y + 2z &= 1 \\ x - 2y - z &= 0 \end{aligned}$$

Solution:

Step 1: As in the previous example, we form the augmented matrix, and try to transform it to reduced row echelon form. We start in the first column.

$$\left[\begin{array}{ccc|c} 1 & 1 & 1 & 2 \\ 2 & 1 & 2 & 1 \\ 1 & -2 & -1 & 0 \end{array}\right] \overset{-R_1 + R_3}{\Rightarrow} \left[\begin{array}{ccc|c} 1 & 1 & 1 & 2 \\ 2 & 1 & 2 & 1 \\ 0 & -3 & -2 & -2 \end{array}\right]$$

$$\overset{-2R_1 + R_2}{\Rightarrow} \left[\begin{array}{ccc|c} 1 & 1 & 1 & 2 \\ 0 & -1 & 0 & -3 \\ 0 & -3 & -2 & -2 \end{array}\right]$$

Step 2: We now work on the second column.

$$\left[\begin{array}{ccc|c} 1 & 1 & 1 & 2 \\ 0 & -1 & 0 & -3 \\ 0 & -3 & -2 & -2 \end{array}\right] \overset{-1 \cdot R_2}{\Rightarrow} \left[\begin{array}{ccc|c} 1 & 1 & 1 & 2 \\ 0 & 1 & 0 & 3 \\ 0 & -3 & -2 & -2 \end{array}\right]$$

$$\left[\begin{array}{ccc|c} 1 & 1 & 1 & 2 \\ 0 & 1 & 0 & 3 \\ 0 & -3 & -2 & -2 \end{array}\right] \underset{3R_2 + R_3}{\overset{-R_2 + R_1}{\Rightarrow}} \left[\begin{array}{ccc|c} 1 & 0 & 1 & -1 \\ 0 & 1 & 0 & 3 \\ 0 & 0 & -2 & 7 \end{array}\right]$$

Step 3: We now work on the third column.

$$\left[\begin{array}{ccc|c} 1 & 0 & 1 & -1 \\ 0 & 1 & 0 & 3 \\ 0 & 0 & -2 & 7 \end{array}\right] \overset{-\frac{1}{2}R_3}{\Rightarrow} \left[\begin{array}{ccc|c} 1 & 0 & 1 & -1 \\ 0 & 1 & 0 & 3 \\ 0 & 0 & 1 & -7/2 \end{array}\right]$$

$$\left[\begin{array}{ccc|c} 1 & 0 & 1 & -1 \\ 0 & 1 & 0 & 3 \\ 0 & 0 & 1 & -7/2 \end{array}\right] \overset{-R_3 + R_1}{\Rightarrow} \left[\begin{array}{ccc|c} 1 & 0 & 0 & 5/2 \\ 0 & 1 & 0 & 3 \\ 0 & 0 & 1 & -7/2 \end{array}\right]$$

Step 4: Since the final form of the augmented matrix is:

$$\left[\begin{array}{ccc|c} 1 & 0 & 0 & 5/2 \\ 0 & 1 & 0 & 3 \\ 0 & 0 & 1 & -7/2 \end{array}\right]$$

we see that the solution is $x = 5/2$, $y = 3$, and $z = -7/2$.

Example-6: *(A linear system with more unknowns than equations)*
Solve the 3×4 linear system:

$$\begin{aligned} x_1 - 2x_2 + x_3 + x_4 &= 1 \\ 2x_1 + x_2 + x_3 &= 0 \\ -x_1 + x_2 + x_3 + 2x_4 &= 2 \end{aligned}$$

Solution:

Step 1: We first write down the augmented matrix as:

$$\left[\begin{array}{cccc|c} 1 & -2 & 1 & 1 & 1 \\ 2 & 1 & 1 & 0 & 0 \\ -1 & 1 & 1 & 2 & 2 \end{array}\right]$$

Step 2: We transform the first column as follows:

$$\left[\begin{array}{cccc|c} 1 & -2 & 1 & 1 & 1 \\ 2 & 1 & 1 & 0 & 0 \\ -1 & 1 & 1 & 2 & 2 \end{array}\right] \begin{array}{c} -2R_1 + R_2 \\ \Rightarrow \\ R_1 + R_3 \end{array} \left[\begin{array}{cccc|c} 1 & -2 & 1 & 1 & 1 \\ 0 & 5 & -1 & -2 & -2 \\ 0 & -1 & 2 & 3 & 3 \end{array}\right]$$

Step 3: We now work on the second column.

$$\left[\begin{array}{cccc|c} 1 & -2 & 1 & 1 & 1 \\ 0 & 5 & -1 & -2 & -2 \\ 0 & -1 & 2 & 3 & 3 \end{array}\right] \begin{array}{c} R_2 \leftrightarrow R_3 \\ \Rightarrow \\ -1 \cdot R_2 \end{array} \left[\begin{array}{cccc|c} 1 & -2 & 1 & 1 & 1 \\ 0 & 1 & -2 & -3 & -3 \\ 0 & 5 & -1 & -2 & -2 \end{array}\right]$$

Step 4: Next, we work on column three.

$$\left[\begin{array}{cccc|c} 1 & -2 & 1 & 1 & 1 \\ 0 & 1 & -2 & -3 & -3 \\ 0 & 5 & -1 & -2 & -2 \end{array}\right] \begin{array}{c} -5R_2 + R_3 \\ \Rightarrow \\ 2R_2 + R_1 \end{array} \left[\begin{array}{cccc|c} 1 & 0 & -3 & -5 & -5 \\ 0 & 1 & -2 & -3 & -3 \\ 0 & 0 & 9 & 13 & 13 \end{array}\right]$$

$$\left[\begin{array}{cccc|c} 1 & 0 & -3 & -5 & -5 \\ 0 & 1 & -2 & -3 & -3 \\ 0 & 0 & 9 & 13 & 13 \end{array}\right] \begin{array}{c} \frac{1}{9}R_3 \\ \Rightarrow \end{array} \left[\begin{array}{cccc|c} 1 & 0 & -3 & -5 & -5 \\ 0 & 1 & -2 & -3 & -3 \\ 0 & 0 & 1 & 13/9 & 13/9 \end{array}\right]$$

$$\left[\begin{array}{cccc|c} 1 & 0 & -3 & -5 & -5 \\ 0 & 1 & -2 & -3 & -3 \\ 0 & 0 & 1 & 13/9 & 13/9 \end{array}\right] \begin{array}{c} 2R_3 + R_2 \\ \Rightarrow \\ 3R_3 + R_1 \end{array} \left[\begin{array}{cccc|c} 1 & 0 & 0 & -2/3 & -2/3 \\ 0 & 1 & 0 & -1/9 & -1/9 \\ 0 & 0 & 1 & 13/9 & 13/9 \end{array}\right]$$

Step 5: The last matrix corresponds to the linear system

$$\begin{aligned} x_1 - \frac{2}{3}x_4 &= -\frac{2}{3} \\ x_2 - \frac{1}{9}x_4 &= -\frac{1}{9} \\ x_3 + \frac{13}{9}x_4 &= \frac{13}{9} \end{aligned}$$

We now solve each of these equations for the unknowns that correspond to the leading ones, to obtain

$$\begin{aligned} x_1 &= -\frac{2}{3} + \frac{2}{3}x_4 \\ x_2 &= -\frac{1}{9} + \frac{1}{9}x_4 \\ x_3 &= \frac{13}{9} - \frac{13}{9}x_4 \end{aligned}$$

Since x_4 is arbitrary, we set $x_4 = s$ where s is an arbitrary real number, and the solution of the system is

$$\begin{aligned} x_1 &= -\frac{2}{3} + \frac{2}{3}s \\ x_2 &= -\frac{1}{9} + \frac{1}{9}s \\ x_3 &= \frac{13}{9} - \frac{13}{9}s \\ x_4 &= s \end{aligned}$$

Since the parameter s is arbitrary, the system has *infinitely many solutions.*

In the previous example, we saw that a system with more unknowns than equations gave us infinitely many solutions. This is no accident, and it turns out to be true in general; this is the content of the following theorem.

Theorem 4.1: *(Linear System with More Unknowns than Equations)*

If the $m \times n$ linear system $A\mathbf{x} = \mathbf{b}$ of m equations in n unknowns, has more unknowns than equations, then it has *infinitely many* solutions.

Proof: Assume that after an appropriate sequence of elementary row operations, the augmented matrix $[A|\mathbf{b}]$ is transformed to the matrix $[A'|\mathbf{b}']$, where A' is in reduced row echelon form. If r is the number of non-zero rows in A', then clearly, $r \leq m$. But since $m < n$ by hypothesis, we are forced to conclude that $r < n$ as well. Consequently, the matrix $[A'|\mathbf{b}']$ represents exactly r equations in n unknowns, and we can simply solve for the r unknowns (corresponding to the leading ones) in terms of the remaining $(n - r)$ unknowns - whose values are arbitrary. Thus, the system has infinitely many solutions. □.

Corollary 4.2: *(Homogeneous System with More Unknowns than Equations)*

If the $m \times n$ linear homogeneous system $A\mathbf{x} = \mathbf{0}$ of m equations in n unknowns, with $m < n$, so it has more unknowns than equations, then it always has at least one *nontrivial solution.*

Proof: Set $\mathbf{b} = \mathbf{0}$ in the previous theorem, so the homogeneous system $A\mathbf{x} = \mathbf{0}$ must have infinitely many solutions, and hence, at least one of those solutions must be nontrivial. □.

From the previous theorem and its corollary, we see that when we are attempting to solve a linear system, then only one of three results are possible:

1. The linear system is *inconsistent* so it has no solution.
2. The linear system has a unique solution, i.e., one and only one solution.
3. The linear system has infinitely many solutions.

This state of affairs is shown in Figure-4.1 below.

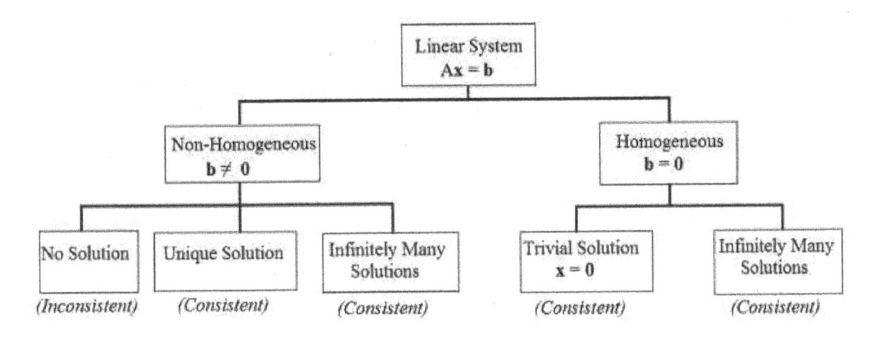

Figure 4.1: Possible solutions of the linear system $A\mathbf{x} = \mathbf{b}$.

A question which naturally arises is how do we determine when a linear is inconsistent? It turns out that when we are attempting to solve the system via the Gauss-Jordan method, and reduce the augmented matrix to reduced row echelon form, *an inconsistent system is signaled by the appearance of a zero row to the left of the vertical bar, and a non-zero constant (in the same row) to the right of the vertical bar.* Take a look at the next example.

Example-7: *(An Inconsistent System)* Solve the linear system:

$$\begin{aligned} 2x + 6y - 4z &= 1 \\ x + 2y - z &= 0 \\ -x - 3y + 2z &= 1 \end{aligned}$$

Solution:

Step 1: We first form the augmented matrix

$$\left[\begin{array}{ccc|c} 2 & 6 & -4 & 1 \\ 1 & 2 & -1 & 0 \\ -1 & -3 & 2 & 1 \end{array}\right]$$

Step 2: Now, we work on the first column.

$$\left[\begin{array}{ccc|c} 2 & 6 & -4 & 1 \\ 1 & 2 & -1 & 0 \\ -1 & -3 & 2 & 1 \end{array}\right] \overset{R_1 \leftrightarrow R_2}{\Rightarrow} \left[\begin{array}{ccc|c} 1 & 2 & -1 & 0 \\ 2 & 6 & -4 & 1 \\ -1 & -3 & 2 & 1 \end{array}\right]$$

$$\left[\begin{array}{ccc|c} 1 & 2 & -1 & 0 \\ 2 & 6 & -4 & 1 \\ -1 & -3 & 2 & 1 \end{array}\right] \begin{array}{c} -2R_1 + R_2 \\ \Rightarrow \\ R_1 + R_2 \end{array} \left[\begin{array}{ccc|c} 1 & 2 & -1 & 0 \\ 0 & 2 & -2 & 1 \\ 0 & -1 & 1 & 1 \end{array}\right]$$

Step 3: Next, we work on column two.

$$\left[\begin{array}{ccc|c} 1 & 2 & -1 & 0 \\ 0 & 2 & -2 & 1 \\ 0 & -1 & 1 & 1 \end{array}\right] \overset{\frac{1}{2}R_2}{\Rightarrow} \left[\begin{array}{ccc|c} 1 & 2 & -1 & 0 \\ 0 & 1 & -1 & 1/2 \\ 0 & -1 & 1 & 1 \end{array}\right]$$

$$\left[\begin{array}{ccc|c} 1 & 2 & -1 & 0 \\ 0 & 1 & -1 & 1/2 \\ 0 & -1 & 1 & 1 \end{array}\right] \begin{array}{c} -2R_2 + R_1 \\ \Rightarrow \\ R_2 + R_3 \end{array} \left[\begin{array}{ccc|c} 1 & 0 & 1 & -1 \\ 0 & 1 & -1 & 1/2 \\ 0 & 0 & 0 & 3/2 \end{array}\right]$$

Step 4: Observe that the last matrix implies that the following equivalent system of equations must hold:

$$\begin{aligned} x + (0)y + z &= -1 \\ (0)x + y - z &= 1/2 \\ (0)x + (0)y + (0)z &= 3/2 \end{aligned}$$

Since the last equation can never be satisfied, we conclude that the linear system is *inconsistent*; it has no solution.

Example-8: *(A Homogeneous System)* Solve the homogeneous system:

$$\begin{bmatrix} 1 & 0 & 2 & 0 \\ 1 & -1 & 1 & 1 \\ 2 & 0 & 1 & 0 \end{bmatrix} \begin{bmatrix} x_1 \\ x_2 \\ x_3 \\ x_4 \end{bmatrix} = \mathbf{0}$$

Solution:

Step 1: We first form the augmented matrix:

$$\left[\begin{array}{cccc|c} 1 & 0 & 2 & 0 & 0 \\ 1 & -1 & 1 & 1 & 0 \\ 2 & 0 & 1 & 0 & 0 \end{array}\right]$$

Step 2: We reduce the augmented matrix to reduced row echelon form to obtain (exercise):

$$\left[\begin{array}{cccc|c} 1 & 0 & 0 & 0 & 0 \\ 0 & 1 & 0 & -1 & 0 \\ 0 & 0 & 1 & 0 & 0 \end{array}\right]$$

Step 3: The last matrix corresponds to the linear system

$$\begin{aligned} x_1 &= 0 \\ x_2 - x_4 &= 0 \\ x_3 &= 0 \end{aligned}$$

Solving each equation for the variables corresponding to the leading ones of the previous matrix, we obtain

$$\begin{aligned} x_1 &= 0 \\ x_2 &= r \\ x_3 &= 0 \\ x_4 &= r \end{aligned}$$

where r is an arbitrary real number.

Now that we have seen numerous examples of applying the Gauss-Jordan method, we summarize the results of our work below.

Summary of the Gauss-Jordan Method

In order to solve the linear system $A\mathbf{x} = \mathbf{b}$ by Gauss-Jordan reduction, we proceed as follows:

1. Write down the augmented matrix $[A|\mathbf{b}]$ corresponding to the system.
2. By using an appropriate sequence of elementary row operations, we transform the augmented matrix to the matrix $[A'|\mathbf{b}']$ which is in *reduced row echelon form.*
3. If a given row of the matrix A' is a zero row, but the *same* row of the column vector $\mathbf{b}'$ is not zero, then the original system is *inconsistent*, and we are done.
4. Otherwise, for each non-zero row of the transformed matrix $[A'|\mathbf{b}']$, we solve for the unknown that corresponds to the leading one of that row.

If a row consists entirely of zeroes (both to the left and right of the vertical bar), then it may be safely ignored.

Another very efficient way of solving a linear system $A\mathbf{x} = \mathbf{b}$, is to use a slight variation of the Gauss-Jordan method. This variation is simply called **Gaussian Elimination.**

In this new procedure, we simply use elementary row operations to transform the augmented matrix $[A|\mathbf{b}]$ to *row echelon form*, and then work backwards, using a method called **back substitution**, to obtain the solution. Let's see how this works.

Example-9: Use Gaussian elimination to solve the system

$$\begin{aligned} x + 2y + z &= 3 \\ 2x + 2y + z &= 2 \\ x + 3y - z &= 0 \end{aligned}$$

Solution:

Step 1: We first form the augmented matrix

$$\left[\begin{array}{ccc|c} 1 & 2 & 1 & 3 \\ 2 & 2 & 1 & 2 \\ 1 & 3 & -1 & 0 \end{array}\right]$$

Step 2: Next, using elementary row operations, we reduce the augmented matrix to *row echelon form* to obtain

$$\left[\begin{array}{ccc|c} 1 & 2 & 1 & 3 \\ 0 & 1 & 1/2 & 2 \\ 0 & 0 & 1 & 2 \end{array}\right]$$

Step 3: The last matrix corresponds to the simplified system of equations

$$\begin{aligned} x + 2y + z &= 3 \\ y + \frac{1}{2}z &= 2 \\ z &= 2 \end{aligned}$$

We now perform *back substitution*, i.e., we substitute the result $z = 2$ into the second equation to get

$$y + \frac{1}{2}(2) = 2$$

so that $y = 1$. Finally, we substitute the values $y = 1$ and $z = 2$ into the first equation to get

$$x + 2(1) + (2) = 3$$

so $x = -1$. Thus, the solution of the system is

$$\begin{aligned} x &= -1 \\ y &= 1 \\ z &= 2 \end{aligned}$$

Exercise Set 4

In Exercises 1-10, use the Gauss-Jordan method or Gaussian elimination (with back substitution) to find the solution of each linear system. If a system is inconsistent then state the same.

1.

$$\begin{aligned} 2x + y &= 2 \\ x + 2y &= 1 \end{aligned}$$

2.

$$\begin{aligned} 3x + y &= 0 \\ 6x + y &= 1 \end{aligned}$$

3.

$$\begin{aligned} x + y + z &= 3 \\ 2x + 3y + z &= 5 \\ x + 2y + z &= 4 \end{aligned}$$

4.

$$\begin{aligned} x + y + z &= 0 \\ x + 3y + 2z &= -1 \\ x + y + 3z &= 2 \end{aligned}$$

5.

$$\begin{aligned} 2x + y + z &= 1 \\ 6x + 3y + z &= 3 \\ x + 2y + z &= 1 \end{aligned}$$

6.

$$\begin{aligned} 3x + y + z &= 8 \\ x + 2y + z &= 8 \\ x - y + z &= 2 \end{aligned}$$

7.

$$\begin{aligned} 2x + 3y + 4z &= 1 \\ 2x - 3y + 8z &= 2 \\ 4x - 3y + 12z &= 2 \end{aligned}$$

8.

$$\begin{aligned} x - 3y + 4z &= 0 \\ 2x - 3y + 8z &= 0 \\ -2x + 6y - 8z &= 0 \end{aligned}$$

9.

$$\begin{aligned} x + y + z - w &= 0 \\ 2x - y + 3z - 3w &= 0 \\ 3x - 3y + 4z - w &= 1 \\ 2x + y + z - 2w &= -2 \end{aligned}$$

10.

$$
\begin{aligned}
x + y + z - w &= 0 \\
2x - 2y + 3z - w &= 0 \\
-x - y + 2z &= 0 \\
3x + 2y + z - 2w &= 0
\end{aligned}
$$

11. *(Application to Electrical Networks)* The analysis of electrical networks often involves the solution of linear systems. The analysis of such networks is based upon **Kirchoff's Laws:**

(1) Kirchoff's Voltage Law: The sum of the voltage drops IR (where I is the current, and R is the resistance) around a closed circuit path is equal to the total voltage applied to that path.

(2) Kirchoff's Current Law: The total current flowing into a junction is equal to the total current flowing out of that junction.

(a) Consider the electrical network in Figure 4.2.

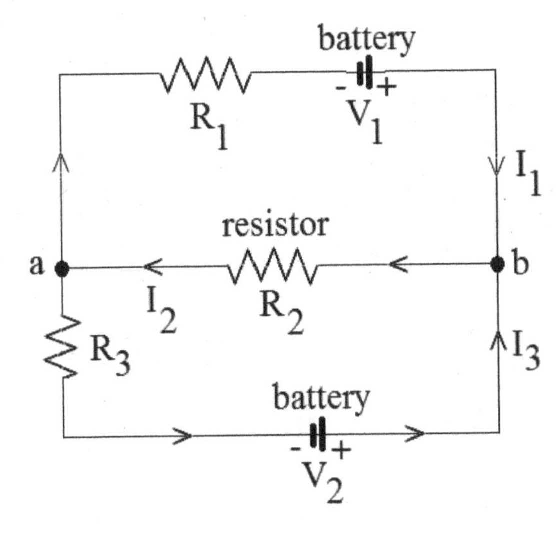

Figure 4.2: A typical electrical network.

Show that an application of Kirchoff's laws to the given network leads to the system of equations:

$$
\begin{aligned}
I_1 - I_2 + I_3 &= 0 \quad \text{(At junction } \mathbf{a} \text{ or junction } \mathbf{b}\text{)} \\
R_1 I_1 + R_2 I_2 &= V_1 \\
R_2 I_2 + R_3 I_3 &= V_2
\end{aligned}
$$

(b) Find the currents I_1, I_2, and I_3 in each branch of the network, given that $R_1 = 4, R_2 = 3, R_3 = 6, V_1 = 3$, and $V_2 = 6$.

5. The Inverse of a Matrix

In Theorem 2.2, we saw that matrices obey many of the same basic properties as real numbers, with the most glaring exception being that matrix multiplication is not commutative. One very important property of real numbers is that every *non-zero* real number has a multiplicative inverse.

That is, for any real number $a \neq 0$, there exists a *unique* real number a^{-1}, called the *multiplicative inverse* of a, such that

$$a^{-1} \cdot a = a \cdot a^{-1} = 1 \tag{5.1}$$

where the number 1 is the *identity element* for multiplication. It turns out that *some* square matrices enjoy a property which is analogous to (5.1). This fact leads us to our first definition.

Definition: *(Inverse of a Matrix)*

Let A be a square matrix of order n. We shall say the A is **invertible** (or **non-singular**) if there exists an $n \times n$ matrix B such that

$$AB = BA = I \tag{5.2}$$

where I is the $n \times n$ identity matrix. The matrix B is called the **inverse** of A, and we agree to write

$$B = A^{-1} \tag{5.3}$$

Clearly, (5.2) is the matrix analogue of the multiplicative inverse property of real numbers; here the $n \times n$ identity matrix plays the role of the identity element for matrix multiplication, and the matrix A^{-1} acts like a multiplicative inverse.

Now, it turns out that not every square matrix has an inverse. In other words, as we shall see in the examples that follow, it may happen that no $n \times n$ matrix B exists which satisfies (5.2); in this case, we say that A is **non-invertible**, or **singular**.

If a square matrix is invertible, however, then its inverse can be found by using a suitable adaptation of the Gauss-Jordan method. To see this, suppose that we want to find the inverse of an $n \times n$ matrix A. This means that we must find an $n \times n$ matrix X such that

$$AX = I \tag{5.4}$$

To this end, let the columns of the unknown matrix X be given by column vectors

$$\mathbf{x}_1 = \begin{bmatrix} x_{11} \\ x_{21} \\ \vdots \\ x_{n1} \end{bmatrix}, \quad \mathbf{x}_2 = \begin{bmatrix} x_{12} \\ x_{22} \\ \vdots \\ x_{n2} \end{bmatrix}, \ldots, \mathbf{x}_n = \begin{bmatrix} x_{1n} \\ x_{2n} \\ \vdots \\ x_{nn} \end{bmatrix}$$

Now, in order to solve (5.4), we need to solve the n linear systems

$$A\mathbf{x}_1 = \mathbf{e}_1, \quad A\mathbf{x}_2 = \mathbf{e}_2, \ldots, A\mathbf{x}_n = \mathbf{e}_n. \tag{5.5}$$

where $\mathbf{e}_1, \mathbf{e}_2, \ldots, \mathbf{e}_n$ are the column vectors:

$$\mathbf{e}_1 = \begin{bmatrix} 1 \\ 0 \\ 0 \\ \vdots \\ 0 \end{bmatrix}, \quad \mathbf{e}_2 = \begin{bmatrix} 0 \\ 1 \\ 0 \\ \vdots \\ 0 \end{bmatrix}, \ldots, \mathbf{e}_n = \begin{bmatrix} 0 \\ 0 \\ 0 \\ \vdots \\ 1 \end{bmatrix}$$

Following the usual Gauss-Jordan procedure, we form the n augmented matrices

$$[A|\mathbf{e}_1], \quad [A|\mathbf{e}_2], \ldots, [A|\mathbf{e}_n] \tag{5.6}$$

and we transform each augmented matrix to reduced row echelon form. Assuming that each of the linear systems in (5.5) has a *unique solution*, then the solution vectors $\mathbf{x}_1, \mathbf{x}_2, \ldots, \mathbf{x}_n$ will appear as column vectors to the right of each vertical bar as follows

$$[I|\mathbf{x}_1], \quad [I|\mathbf{x}_2], \ldots, [I|\mathbf{x}_n]$$

We can then reconstruct the columns of the solution matrix X as

$$X = \begin{bmatrix} \mathbf{x}_1 & \mathbf{x}_2 & \mathbf{x}_3 & \cdots & \mathbf{x}_n \end{bmatrix}$$

But since the *same matrix* appears to the left of each vertical bar in (5.6), a more efficient procedure is to form only one augmented matrix

$$[A|I] = \begin{bmatrix} A|\mathbf{e}_1 & \mathbf{e}_2 & \mathbf{e}_3 & \cdots & \mathbf{e}_n \end{bmatrix}$$

and then use elementary row operations to transform A to reduced row echelon form. If this can be done, we will obtain an augmented matrix $[I|B]$, and the resulting matrix B will be the inverse of A; that is, $B = A^{-1}$.

Symbolically, the procedure for finding the inverse of A is

$$[A|I] \overset{\text{elementary row operations}}{\Rightarrow} [I|B] \tag{5.7}$$

This general procedure can be summarized as follows:

Finding the Inverse of a Matrix by the Gauss-Jordan Method

Let A be a square matrix of order n. To find its inverse, we proceed as follows:
1. Form the augmented matrix $[A|I]$ which consists of the original matrix A to the left of the vertical bar, and the identity matrix of order n to its right.
2. As shown in (5.7), perform a sequence of elementary row operations to transform the augmented matrix to the form $[I|B]$. Remember that any elementary row operation that we perform to the left of the vertical bar must also be performed to the right of the vertical bar.
3. The required inverse of A is then $A^{-1} = B$.
4. If the augmented matrix cannot be reduced to the form $[I|B]$, then A is not invertible; i.e., A^{-1} does not exist.

Let's see how this algorithm actually works.

Example-1: *(Non-Singular* 2 × 2 *matrix)* Find the inverse of the matrix

$$A = \begin{bmatrix} 1 & 2 \\ 2 & 5 \end{bmatrix}$$

Solution:

Step 1. We form the augmented matrix $[A|I]$ or,

$$\left[\begin{array}{cc|cc} 1 & 2 & 1 & 0 \\ 2 & 5 & 0 & 1 \end{array}\right]$$

Step 2. Using elementary row operations, we obtain the matrix $[I|A^{-1}]$ or

$$\left[\begin{array}{cc|cc} 1 & 2 & 1 & 0 \\ 2 & 5 & 0 & 1 \end{array}\right] \overset{-2R_1 + R_2}{\Rightarrow} \left[\begin{array}{cc|cc} 1 & 2 & 1 & 0 \\ 0 & 1 & -2 & 1 \end{array}\right]$$

$$\overset{-2R_2 + R_1}{\Rightarrow} \left[\begin{array}{cc|cc} 1 & 0 & 5 & -2 \\ 0 & 1 & -2 & 1 \end{array}\right]$$

Step 3. By inspection of the last matrix, we find that

$$A^{-1} = \begin{bmatrix} 5 & -2 \\ -2 & 1 \end{bmatrix}$$

Step 4. We check our work:

$$AA^{-1} = \begin{bmatrix} 1 & 2 \\ 2 & 5 \end{bmatrix}\begin{bmatrix} 5 & -2 \\ -2 & 1 \end{bmatrix} = \begin{bmatrix} 1 & 0 \\ 0 & 1 \end{bmatrix} = I$$

$$A^{-1}A = \begin{bmatrix} 5 & -2 \\ -2 & 1 \end{bmatrix}\begin{bmatrix} 1 & 2 \\ 2 & 5 \end{bmatrix} = \begin{bmatrix} 1 & 0 \\ 0 & 1 \end{bmatrix} = I$$

Example-2: *(Non-Singular* 3×3 *matrix)* Find the inverse of the matrix

$$A = \begin{bmatrix} 1 & 0 & 1 \\ 2 & 1 & 2 \\ 0 & 1 & 1 \end{bmatrix}$$

Solution:

Step 1. We first form the augmented matrix

$$\left[\begin{array}{ccc|ccc} 1 & 0 & 1 & 1 & 0 & 0 \\ 2 & 1 & 2 & 0 & 1 & 0 \\ 0 & 1 & 1 & 0 & 0 & 1 \end{array}\right]$$

Step 2. After an appropriate sequence of elementary row operations, we obtain the augmented matrix

$$\left[\begin{array}{ccc|ccc} 1 & 0 & 0 & -1 & 1 & -1 \\ 0 & 1 & 0 & -2 & 1 & 0 \\ 0 & 0 & 1 & 2 & -1 & 1 \end{array}\right]$$

which is in the form $[I|A^{-1}]$. We conclude that the inverse is

$$A^{-1} = \begin{bmatrix} -1 & 1 & -1 \\ -2 & 1 & 0 \\ 2 & -1 & 1 \end{bmatrix}$$

We leave it to the reader to check our work.

Example-3: *(A Singular* 3×3 *Matrix)* Find the inverse - if it exists - of the matrix:

$$A = \begin{bmatrix} 1 & 2 & 1 \\ 0 & 1 & 2 \\ 1 & 3 & 3 \end{bmatrix}$$

Solution:

Step 1. We form the augmented matrix $[A|I]$ or

$$\left[\begin{array}{ccc|ccc} 1 & 2 & 1 & 1 & 0 & 0 \\ 0 & 1 & 2 & 0 & 1 & 0 \\ 1 & 3 & 3 & 0 & 0 & 1 \end{array}\right]$$

Step 2. After a sequence of elementary row operations, we obtain the following matrix (the reader should verify this) in reduced row echelon form:

$$\left[\begin{array}{ccc|ccc} 1 & 0 & -3 & 0 & -3 & 1 \\ 0 & 1 & 2 & 0 & 1 & 0 \\ 0 & 0 & 0 & 1 & 1 & -1 \end{array}\right]$$

Since this matrix is *not* in the form $[I|B]$ where the identity matrix is to the left of the vertical bar, we conclude that A is not invertible, or equivalently, A^{-1} does not exist.

Next, we explore some of the useful properties of matrix inversion.

Theorem 5.1: *(Uniqueness of an Inverse)*

If A is an invertible matrix, then its inverse is *unique*.

Proof: Assume that A is non-singular, and suppose that B and B' are both inverses of A. Consequently, we must have

$$AB = BA = I \text{ and } AB' = B'A = I$$

Now, observe that $AB = I$ implies that

$$B'(AB) = B'I = B'$$

But, since matrix multiplication is associative, we can also write,

$$B'(AB) = (B'A)B = IB = B$$

Comparing the last two expressions, we see that $B' = B$. □.

Theorem 5.2: *(Inverse of a Product)*

If A and B are invertible matrices of order n, then their product AB is also invertible, and

$$(AB)^{-1} = B^{-1}A^{-1} \tag{5.8}$$

In other words, the inverse of the product of two invertible matrices is equal to the product of their inverses, taken in reverse order.

Proof: Observe that

$$AB \cdot (B^{-1}A^{-1}) = A(B \cdot B^{-1})A^{-1} = A(I)A^{-1} = (AI)A^{-1} = AA^{-1} = I$$
$$(B^{-1}A^{-1}) \cdot AB = B^{-1}(A^{-1} \cdot A)B = B^{-1}(I)B = B^{-1}(IB) = B^{-1}B = I$$

So, we have shown that

$$AB \cdot (B^{-1}A^{-1}) = I \text{ and } (B^{-1}A^{-1}) \cdot AB = I$$

and by the *definition* of matrix inverse, the result follows. □.

Corollary 5.3: *(Inverse of the Product of Finitely Many Matrices)*

If $A_1, A_2, \ldots, A_n$ are all invertible $n \times n$ matrices, then their product $A_1 A_2 \cdots A_n$ is also invertible, and

$$(A_1 A_2 \cdots A_n)^{-1} = A_n^{-1} A_{n-1}^{-1} \cdots A_1^{-1} \tag{5.9}$$

Proof: Left as an exercise.

Theorem 5.4: *(Some Useful Properties of Inverses)*

If A is an invertible matrix, then the following statements are true:
(a)

$$(A^{-1})^{-1} = A \tag{5.10}$$

(b) For any positive integer n,

$$(A^n)^{-1} = (A^{-1})^n \tag{5.11}$$

(c) The inverse of the transpose of A is equal to the transpose of its inverse:

$$(A^T)^{-1} = (A^{-1})^T \tag{5.12}$$

Proof: We prove only (c), and leave the remaining properties as exercises. To establish (c), observe that we have both

$$A^T \cdot (A^{-1})^T = (A^{-1}A)^T = I^T = I$$
$$(A^{-1})^T \cdot A^T = (AA^{-1})^T = I^T = I$$

so evidently, $(A^{-1})^T = (A^T)^{-1}$. $\square$.

We now turn our attention towards investigating the connection between the operation of matrix inversion and the solution of linear systems. From the previous lesson, we know that a general $m \times n$ linear system consisting of m equations in n unknowns, can either have a unique solution, no solution, or infinitely many solutions.

However, for an $n \times n$ linear system $A\mathbf{x} = \mathbf{b}$, where A is non-singular, we can always find its solution via the operation of matrix inversion, and this solution will be unique. This is the content of the next theorem.

Theorem 5.4: *(Solution of n × n Linear System)*

Let $A\mathbf{x} = \mathbf{b}$ be an $n \times n$ linear system. If the matrix A is invertible, then the linear system has the *unique* solution

$$\mathbf{x} = A^{-1}\mathbf{b} \tag{5.13}$$

Proof: Observe that

$$\begin{aligned} A\mathbf{x} &= \mathbf{b} \\ &\Rightarrow A^{-1}(A\mathbf{x}) = A^{-1}\mathbf{b} \\ &\Rightarrow (A^{-1}A)\mathbf{x} = A^{-1}\mathbf{b} \\ &\Rightarrow I\mathbf{x} = A^{-1}\mathbf{b} \\ &\Rightarrow \mathbf{x} = A^{-1}\mathbf{b} \end{aligned}$$

as claimed. To see that this solution is unique, assume that the linear system has two solutions, say $\mathbf{x}_1$ and $\mathbf{x}_2$. Consequently, we must have both $A\mathbf{x}_1 = \mathbf{b}$ and $A\mathbf{x}_2 = \mathbf{b}$. Subtracting the second equation from the first one, we get

$$A(\mathbf{x}_1 - \mathbf{x}_2) = A\mathbf{x}_1 - A\mathbf{x}_2 = \mathbf{0}$$

Next, we multiply on the left by A^{-1} to get

$$\begin{aligned} A^{-1}A(\mathbf{x}_1 - \mathbf{x}_2) &= A^{-1} \cdot \mathbf{0} = \mathbf{0} \\ &\Rightarrow I(\mathbf{x}_1 - \mathbf{x}_2) = \mathbf{0} \\ &\Rightarrow \mathbf{x}_1 - \mathbf{x}_2 = \mathbf{0} \end{aligned}$$

so $\mathbf{x}_1 = \mathbf{x}_2$ and the solution must be unique. $\square$.

Example-4: Use the previous theorem to solve the linear system

$$
\begin{aligned}
x_1 + x_2 + 3x_3 &= 1 \\
x_1 + 2x_2 - x_3 &= 0 \\
x_1 - x_2 + x_3 &= 3
\end{aligned}
$$

Solution:

Step 1. We first write the system in the form $A\mathbf{x} = \mathbf{b}$:

$$
\begin{bmatrix} 1 & 1 & 3 \\ 1 & 2 & -1 \\ 1 & -1 & 1 \end{bmatrix} \begin{bmatrix} x_1 \\ x_2 \\ x_3 \end{bmatrix} = \begin{bmatrix} 1 \\ 0 \\ 3 \end{bmatrix}
$$

Step 2. Next, we find the inverse of A to be

$$
A^{-1} = \begin{bmatrix} -\frac{1}{10} & \frac{2}{5} & \frac{7}{10} \\ \frac{1}{5} & \frac{1}{5} & -\frac{2}{5} \\ \frac{3}{10} & -\frac{1}{5} & -\frac{1}{10} \end{bmatrix}
$$

Step 3. According to (5.13), the solution vector is then

$$
\mathbf{x} = A^{-1}\mathbf{b} = \begin{bmatrix} -\frac{1}{10} & \frac{2}{5} & \frac{7}{10} \\ \frac{1}{5} & \frac{1}{5} & -\frac{2}{5} \\ \frac{3}{10} & -\frac{1}{5} & -\frac{1}{10} \end{bmatrix} \begin{bmatrix} 1 \\ 0 \\ 3 \end{bmatrix} = \begin{bmatrix} 2 \\ -1 \\ 0 \end{bmatrix}
$$

so that

$$
\begin{aligned}
x_1 &= 2 \\
x_2 &= -1 \\
x_3 &= 0
\end{aligned}
$$

The previous example illustrates, that while the previous theorem is of theoretical interest, the use of (5.13) to solve a linear system is usually quite tedious; first we have to find A^{-1} and then carry out matrix multiplication. Clearly, the Gauss-Jordan method, or Gauss elimination with back substitution are more efficient methods of solution.

Exercise Set 5

In Exercises 1-10, find the inverse of each matrix - if it exists.

1. $\begin{bmatrix} -1 & 1 \\ 1 & 1 \end{bmatrix}$

2. $\begin{bmatrix} 1 & 2 \\ 3 & 4 \end{bmatrix}$

3. $\begin{bmatrix} 1 & 0 & 1 \\ 1 & -1 & 0 \\ 2 & 0 & 1 \end{bmatrix}$

4. $\begin{bmatrix} -1 & 2 & 3 \\ 0 & 1 & 1 \\ -2 & 1 & 1 \end{bmatrix}$

5. $\begin{bmatrix} 1 & 0 & 0 \\ 4 & 1 & 0 \\ 2 & 1 & 1 \end{bmatrix}$

6. $\begin{bmatrix} 1 & 2 & 1 \\ 0 & 1 & 1 \\ 0 & -1 & 1 \end{bmatrix}$

7. $\begin{bmatrix} 1 & -1 & 1 \\ 2 & 0 & 1 \\ 3 & -1 & 2 \end{bmatrix}$

8. $\begin{bmatrix} 2 & 1 & 0 \\ 0 & 2 & -1 \\ 2 & 1 & 1 \end{bmatrix}$

9. $\begin{bmatrix} 1 & 0 & 1 & 1 \\ 0 & 2 & 1 & 1 \\ 0 & 0 & 1 & -1 \\ 0 & 0 & 0 & 3 \end{bmatrix}$

10. $\begin{bmatrix} 1 & 0 & 0 & -1 \\ 2 & 1 & 0 & 0 \\ -1 & 0 & 1 & 0 \\ 0 & 0 & 0 & 2 \end{bmatrix}$

11. Given that

$$A^{-1} = \begin{bmatrix} -1 & 3 \\ 2 & 1 \end{bmatrix}$$

find A.

12. Use the inverse matrix and (5.13) to solve the linear systems:
(a)

$$\begin{aligned} 2x + y &= -1 \\ x + 2y &= 1 \end{aligned}$$

(b)

$$\begin{aligned} x + 2y + 1 &= 2 \\ 2x + y + z &= 0 \\ x + 3y + z &= 3 \end{aligned}$$

13. Show that the inverse of the matrix

$$A = \begin{bmatrix} a & b \\ c & d \end{bmatrix}$$

is given by

$$A^{-1} = \frac{1}{D}\begin{bmatrix} d & -b \\ -c & a \end{bmatrix}$$

provided the quantity $D = ad - bc \neq 0$

14. Given that a square matrix A satisfies the matrix equation $A^2 - A - I = 0$, show that A is invertible, and find A^{-1}.

15. If A is invertible, then show that for any scalar $\lambda \neq 0$,

$$(\lambda A)^{-1} = \frac{1}{\lambda}A^{-1}$$

16. Given the $n \times n$ diagonal matrix

$$A = \begin{bmatrix} a_{11} & 0 & 0 & 0 \\ 0 & a_{22} & 0 & 0 \\ \vdots & \vdots & \vdots & \vdots \\ 0 & 0 & 0 & a_{nn} \end{bmatrix}$$

show that A is invertible provided $a_{11}a_{22} \cdots a_{nn} \neq 0$, and find A^{-1}.

17. *(Cancellation Laws)* Prove that if X is an invertible matrix, then
(a) $AX = BX$ implies $A = B$
(b) $XA = XB$ implies $A = B$

18. Prove that if the $n \times n$ homogeneous system $A\mathbf{x} = \mathbf{0}$ has at least one non-trivial solution, then A must be singular.

Unit II: Determinants

Lesson 6 Introduction to Determinants

Lesson 7 Properties of Determinants

Lesson 8 Applications of Determinants

Colin Maclaurin (1698-1746)

The theory of determinants gradually developed during the 17th and 18th centuries. This development was due to the diligent work of many famous mathematicians including Leibnitz, Laplace, Euler, and Vandermonde. It is curious to note that although matrices are the logical predecessors of determinants, the development of determinant theory actually preceded the formal birth of matrix theory.

The use of determinants in the solution of linear systems in two, three, and four unknowns was first elucidated by the notable Scottish mathematician Colin Maclaurin. Like many of his mathematical cohorts, Maclaurin was a child prodigy. He entered the University of Glasgow at just eleven years of age, and completed his MA degree just three years later.

Maclaurin developed a method of solving linear systems with determinants in 1729; however, it wasn't until 1748 that this important work was formally published in his *Treatise of Algebra.* Maclaurin's method was later generalized by the mathematician Gabriel Cramer (1704-1752), and today it is known as Cramer's Rule.

6. Introduction to Determinants

In this lesson, we introduce a special type of mapping, called the *determinant function*, which maps each *square* matrix into a unique scalar. In our future work, the determinant function will be instrumental to our understanding of systems of linear equations, and in exploring the properties of matrices.

Before we can do so, however, we need to introduce the concept of a *permutation*, and what is commonly called the *Levi-Civita permutation symbol*.

Definition: *(Permutation)*

A **permutation** π of a set S is a one-to-one mapping of S onto itself; that is, it is some *arrangement* of *all* of the elements in S, without any repetitions. In particular, a permutation of the set of integers $S_n = \{1,2,3,\ldots,n\}$ is an arrangement of all of these integers, without any repetitions.

Example-1: Find all permutations of the set $S_3 = \{1,2,3\}$.

Solution: By inspection, there are six permutations of S_3; they are:

$$(1,2,3), (3,1,2), (2,3,1)$$
$$(3,2,1), (1,3,2), (2,1,3)$$

By the Fundamental Counting Principle, it's easy to see that there are exactly $n!$ permutations of the elements in the set S_n. As in the above example, to denote any permutation π of S_n, we shall use the shorthand notation

$$\pi = (i_1, i_2, \ldots, i_n) \tag{6.1a}$$

where i_1 is the first integer in the permutation, i_2 is the second integer in the permutation, and so on.

We now introduce the concept of *inversion.* Given any permutation such as (6.1a), we say that an **inversion** occurs in the given permutation whenever a larger integer *precedes* a smaller integer. Now, we can find the total number of inversions in the above permutation in the following manner: let k_1 denote the number of integers less than i_1 and that follow i_1 in π, let k_2 denote the number of integers less than i_2 and that follow i_2 in π, and so on. Then the total number of inversions in π is just the *sum* of these numbers, i.e.,

$$\text{Total Number of Inversions} = k_1 + k_2 + \ldots + k_{n-1} \tag{6.1b}$$

Example-2: Find the number of inversions in each of the following permutations:
(a) $(4,2,3,1)$
(b) $(3,5,4,2,1)$
(c) $(1,3,6,2,4,5)$

Solution:
(a) The total number of inversions is $3+1+1=5$.
(b) The total number of inversions is $2+3+2+1=8$.
(c) The total number of inversions is $0+1+3+0+0=4$.

Definition: *(Parity of a Permutation)*

Given any permutation $\pi = (i_1, i_2, \ldots, i_n)$ of the set of integers $S_n = \{1,2,3,\ldots,n\}$, we shall say that π is **even** (or has **even parity**) if the total number of inversions in π is an *even* number, and say that π is **odd** (or has **odd parity**) if the total number of inversions in π is an odd number.

Example-3: We list each permutation of $S_3 = \{1,2,3\}$, along with its parity:

Permutation	Total Number of Inversions	Parity
$(1,2,3)$	0	even
$(3,1,2)$	2	even
$(2,3,1)$	2	even
$(3,2,1)$	3	odd
$(1,3,2)$	1	odd
$(2,1,3)$	1	odd

Definition: *(Levi-Civita Permutation Symbol)*

The (Levi-Civita) **permutation symbol** $\varepsilon_{i_1 i_2 i_3 \ldots i_n}$ **of order** n is defined by:

$$\varepsilon_{i_1 i_2 i_3 \ldots i_n} = \begin{cases} +1, \text{ if } (i_i, i_2, i_3, \ldots, i_n) \text{ is an even permutation of } S_n \\ -1, \text{ if } (i_i, i_2, i_3, \ldots, i_n) \text{ is an odd permutation of } S_n \\ 0, \text{ otherwise} \end{cases} \tag{6.2}$$

Example-4: Evaluate each of the following permutation symbols:
(a) ε_{312}

(b) ε_{3142}
(c) ε_{1342}
(d) ε_{1123}

Solution:
(a) $\varepsilon_{312} = +1$ since $(3,1,2)$ is an even permutation of S_3.
(b) $\varepsilon_{3142} = -1$ since $(3,1,4,2)$ is an odd permutation of S_4.
(c) $\varepsilon_{1342} = +1$ since $(1,3,4,2)$ is an even permutation of S_4.
(d) $\varepsilon_{1123} = 0$ since $(1,1,2,3)$ is **not** a permutation of S_4 (since the element 1 is repeated).

From the previous example, it's easy to see that the permutation symbol is **anti-symmetric** (or **skew symmetric**); that is, it *changes its sign whenever any two of its indices are interchanged; and it is identically zero if any two of its indices are equal.*

Finally, we are now in a position to define the determinant of a square matrix.

Definition: *(Determinant)*

Given a square matrix $A = [a_{ij}]$ of size n, the **determinant** of A, denoted by $\det(A)$ or $|A|$, is given by

$$\det(A) = \sum_{i_1,i_2,i_3,\ldots,i_n=1}^{n} \varepsilon_{i_1 i_2 i_3 \ldots i_n} a_{1i_1} a_{2i_2} a_{3i_3} \cdots a_{ni_n} \tag{6.3a}$$

or equivalently, by

$$\det(A) = \sum_{i_1,i_2,i_3,\ldots,i_n=1}^{n} \varepsilon_{i_1 i_2 i_3 \ldots i_n} a_{i_1 1} a_{i_2 2} a_{i_3 3} \cdots a_{i_n n} \tag{6.3b}$$

In this definition, it is important to note that the sum which appears in both (6.3a) and (6.3b) denotes and the sum over all of the indices; thus, when we evaluate the determinant of a square matrix of size n, the sum will necessarily involve a total of $n!$ terms.

In particular, for the special case of a 1×1 matrix, say $A = [a_{11}]$, we simply define $\det(A) = a_{11}$. Furthermore, given a square matrix $A = [a_{ij}]$ of size n, it is customary to write

$$|A| \equiv \det(A) = \begin{vmatrix} a_{11} & a_{12} & \cdots & a_{1n} \\ a_{21} & a_{22} & \cdots & a_{2n} \\ \cdots & \cdots & \cdots & \cdots \\ a_{n1} & a_{n2} & \cdots & a_{nn} \end{vmatrix} \tag{6.4}$$

where vertical bars are used to denote the determinant of A. Also, it is customary to refer to (6.4) as a determinant of **order** n.

Example-5: Show that

$$\begin{vmatrix} a_{11} & a_{12} \\ a_{21} & a_{22} \end{vmatrix} = a_{11}a_{22} - a_{12}a_{21} \tag{6.5}$$

Solution: From (6.3a), we can write

$$\begin{aligned} \det(A) &= \sum_{i,j=1}^{2} \varepsilon_{ij}a_{1i}a_{2j} = \varepsilon_{12}a_{11}a_{22} + \varepsilon_{21}a_{12}a_{21} \\ &= (+1)a_{11}a_{22} + (-1)a_{12}a_{21} \\ &= a_{11}a_{22} - a_{12}a_{21} \end{aligned}$$

The reader should verify that one obtains the same result by (6.3b).

Example-6: Evaluate the determinant

$$\begin{vmatrix} a_{11} & a_{12} & a_{13} \\ a_{21} & a_{22} & a_{23} \\ a_{31} & a_{32} & a_{33} \end{vmatrix}$$

Solution: From (6.3a), we have

$$\begin{aligned} \det(A) &= \sum_{i,j,k=1}^{3} \varepsilon_{ijk}a_{1i}a_{2j}a_{3k} \\ &= \varepsilon_{123}a_{11}a_{22}a_{33} + \varepsilon_{312}a_{13}a_{21}a_{32} + \varepsilon_{231}a_{12}a_{23}a_{31} \\ &\quad + \varepsilon_{321}a_{13}a_{22}a_{31} + \varepsilon_{132}a_{11}a_{23}a_{32} + \varepsilon_{213}a_{12}a_{21}a_{33} \\ &= a_{11}a_{22}a_{33} + a_{13}a_{21}a_{32} + a_{12}a_{23}a_{31} - a_{13}a_{22}a_{31} \\ &\quad - a_{11}a_{23}a_{32} - a_{12}a_{21}a_{33} \end{aligned}$$

Consequently,

$$\begin{vmatrix} a_{11} & a_{12} & a_{13} \\ a_{21} & a_{22} & a_{23} \\ a_{31} & a_{32} & a_{33} \end{vmatrix} = a_{11}(a_{22}a_{33} - a_{23}a_{32}) - a_{12}(a_{21}a_{33} - a_{23}a_{31}) + a_{13}(a_{21}a_{32} - a_{22}a_{31}) \tag{6.6}$$

The last example shows that evaluating the determinant of order n (where $n \geq 3$) can be quite tedious, and is easily prone to error. Consequently, we now turn our attention to the various ways that we can simplify such computations.

Definition: *(Minors and Cofactors)*

Given any square matrix A of size n (where $n \geq 2$), the **minor** M_{ij} of the element a_{ij} is defined to be the determinant of the $(n-1) \times (n-1)$ matrix which results from deleting the i-th row and j-th column of A.

Furthermore, the **cofactor** C_{ij} of the element a_{ij} is given by

$$C_{ij} = (-1)^{i+j} M_{ij} \tag{6.7}$$

Example-7: Given the matrix

$$A = \begin{bmatrix} 4 & 3 & 2 \\ 5 & 1 & 7 \\ 8 & 4 & 9 \end{bmatrix}$$

find the cofactors of the elements a_{11} and a_{31}.

Solution:

$$C_{11} = (-1)^{1+1} M_{11} = \begin{vmatrix} 1 & 7 \\ 4 & 9 \end{vmatrix} = (1 \cdot 9) - (7 \cdot 4) = -19$$

$$C_{31} = (-1)^{3+1} M_{31} = \begin{vmatrix} 3 & 2 \\ 1 & 7 \end{vmatrix} = (3 \cdot 7) - (2 \cdot 1) = 19$$

Now, observe that we can use cofactors to rewrite (6.6) in the form

$$\begin{vmatrix} a_{11} & a_{12} & a_{13} \\ a_{21} & a_{22} & a_{23} \\ a_{31} & a_{32} & a_{33} \end{vmatrix} = a_{11} \begin{vmatrix} a_{22} & a_{23} \\ a_{32} & a_{33} \end{vmatrix} - a_{12} \begin{vmatrix} a_{21} & a_{23} \\ a_{31} & a_{33} \end{vmatrix} + a_{13} \begin{vmatrix} a_{21} & a_{22} \\ a_{31} & a_{32} \end{vmatrix}$$

$$= a_{11}C_{11} + a_{12}C_{12} + a_{13}C_{13}$$

In other words, we can calculate $|A|$ by simply multiplying each element in the first row by its corresponding cofactor, and then adding the resulting products.

Similarly, by rearranging terms in (6.6), it's easy to show that

$$
\begin{aligned}
|A| &= a_{21}C_{21} + a_{22}C_{22} + a_{23}C_{23} \\
&= a_{31}C_{31} + a_{32}C_{32} + a_{33}C_{33}
\end{aligned}
$$

as well. So, we see that we can calculate $|A|$ by simply multiplying each element in *any* row we like by its corresponding cofactor, and then adding the resulting products. The reader should also verify that this procedure works equally well for any given *column*; that is,

$$
\begin{aligned}
|A| &= a_{11}C_{11} + a_{12}C_{12} + a_{13}C_{13} \\
&= a_{21}C_{21} + a_{22}C_{22} + a_{23}C_{23} \\
&= a_{31}C_{31} + a_{32}C_{12} + a_{33}C_{13}
\end{aligned}
$$

It may be shown that this computational scheme works for any square matrix of size $n \geq 2$. We summarize this general procedure, called **expansion by cofactors**, in the following theorem, whose proof we omit.

Theorem-6.1: *(Expansion by Cofactors)*

If A is any square matrix of size $n \geq 2$, then its determinant can be calculated by multiplying each element in any given row (or column) by its corresponding cofactor, and then adding the resulting products. That is, we have both:

$$|A| = \sum_{j=1}^{n} a_{ij}C_{ij} = a_{i1}C_{i1} + a_{i2}C_{i2} + \ldots + a_{in}C_{in} \quad \text{(expansion on } i\text{-th row)} \tag{6.8a}$$

$$|A| = \sum_{i=1}^{n} a_{ij}C_{ij} = a_{1j}C_{1j} + a_{2j}C_{2j} + \ldots + a_{nj}C_{nj} \quad \text{(expansion on } j\text{-th column)} \tag{6.8b}$$

Example-8: Compute the determinant

$$
\begin{vmatrix}
4 & 3 & 2 \\
5 & 1 & 7 \\
8 & 4 & 9
\end{vmatrix}
$$

in two different ways, i.e., by expanding it along the first row, and by expanding it along the second column.

Solution: Expanding along the first row, we obtain:

$$\begin{vmatrix} 4 & 3 & 2 \\ 5 & 1 & 7 \\ 8 & 4 & 9 \end{vmatrix} = 4\begin{vmatrix} 1 & 7 \\ 4 & 9 \end{vmatrix} - 3\begin{vmatrix} 5 & 7 \\ 8 & 9 \end{vmatrix} + 2\begin{vmatrix} 5 & 1 \\ 8 & 4 \end{vmatrix}$$

$$= 4(-19) - 3(-11) + 2(12) = -19$$

Similarly, expanding along the second column, we get

$$\begin{vmatrix} 4 & 3 & 2 \\ 5 & 1 & 7 \\ 8 & 4 & 9 \end{vmatrix} = -3\begin{vmatrix} 5 & 7 \\ 8 & 9 \end{vmatrix} + 1\begin{vmatrix} 4 & 2 \\ 8 & 9 \end{vmatrix} - 4\begin{vmatrix} 4 & 2 \\ 5 & 7 \end{vmatrix}$$

$$= -3(-11) + 1(20) - 4(18)$$

$$= -19$$

Clearly, other expansions are possible.

Exercise Set 6

In exercises 1-5, use expansion by minors to evaluate each of the following determinants:

1. $\begin{vmatrix} 4 & 3 \\ 1 & 3 \end{vmatrix}$

2 $\begin{vmatrix} 0 & 1 & 4 \\ -1 & 3 & 2 \\ 3 & 1 & 4 \end{vmatrix}$

3. $\begin{vmatrix} 1 & x & x^2 \\ 1 & x & x^2 \\ 1 & x & x^2 \end{vmatrix}$

4. $\begin{vmatrix} 0 & -1 & 8 \\ 2 & 1 & 3 \\ 1 & 0 & 1 \end{vmatrix}$

5. $\begin{vmatrix} 1 & 0 & 0 \\ 0 & \sin t & -\cos t \\ 0 & \cos t & \sin t \end{vmatrix}$

6. Given the matrix

$$A = \begin{bmatrix} 1 & 0 & 1 \\ -1 & 1 & 0 \\ 0 & -1 & 1 \end{bmatrix}$$

Verify by direct computation that $\det(A^T) = \det(A)$.

7. Let A be the 2×2 matrix

$$A = \begin{bmatrix} a & b \\ c & d \end{bmatrix}$$

where $\det(A) \neq 0$. Show each of the following:

(a)

$$A^{-1} = \frac{1}{\det(A)} \begin{bmatrix} d & -b \\ -c & a \end{bmatrix}$$

(b)

$$\det(A^{-1}) = \frac{1}{\det(A)} = \frac{1}{ad - bc}$$

8. If A is any 2×2 matrix, then show that

$$\det(A) = \det(A^T)$$

9. If A and B are arbitrary 2×2 matrices, then show that

$$\det(AB) = \det(A) \cdot \det(B)$$

10. Evaluate the determinant

$$\begin{vmatrix} 1 & 0 & 0 & 0 \\ 0 & 2 & 1 & 1 \\ 0 & 0 & 3 & 1 \\ 0 & 0 & 0 & 4 \end{vmatrix}$$

11. Solve the following determinantal equation for λ:

$$\begin{vmatrix} \lambda & 0 & 1 \\ 0 & (\lambda - 1) & 1 \\ 0 & -1 & (\lambda - 2) \end{vmatrix} = 0$$

12. Show that

$$\begin{vmatrix} 1 & a & a^2 \\ 1 & b & b^2 \\ 1 & c & c^2 \end{vmatrix} = (b - a)(c - a)(c - b)$$

7. Properties of Determinants

In this lesson, we explore some of the important properties of determinants. As we might expect, most of these properties are a direct result of the very definition of the determinant, along with the properties of the Levi-Civita permutation symbol.

Theorem-7.1: *(Determinant of a Transpose)*

If A is any $n \times n$ matrix, then

$$\det(A) = \det(A^T) \tag{7.1}$$

Proof: Let $B = A^T$, so that by the definition of a determinant, we have

$$\begin{aligned}\det(B) &= \sum_{i_1,i_2,i_3,\ldots,i_n=1}^{n} \varepsilon_{i_1 i_2 i_3 \ldots i_n} b_{1i_1} b_{2i_2} b_{3i_3} \cdots b_{ni_n} \\ &= \sum_{i_1,i_2,i_3,\ldots,i_n=1}^{n} \varepsilon_{i_1 i_2 i_3 \ldots i_n} a_{i_1 1} a_{i_2 2} a_{i_3 3} \cdots a_{i_n n} \\ &= \det(A)\end{aligned}$$

by virtue of (6.3b). □.

Theorem-7.2: *(The Result of Interchanging the Rows of a Determinant)*

Let A be any $n \times n$ matrix. If we interchange any two rows (or columns) of A to obtain a matrix B, then $\det(B) = -\det(A)$.

Proof: For definiteness, let B be the matrix that results from interchanging the first and second rows of A. Consequently, we have

$$\begin{aligned}\det(B) &= \sum_{i_1,i_2,i_3,\ldots,i_n=1}^{n} \varepsilon_{i_1 i_2 i_3 \ldots i_n} b_{1i_1} b_{2i_2} b_{3i_3} \cdots b_{ni_n} \\ &= \sum_{i_1,i_2,i_3,\ldots,i_n=1}^{n} \varepsilon_{i_1 i_2 i_3 \ldots i_n} a_{2i_1} a_{1i_2} a_{3i_3} \cdots a_{ni_n} \\ &= -\sum_{i_1,i_2,i_3,\ldots,i_n=1}^{n} \varepsilon_{i_2 i_1 i_3 \ldots i_n} a_{1i_2} a_{2i_1} a_{3i_3} \cdots a_{ni_n} \\ &= -\det(A)\end{aligned}$$

where, in the next to the last step, we used the skew-symmetry of the permutation symbol. Clearly, the same result holds true if we interchange any other two rows of A. Finally, if we interchange any two *columns* the same result follows by virtue of Theorem-7.1. □.

This theorem implies that if we perform the elementary row operation that consists of interchanging any two rows of a determinant, then we can do so along as we remember to change the sign of the resulting determinant.

Lemma 7.3: *(Determinant of a Matrix with Two Rows Equal)*

If A is an $n \times n$ matrix such that any two of its rows (or columns) are equal, then $\det(A) = 0$.

Proof: For brevity of notation, suppose that the first and second rows of A are equal. Then, we have

$$\begin{aligned}\det(A) &= \sum_{j_1, j_2, j_3, \ldots, j_n = 1}^{n} \varepsilon_{j_1 j_2 j_3 \ldots j_n} a_{1j_1} a_{2j_2} a_{3j_3} \cdots a_{nj_n} \\ &= \sum_{j_1, j_2, j_3, \ldots, j_n = 1}^{n} \varepsilon_{j_1 j_2 j_3 \ldots j_n} (a_{1j_1} a_{1j_2}) a_{3j_3} \cdots a_{nj_n} \\ &= 0\end{aligned}$$

where the last step follows from the fact that the expression $a_{1j_1} a_{1j_2}$ is symmetric in the indices $j_1 j_2$ while the permutation symbol is skew symmetric in each of its indices. Clearly, the same result follows for any other two rows of A. The fact that the theorem also holds when any two columns are equal follows immediately from Theorem 7.1. □.

Next, we state and prove a useful alternative expression for the determinant of a square matrix. As we will see, this expression is useful primarily for theoretical purposes.

Theorem 7.4: *(A Useful Expression for* $\det(A)$*)*

If A is any $n \times n$ matrix, then

$$\varepsilon_{i_1 i_2 i_3 \ldots i_n} \det(A) = \sum_{j_1, j_2, j_3, \ldots, j_n = 1}^{n} \varepsilon_{j_1 j_2 j_3 \ldots j_n} a_{i_1 j_1} a_{i_2 j_2} a_{i_3 j_3} \cdots a_{i_n j_n} \tag{7.2}$$

Proof: Observe that for any specific choice of *distinct* values for the indices $i_1, i_2, \ldots, i_n$, we obtain $\pm\det(A)$ on both sides of (7.2). If any two of the indices $i_1, i_2, \ldots, i_n$ are equal, then the lefthand side is zero, and at the same time, the right-hand side is zero due to Lemma 7.3. □.

Theorem 7.5

Let A be any $n \times n$ matrix. If we add a scalar multiple of any row of A to a different row of A to obtain a matrix B, then $\det(A) = \det(B)$.

Proof: For definiteness, assume that we obtain the matrix B by multiplying the second row of A by a scalar λ, and then adding the result to the third row of A. Then,

$$\det(B) = \sum_{j_1, j_2, j_3, \ldots, j_n = 1}^{n} \varepsilon_{j_1 j_2 j_3 \ldots j_n} b_{1j_1} b_{2j_2} b_{3j_3} \cdots b_{nj_n}$$

and consequently,

$$\begin{aligned}\det(B) &= \sum_{j_1, j_2, j_3, \ldots, j_n = 1}^{n} \varepsilon_{j_1 j_2 j_3 \ldots j_n} a_{1j_1} a_{2j_2} (a_{3j_3} + \lambda a_{2j_3}) \cdots a_{nj_n} \\ &= \det(A) + \lambda \left(\sum_{j_1, j_2, j_3, \ldots, j_n = 1}^{n} \varepsilon_{j_1 j_2 j_3 \ldots j_n} a_{1j_1} a_{2j_2} a_{2j_3} \cdots a_{nj_n} \right) \\ &= \det(A)\end{aligned}$$

since the second term is zero due to the fact that the second and third rows of determinant are identical. Clearly, the same result will be obtained for any other two distinct rows of A. □.

Observe that if we are given any determinant, and we perform the elementary row operation which involves multiplying a given row by a non-zero constant, and adding the result to another row, then Theorem 7.5 guarantees that the value of the determinant is unaffected.

The next theorem shows that the determinant of the product of any two square matrices (of the same size) is equal to the product of their respective determinants.

Theorem 7.6: *(Determinant of a Product)*

If A and B are both $n \times n$ matrices, then

$$\det(AB) = \det(A) \cdot \det(B) \tag{7.3}$$

In other words, the determinant of a product is equal to the product of the determinants.

Proof: Let $C = [c_{ij}] = AB$ so that

$$
\begin{aligned}
\det(C) &= \sum_{j_1,j_2,j_3,\ldots,j_n=1}^{n} \varepsilon_{j_1j_2j_3\ldots j_n} c_{1j_1} c_{2j_2} c_{3j_3} \cdots c_{nj_n} \\
&= \sum_{j_1,j_2,j_3,\ldots,j_n=1}^{n} \varepsilon_{j_1j_2j_3\ldots j_n} \left(\sum_{p_1=1}^{n} a_{1p_1} b_{p_1j_1} \right) \left(\sum_{p_2=1}^{n} a_{2p_2} b_{p_2j_2} \right) \cdots \left(\sum_{p_n=1}^{n} a_{np_n} b_{p_nj_n} \right) \\
&= \sum_{p_1,p_2,p_3,\ldots,p_n=1}^{n} a_{1p_1} a_{2p_2} \cdots a_{np_n} \left(\sum_{j_1,j_2,j_3,\ldots,j_n=1}^{n} \varepsilon_{j_1j_2j_3\ldots j_n} b_{p_1j_1} b_{p_2j_2} \cdots b_{p_nj_n} \right) \\
&= \sum_{p_1,p_2,p_3,\ldots,p_n=1}^{n} a_{1p_1} a_{2p_2} \cdots a_{np_n} \left(\varepsilon_{p_1p_2p_3\ldots p_n} \cdot \det(B) \right) \\
&= \left(\sum_{p_1,p_2,p_3,\ldots,p_n=1}^{n} \varepsilon_{p_1p_2p_3\ldots p_n} a_{1p_1} a_{2p_2} \cdots a_{np_n} \right) \cdot \det(B) \\
&= \det(A) \cdot \det(B)
\end{aligned}
$$

where (6.3a) was used in the last step. □.

Finally, the last important property of determinants we shall discuss is the determinant of a triangular matrix. As the next theorem shows, if we are given an $n \times n$ upper triangular (or lower triangular) matrix, and we want to find its determinant, we only have to multiply the elements that lie on its main diagonal. This property is often helpful in evaluating determinants, since if we apply elementary row operations to a messy determinant, we can always put it into triangular form.

Theorem 7.7: *(Determinant of a Triangular Matrix)*

If A is an $n \times n$ upper triangular matrix, i.e.,

$$
A = \begin{bmatrix} a_{11} & a_{12} & a_{13} & \cdots & a_{1n} \\ 0 & a_{22} & a_{23} & \cdots & a_{2n} \\ 0 & 0 & a_{33} & \cdots & a_{3n} \\ \vdots & \vdots & \vdots & \vdots & \vdots \\ 0 & 0 & 0 & 0 & a_{nn} \end{bmatrix}
$$

then

$$
\det(A) = a_{11} a_{22} a_{33} \cdots a_{nn} \tag{7.4}
$$

Also, the same result holds true if A is a lower triangular matrix.

Proof: We repeatedly expand $\det(A)$, along with its cofactors, along the first column, to obtain

$$\det(A) = a_{11} \cdot \begin{vmatrix} a_{22} & a_{23} & a_{24} & \cdots & a_{2n} \\ 0 & a_{33} & a_{34} & \cdots & a_{3n} \\ 0 & 0 & a_{44} & \cdots & a_{4n} \\ \vdots & \vdots & 0 & \vdots & \vdots \\ 0 & 0 & 0 & 0 & a_{nn} \end{vmatrix}$$

$$= a_{11}a_{22} \cdot \begin{vmatrix} a_{33} & a_{34} & a_{35} & \cdots & a_{3n} \\ 0 & a_{44} & a_{45} & \cdots & a_{4n} \\ 0 & 0 & a_{55} & \cdots & a_{5n} \\ \vdots & \vdots & \vdots & \vdots & \vdots \\ 0 & 0 & 0 & 0 & a_{nn} \end{vmatrix}$$

$$\vdots$$

$$= a_{11}a_{22} \cdots a_{nn}$$

Similarly, we can show that the same result holds for a lower triangular matrix. □.

Example-1: Evaluate the determinant:

$$\begin{vmatrix} 2 & 0 & 4 \\ 1 & 2 & 8 \\ 1 & 1 & 3 \end{vmatrix}$$

Solution: We use a sequence of elementary row operations to reduce the determinant to upper triangular form:

$$\begin{vmatrix} 2 & 0 & 4 \\ 1 & 2 & 8 \\ 1 & 1 & 3 \end{vmatrix} = 2 \cdot \begin{vmatrix} 1 & 0 & 2 \\ 1 & 2 & 8 \\ 1 & 1 & 3 \end{vmatrix} \quad \text{(Factor out 2 from row 1)}$$

$$= 2 \cdot \begin{vmatrix} 1 & 0 & 2 \\ 0 & 2 & 6 \\ 0 & 1 & 1 \end{vmatrix} \quad \text{(Take } -R_1 + R_2 \text{ and } -R_1 + R_3\text{)}$$

$$= 2 \cdot 2 \cdot \begin{vmatrix} 1 & 0 & 2 \\ 0 & 1 & 3 \\ 0 & 1 & 1 \end{vmatrix} \quad \text{(Factor out 2 from row 2)}$$

$$= 2 \cdot 2 \cdot \begin{vmatrix} 1 & 0 & 2 \\ 0 & 1 & 3 \\ 0 & 0 & -2 \end{vmatrix} \quad (\text{Take } -R_2 + R_3)$$

$$= 2 \cdot 2 \cdot 1 \cdot 1 \cdot (-2) \quad (\text{By } (7.4))$$

$$= -8$$

Exercise Set 7

---

1. Verify that $\det(A) = \det(A^T)$ for the matrix:

$$A = \begin{bmatrix} 1 & 0 & 3 \\ 4 & 2 & 1 \\ -1 & 2 & 1 \end{bmatrix}$$

2. If A is an $n \times n$ matrix, and λ is a scalar, show that

$$\det(\lambda A) = \lambda^n \det(A)$$

3. If $A_1, A_2, \ldots, A_n$ are $n \times n$ matrices, show that

$$\det(A_1 A_2 \cdots A_n) = \det(A_1)\det(A_2) \cdots \det(A_n)$$

4. If A is an $n \times n$ matrix, then show that

$$\det(A^k) = [\det(A)]^k$$

where k is any positive integer.

5. Given the matrices

$$A = \begin{bmatrix} 1 & 2 & 1 \\ 0 & 1 & 3 \\ 1 & -1 & 2 \end{bmatrix} \quad \text{and} \quad B = \begin{bmatrix} -1 & 0 & 1 \\ 2 & -1 & 3 \\ 1 & 1 & 2 \end{bmatrix}$$

verify that $\det(AB) = \det(A) \cdot \det(B)$.

6. Evaluate each of the given determinants by reducing them to triangular form, and then using equation (7.4):

(a)

$$\begin{vmatrix} 1 & 2 & 3 \\ 2 & 0 & 1 \\ 1 & -1 & 2 \end{vmatrix}$$

(b)

$$\begin{vmatrix} 2 & 0 & 6 \\ 1 & 3 & 2 \\ 4 & 1 & 1 \end{vmatrix}$$

(c)

$$\begin{vmatrix} 1 & 2 & 0 & 1 \\ 1 & -1 & 1 & 0 \\ 2 & 1 & 3 & 1 \\ 0 & 1 & 2 & 3 \end{vmatrix}$$

7. If A is an $n \times n$ anti-symmetric matrix, then show that

$$\det(A) = (-1)^n \det(A)$$

8. If A is an $n \times n$ anti-symmetric matrix where n is an odd counting number, then we must have $\det(A) = 0$.

9. If A is an invertible $n \times n$ matrix, then show that

$$\det(A^{-1}) = \frac{1}{\det(A)}$$

10. *(Derivative of a Determinant)* If A is a 2×2 matrix of the form

$$A = \begin{bmatrix} a_{11}(t) & a_{12}(t) \\ a_{21}(t) & a_{22}(t) \end{bmatrix}$$

whose elements are differentiable functions of a parameter t, show that

$$\frac{d}{dt}(\det A) = \begin{vmatrix} \frac{d}{dt}a_{11}(t) & \frac{d}{dt}a_{12}(t) \\ a_{21}(t) & a_{22}(t) \end{vmatrix} + \begin{vmatrix} a_{11}(t) & a_{12}(t) \\ \frac{d}{dt}a_{21}(t) & \frac{d}{dt}a_{22}(t) \end{vmatrix}$$

8. Applications of Determinants

In this lesson, we examine some of the applications of determinants. In particular, we will see how determinants can be useful in finding the inverses of square matrices and in solving linear systems. But first, we need to discuss some preliminary items.

Definition: *(Cofactor Matrix)*

Let $A = [a_{ij}]$ be any a square matrix of order n, and construct the square matrix $C = [C_{ij}]$ such that each element C_{ij} is the *cofactor* of the element a_{ij} of A, then C is called the **cofactor matrix** of A.

Example-1: Given the matrix

$$B = \begin{bmatrix} 1 & 2 \\ 0 & 4 \end{bmatrix}$$

find the cofactor matrix of B.

Solution: Let C denote the matrix of cofactors of B. The cofactors of B are

$$C_{11} = 4, \quad C_{12} = 0,$$
$$C_{21} = -2, \quad C_{22} = 1$$

so the required cofactor matrix is

$$C = \begin{bmatrix} 4 & 0 \\ -2 & 1 \end{bmatrix}$$

Example-2: Find the cofactor matrix of

$$A = \begin{bmatrix} 1 & 0 & 1 \\ 2 & 1 & 3 \\ 4 & 1 & 2 \end{bmatrix}$$

Solution: The cofactors of A are:

$$C_{11} = -1, \quad C_{12} = 8, \quad C_{13} = -2$$
$$C_{21} = 1, \quad C_{22} = -2, \quad C_{23} = -1$$
$$C_{31} = -1, \quad C_{32} = -1, \quad C_{33} = 1$$

so that the cofactor matrix of A is

$$C = \begin{bmatrix} -1 & 8 & -2 \\ 1 & -2 & -1 \\ -1 & -1 & 1 \end{bmatrix}$$

Theorem 8.1: *(A Useful Relation)*

If A is any square matrix of order n, and let $C = [C_{ij}]$ be the matrix of cofactors of A, then

$$\sum_{k=1}^{n} a_{ik} C_{jk} = \delta_{ij} \cdot \det(A) \tag{8.1}$$

or equivalently,

$$A \cdot C^T = I \cdot \det(A) \tag{8.2}$$

where I is the identity matrix of order n.

Proof: We first show that

$$\sum_{k=1}^{n} a_{ik} C_{jk} = 0 \quad \text{for } i \neq j$$

Let's construct the matrix $A' = \left[a'_{ij}\right]$ which is identical to A, but with the exception that the j-th row of A' is identical to the i-th row of A. Since the i-th and j-th rows of A' are identical, then $\det(A') = 0$. In particular, if we expand A' along j-th row, then we must have

$$\sum_{k=1}^{n} a_{jk} C'_{jk} = 0$$

But by our construction, $a'_{jk} = a_{ik}$ and $C'_{jk} = C_{jk}$, so the last expression becomes

$$\sum_{k=1}^{n} a_{ik} C_{jk} = 0$$

which establishes our claim. On the other hand, if we simply expand A along its i-th row, we have

$$\det(A) = \sum_{k=1}^{n} a_{ik} C_{ik}$$

We can now combine the two previous results by writing

$$\sum_{k=1}^{n} a_{ik} C_{jk} = \delta_{ij} \cdot \det(A)$$

as claimed. □.

Theorem 8.2: *(Inverse of a Square Matrix)*

If A is any square matrix of order n such that $\det(A) \neq 0$, then A is *invertible*, and furthermore,

$$A^{-1} = \frac{1}{\det(A)} C^T \tag{8.3}$$

Proof: Observe that

$$A \cdot \left(\frac{1}{\det(A)} C^T \right) = \frac{1}{\det(A)} (A \cdot C^T) = \frac{1}{\det(A)} (I \cdot \det(A)) = I$$

as required. □.

Example-3: Find the inverse of the matrix A in the previous example.

Solution: We find that $\det(A) = -3$, so A^{-1} exists, and is given by

$$A^{-1} = \frac{1}{\det(A)} C^T = -\frac{1}{3} \begin{bmatrix} -1 & 8 & -2 \\ 1 & -2 & -1 \\ -1 & -1 & 1 \end{bmatrix}^T = \begin{bmatrix} \frac{1}{3} & -\frac{1}{3} & \frac{1}{3} \\ -\frac{8}{3} & \frac{2}{3} & \frac{1}{3} \\ \frac{2}{3} & \frac{1}{3} & -\frac{1}{3} \end{bmatrix}$$

While equation (8.3) provides us with a method of finding the inverse of any non-singular square matrix, this formula is primarily of *theoretical interest* since it is quite tedious to use for square matrices of order $n > 3$.

In fact, even in machine-based computations, when matrices are often used to model physical problems, it is often necessary to compute the inverses of relatively large square matrices of order $n \geq 64$. Clearly, any attempt to use equation (8.3) for such relatively large matrices would involve at least 64! or about 10^{89} floating-point operations! So, we see that the Gauss-Jordan method is clearly superior.

The previous theorem, does however, enable us to write down a general formula for the solution of linear systems in n-equations and n-unknowns; the formula is known as **Cramer's Rule**. The same caveat, however, applies to its use; this formula is primarlily of theoretical interest, as it is tedious to use when $n > 3$.

Theorem 8.3: *(Cramer's Rule)*

Consider the linear system $A\mathbf{x} = \mathbf{b}$ of n equations in n unknowns, i.e.,

$$\begin{bmatrix} a_{11} & a_{12} & \cdots & a_{1n} \\ a_{21} & a_{22} & \cdots & a_{2n} \\ \vdots & \vdots & \vdots & \vdots \\ a_{n1} & a_{n2} & \cdots & a_{22} \end{bmatrix} \begin{bmatrix} x_1 \\ x_2 \\ \vdots \\ x_n \end{bmatrix} = \begin{bmatrix} b_1 \\ b_2 \\ \vdots \\ b_n \end{bmatrix} \tag{8.4}$$

If $\det(A) \neq 0$, then this system has the *unique solution* given by

$$x_k = \frac{\det(A_k)}{\det(A)} \quad \text{for } k = 1,2,3,\ldots,n \tag{8.5}$$

where A_k is the matrix obtained from A by replacing the k-th column of A with the column vector $\mathbf{b}$.

Proof: Since $\det(A) \neq 0$, then A^{-1} exists. Consequently, $\mathbf{x} = A^{-1}\mathbf{b}$ so that

$$\begin{bmatrix} x_1 \\ x_2 \\ \vdots \\ x_n \end{bmatrix} = \frac{C^T}{\det(A)}\mathbf{b} = \frac{1}{\det(A)} \begin{bmatrix} C_{11} & C_{21} & \cdots & C_{n1} \\ C_{12} & C_{22} & \cdots & C_{2n} \\ \vdots & \vdots & \vdots & \vdots \\ C_n & C_{2n} & \vdots & C_{22} \end{bmatrix} \begin{bmatrix} b_1 \\ b_2 \\ \vdots \\ b_n \end{bmatrix}$$

Equating corresponding elements, we obtain

$$\begin{aligned} x_1 &= \frac{1}{\det(A)}(b_1C_{11} + b_2C_{21} + \ldots + b_nC_{n1}) \\ x_2 &= \frac{1}{\det(A)}(b_1C_{12} + b_2C_{22} + \ldots + b_nC_{n2}) \\ &\vdots \\ x_n &= \frac{1}{\det(A)}(b_1C_{1n} + b_2C_{2n} + \ldots + b_nC_{nn}) \end{aligned}$$

In other words,

$$x_k = \frac{1}{\det(A)}(b_1C_{1k} + b_2C_{2k} + \ldots + b_nC_{nk}) \quad \text{for } k = 1,2,3,\ldots,n$$

But for each fixed k, the expression in the parenthesis is just the $\det(A_k)$ which is obtained by expanding A_k along its k-th column.

This completes the proof of the theorem . □.

Example-4: Use Cramer's Rule to find the solution of the linear system

$$\begin{aligned} 2x_1 + 2x_2 + x_3 &= 2 \\ x_1 + 3x_2 + x_3 &= 5 \\ -x_1 + 2x_2 + 2x_3 &= 5 \end{aligned}$$

Solution: We find that

$$\det(A) = \begin{vmatrix} 2 & 2 & 1 \\ 1 & 3 & 1 \\ -1 & 2 & 2 \end{vmatrix} = 7, \quad \det(A_1) = \begin{vmatrix} 2 & 2 & 1 \\ 5 & 3 & 1 \\ 5 & 2 & 2 \end{vmatrix} = -7$$

$$\det(A_2) = \begin{vmatrix} 2 & 2 & 1 \\ 1 & 5 & 1 \\ -1 & 5 & 2 \end{vmatrix} = 14, \quad \det(A_3) = \begin{vmatrix} 2 & 2 & 2 \\ 1 & 3 & 5 \\ -1 & 2 & 5 \end{vmatrix} = 0$$

so that

$$x_1 = \frac{\det(A_1)}{\det(A)} = \frac{-7}{7} = -1,$$

$$x_2 = \frac{\det(A_2)}{\det(A)} = \frac{14}{7} = 2,$$

$$x_3 = \frac{\det(A_3)}{\det(A)} = \frac{0}{7} = 0$$

Exercise Set 8

---

1. *(A Useful Formula)* If A is the 2×2 matrix

$$A = \begin{bmatrix} a_{11} & a_{12} \\ a_{21} & a_{22} \end{bmatrix}$$

where $\det(A) \neq 0$, then show that

$$A^{-1} = \frac{1}{\det(A)} \begin{bmatrix} a_{22} & -a_{12} \\ -a_{21} & a_{11} \end{bmatrix} \tag{8.6}$$

2. Use the result of the previous problem to find the inverses of the 2×2

matrices:

(a)

$$\begin{bmatrix} \cos\theta & \sin\theta \\ -\sin\theta & \cos\theta \end{bmatrix}$$

(b)

$$\begin{bmatrix} 1 & 2 \\ 3 & 4 \end{bmatrix}$$

3. Use equation (8.3) to find the inverses of each of the following matrices:

(a)

$$\begin{bmatrix} 2 & 0 & 1 \\ -1 & 2 & 1 \\ 1 & 0 & 1 \end{bmatrix}$$

(b)

$$\begin{bmatrix} -1 & 2 & 0 \\ 1 & 1 & 3 \\ 2 & 2 & 4 \end{bmatrix}$$

(c)

$$\begin{bmatrix} 1 & 2 & 3 \\ 3 & 2 & 0 \\ -1 & 0 & 2 \end{bmatrix}$$

4. Use Cramer's rule to find the solutions of the following linear systems:

(a)

$$\begin{aligned} 2x_1 + 3x_2 &= -1 \\ 3x_1 + 2x_2 &= 1 \end{aligned}$$

(b)

$$\begin{aligned} x_1 + 2x_2 + x_3 &= 0 \\ 2x_1 + 3x_2 + x_3 &= 1 \\ x_1 + x_2 - x_3 &= 0 \end{aligned}$$

(c)

$$\begin{aligned} 2x_1 + x_2 - x_3 + x_4 &= 3 \\ x_1 + x_2 + x_3 - x_4 &= 1 \\ x_1 + x_2 &= 2 \\ x_2 + 2x_3 + x_4 &= 6 \end{aligned}$$

5. Find the values of λ for which the following matrix is *not* invertible:

$$\begin{bmatrix} 1-\lambda & 0 & 1 \\ 2 & 3-\lambda & -1 \\ 2 & 1 & 2-\lambda \end{bmatrix}$$

6. Let $A\mathbf{x} = \mathbf{b}$ be a linear system of n equations in n unknowns where the elements of matrix A and the column vector $\mathbf{b}$ are all integers. Show that if $\det(A) = 1$, then the system has a *unique* solution vector $\mathbf{x} = \mathbf{x}_0$ where the elements of $\mathbf{x}_0$ are all integers.

7. Let $A\mathbf{x} = \mathbf{0}$ be a linear homogeneous system of n equations in n unknowns. Prove that this system has a *non-trivial* solution if and only if $\det(A) = 0$.

8. Let A be an $n \times n$ matrix whose elements are all integers. Show that if $\det(A) = 1$, then A^{-1} has all integer elements as well.

9. Given the matrix

$$R = \begin{bmatrix} 1 & 0 & 0 \\ 0 & \cos\theta & -\sin\theta \\ 0 & \sin\theta & \cos\theta \end{bmatrix}$$

(a) By direct calculation, show that $\det(R) = 1$.

(b) Find its inverse, R^{-1}.

(c) Any square matrix A which satisfies the property

$$A^{-1} = A^T$$

is called an **orthogonal matrix**. Show that R is an orthogonal matrix.

10. If R is an orthogonal matrix, then show that $\det(R) = \pm 1$.

Unit III: A First Look at Vector Spaces

Giuseppe Peano (1858-1932)
(Photo taken circa 1910 by unknown photographer)

The first modern conception of the notion of a vector space is generally attributed to the Italian mathematician, Giuseppe Peano. Peano was born on a farm in a small village in Piedmont, Italy and later attended the Liceo Classico Cavour in Turin.

In 1876, he entered the University of Turin and completed his studies in 1880, graduating with honors. Later on, he was granted a full professorship at University of Turin, where he maintained this post until his death.

In his important book *Calcolo geometrico secondo l'Ausdehnungslere di H. Grassmann prededuto dalle operazioni della logica deduttiva*, which first appeared in 1888, Peano not only introduced modern set notation into mathematical logic, but in Chapter IX of this book, he furnished the first modern set of axioms for a vector space.

During his productive career, Peano published more than 200 papers and books on mathematics. He made major contributions to modern mathematical logic, the axiomatization of the natural numbers, analysis, differential equations, linear algebra and vector analysis.

9. Introduction to Vector Spaces

The reader is no doubt familiar with ordinary plane vectors and space vectors. In this section, we generalize the properties of such vectors by introducing the idea of an abstract *vector space*. Roughly speaking, a vector space is a set of objects, called *vectors*, on which two basic operations have been defined. These two operations are called *vector addition*, and *scalar multiplication*.

Definition: *(Vector Space)*

Let V be a non-empty set of objects on which two operations, called **vector addition** and **scalar multiplication**, have been specified. We shall say that V is a **vector space** if each of the following *axioms* are satisfied:

(1) The set V is *closed* under the operation of vector addition:

$$\mathbf{u} + \mathbf{v} \in V \text{ for all } \mathbf{u}, \mathbf{v} \in V$$

(2) Vector addition is *commutative*:

$$\mathbf{u} + \mathbf{v} = \mathbf{v} + \mathbf{u} \text{ for all } \mathbf{u}, \mathbf{v} \in V$$

(3) Vector addition is *associative*:

$$\mathbf{u} + (\mathbf{v} + \mathbf{w}) = (\mathbf{u} + \mathbf{v}) + \mathbf{w} \text{ for all } \mathbf{u}, \mathbf{v}, \mathbf{w} \in V$$

(4) There exists a vector $\mathbf{0} \in V$ called a **zero vector**, that serves as the *identity element* for vector addition:

$$\mathbf{0} + \mathbf{u} = \mathbf{u} \text{ for every } \mathbf{u} \in V$$

(5) For each $\mathbf{u} \in V$, there exists a vector $-\mathbf{u}$, called the **negative** of $\mathbf{u}$, such that:

$$\mathbf{u} + (-\mathbf{u}) = \mathbf{0}$$

(6) The set V is *closed* under the operation of scalar multiplication:

$$a\mathbf{u} \in V \text{ for all } a \in \mathbb{R},\ \mathbf{u} \in V$$

(7) Scalar multiplication *distributes* over vector addition:

$$a(\mathbf{u} + \mathbf{v}) = a\mathbf{u} + a\mathbf{v} \text{ for all } a \in \mathbb{R},\ \mathbf{u}, \mathbf{v} \in V$$

(8) Scalar multiplication also satisfies the distributive property:

$$(a + b)\mathbf{u} = a\mathbf{u} + b\mathbf{u} \text{ for all } a, b \in \mathbb{R},\ \mathbf{u} \in V$$

(9) Scalar multiplication is *associative*:

$$a(b\mathbf{u}) = (ab)\mathbf{u} \quad \text{for all } a, b \in \mathbb{R},\ \mathbf{u} \in V$$

(10) The real number $1 \in \mathbb{R}$ is the *identity element* for scalar multiplication:

$$1(\mathbf{u}) = \mathbf{u} \quad \text{for all } \mathbf{u} \in V$$

From this definition, we see that if we want to build a vector space, then we must specify four things: a set of vectors, a set of scalars, how vectors are to be added, and how they are to be multiplied by scalars.

Strictly speaking, if we specify that our scalars are real numbers, then we have a **real vector space**. On the other hand, in some applications such as quantum mechanics, where it is convenient to allow our scalars to be complex numbers, then we have a **complex vector space**. For the time being, we shall assume that we are dealing with real vector spaces. Let's take a look at some simple examples.

Example-1: *(Plane Vectors)*
The set of plane vectors in $\mathbb{R}^2$ is a vector space. Let the point $(x_1, x_2) \in \mathbb{R}^2$. Recall that a plane vector $\mathbf{x} \in \mathbb{R}^2$ can be viewed as a directed line segment whose *initial point* is at the origin $(0, 0)$ and whose *terminal point* is (x_1, x_2).

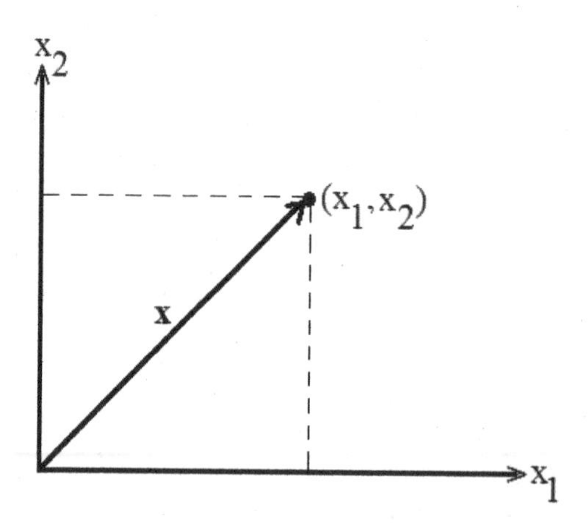

Figure 9.1: A typical vector in $\mathbb{R}^2$.

This state of affairs is shown in Figure 9.1. We shall agree to represent any such vector by

$$\mathbf{x} = \begin{bmatrix} x_1 \\ x_2 \end{bmatrix} \equiv [x_1, x_2]^T$$

We define the operations of vector addition and scalar multiplication according to the rules:

$$\mathbf{x}+\mathbf{y} = \begin{bmatrix} x_1 \\ x_2 \end{bmatrix} + \begin{bmatrix} y_1 \\ y_2 \end{bmatrix} = \begin{bmatrix} x_1+y_1 \\ x_2+y_2 \end{bmatrix} \quad \text{for all } \mathbf{x},\mathbf{y} \in \mathbb{R}^2$$

$$a\mathbf{x} = a\begin{bmatrix} x_1 \\ x_2 \end{bmatrix} = \begin{bmatrix} ax_1 \\ ax_2 \end{bmatrix} \quad \text{for all } a \in \mathbb{R}, \mathbf{x} \in \mathbb{R}^2$$

It's easy to verify that the space of plane vectors satisfies all of the vector space axioms, and consequently, it is a bonefide vector space.

Example-2: *(Euclidean n-Space)*
The set of n-dimensional vectors in $\mathbb{R}^n$ is a vector space. To see this, we just generalize results of the previous example. Consider the abitrary point $(x_1,x_2,\ldots,x_n)$ in $\mathbb{R}^n$. We can imagine a vector in $\mathbb{R}^n$ to be a directed line segment whose initial point is at $(0,0,\ldots,0)$ and whose terminal point is $(x_1,x_2,\ldots,x_n)$. We can denote a typical vector $\mathbf{x} \in \mathbb{R}^n$ by

$$\mathbf{x} = \begin{bmatrix} x_1 \\ x_2 \\ \vdots \\ x_n \end{bmatrix} \equiv [x_1,x_2,\ldots,x_n]^T$$

Similarly, we define vector addition and scalar multiplication by the rules:

$$\mathbf{x}+\mathbf{y} = \begin{bmatrix} x_1 \\ x_2 \\ \vdots \\ x_n \end{bmatrix} + \begin{bmatrix} y_1 \\ y_2 \\ \vdots \\ y_n \end{bmatrix} = \begin{bmatrix} x_1+y_1 \\ x_2+y_2 \\ \vdots \\ x_n+y_n \end{bmatrix} \quad \text{for all } \mathbf{x},\mathbf{y} \in \mathbb{R}^n$$

$$a\mathbf{x} = \begin{bmatrix} ax_1 \\ ax_2 \\ \vdots \\ ax_n \end{bmatrix} \quad \text{for all } a \in \mathbb{R}, \mathbf{x} \in \mathbb{R}^n$$

The vector space $\mathbb{R}^n$ is often called **Euclidean n-space.**

Example-3: *(Vector Space of Continuous Functions)*
Let $C[a,b]$ denote the set of all real-valued functions that are continuous on the closed interval $[a,b]$. It's easy to verify that $C[a,b]$ is a vector space. Here, we define vector addition and scalar multiplication in a pointwise fashion. That is, for each point $x \in [a,b]$, we define:

$$(\mathbf{f}+\mathbf{g})(x) = \mathbf{f}(x)+\mathbf{g}(x) \quad \text{for all } \mathbf{f}(x), \mathbf{g}(x) \in C[a,b]$$
$$(c\mathbf{f})(x) = c[\mathbf{f}(x)] \quad \text{for all } c \in \mathbb{R},\ \mathbf{f}(x) \in C[a,b]$$

From elementary calculus we know that the sum of two continuous functions is a continuous function, and the product of a real number and a continuous function is again a continuous function; so the above operations are closed.

Also, we can take as our zero vector any function $\mathbf{f}_0(x)$ which satisfies the property $\mathbf{f}_0(x) = 0$ for all $x \in [a,b]$. We will then have

$$(\mathbf{f}_0+\mathbf{f})(x) = \mathbf{0}+\mathbf{f}(x) = \mathbf{f}(x)$$

for all $\mathbf{f}(x) \in C[a,b]$, so that axiom (4) is satisfied. It's left to the reader to verify the remaining vector space axioms.

Example-4: *(Simple Harmonic Motion)*
From elementary mechanics, we know that the motion of a simple pendulum is governed by the differential equation

$$\frac{d^2\theta}{dt^2} + \left(\frac{g}{L}\right)\theta = 0$$

As shown in Figure 9.2, θ represents the deflection of the pendulum from the vertical (which is assumed to be small), L is the length of the pendulum arm, g is the acceleration due to gravity, and t is the time.

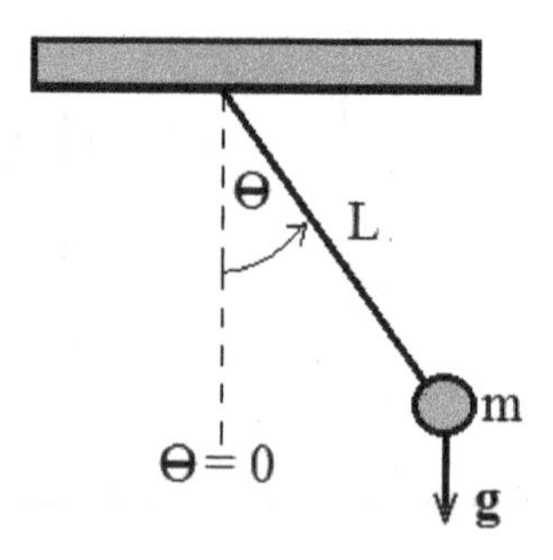

Figure 9.2: A simple pendulum

Now let H denote the set of solutions to this equation. We claim that H is a vector space. To see this, we again define vector addition and scalar multiplication in a pointwise fashion; for each real number $t \geq 0$, we define:

$$(\theta_1+\theta_2)(t) = \theta_1(t)+\theta_2(t) \quad \text{for all } \theta_1(t), \theta_2(t) \in H$$
$$(a\theta)(t) = a[\theta(t)] \quad \text{for all } a \in \mathbb{R},\ \theta(t) \in H$$

These operations are closed since the sum of any two solutions is again a solution, and any scalar multiple of a solution is also a solution - the reader should verify this claim. Note also that $\boldsymbol{\theta}(t) \equiv 0$ is also a solution and plays the role of the zero vector.

From the above examples, it's easy to see that the concept of a vector space is ubiquitous in mathematics. Indeed, a vector space is like an old friend, who pops up his head in some of the most unlikely places. Additional appearances of this important concept are given in the exercises that follow. Before leaving this section, however, we state a simple but useful theorem.

Theorem 9.1: *(Properties of Scalar Multiplication)*

For any vector space V, the operation of scalar multiplication obeys each of the following properties:
(1) $0\mathbf{u} = \mathbf{0}$ for all $\mathbf{u} \in V$
(2) $(-1)\mathbf{u} = -\mathbf{u}$ for all $\mathbf{u} \in V$
(3) If $a\mathbf{u} = \mathbf{0}$ then either $a = 0$ or $\mathbf{u} = \mathbf{0}$.
(4) $a\mathbf{0} = \mathbf{0}$ for all $a \in \mathbb{R}$

Proof: To prove (1), observe that by Axiom (8) we can write

$$0\mathbf{u} = (0+0)\mathbf{u} = 0\mathbf{u} + 0\mathbf{u}$$

so that $0\mathbf{u} = 0\mathbf{u} + 0\mathbf{u}$. By Axiom (5), the vector $-0\mathbf{u}$ must exist, and upon adding this vector to both sides of the last equation we find:

$$\begin{aligned} 0\mathbf{u} + [-0\mathbf{u}] &= 0\mathbf{u} + 0\mathbf{u} + [-0\mathbf{u}] \\ &\Rightarrow \mathbf{0} = 0\mathbf{u} + \mathbf{0} \\ &\Rightarrow \mathbf{0} = 0\mathbf{u} \end{aligned}$$

as claimed. The proofs of the remaining properties are left as exercises. □

Exercise Set 9

In Exercises 1-5, given the vectors

$$\mathbf{u} = \begin{bmatrix} -1 \\ 2 \\ 0 \end{bmatrix}, \quad \mathbf{v} = \begin{bmatrix} 3 \\ 0 \\ 4 \end{bmatrix}, \quad \text{and } \mathbf{w} = \begin{bmatrix} 5 \\ 3 \\ 1 \end{bmatrix}$$

in the vector space $\mathbb{R}^3$, calculate each of the following:

1. $2\mathbf{u} + 4\mathbf{v}$

2. $\mathbf{v} + 3\mathbf{w}$

3. $\mathbf{u} + \mathbf{v} - \mathbf{w}$

4. $\mathbf{u} + 2\mathbf{v} + 3\mathbf{w}$

5. $\mathbf{u} - \mathbf{w}$

6. Find a vector $\mathbf{x} \in \mathbb{R}^3$ such that

$$2\mathbf{x} + \begin{bmatrix} -2 \\ 2 \\ 3 \end{bmatrix} = \mathbf{x} + \begin{bmatrix} 0 \\ 4 \\ 7 \end{bmatrix}$$

7. (*Norm of a Vector in* $\mathbb{R}^n$) Given any vector $\mathbf{x} \in \mathbb{R}^n$, where

$$\mathbf{x} = \begin{bmatrix} x_1 \\ x_2 \\ \vdots \\ x_n \end{bmatrix}$$

we define its **length** or **norm** to be the *non-negative* real number

$$\|\mathbf{x}\| = \sqrt{x_1^2 + x_2^2 + \ \ldots \ + x_n^2}$$

Show that for any real scalar a, we have

$$\|a\mathbf{x}\| = |a| \cdot \|\mathbf{x}\|$$

8. (*Euclidean inner product*) Given the vectors $\mathbf{x} = [x_1, x_2, \ldots, x_n]^T$ and $\mathbf{y} = [y_1, y_2, \ldots, y_n]^T$ in $\mathbb{R}^n$ we define their **Euclidean inner product** as the real number:

$$\langle \mathbf{x}, \mathbf{y} \rangle = x_1 y_1 + x_2 y_2 + \ \ldots \ + x_n y_n$$

Show that:

(a) $\langle \mathbf{x}, \mathbf{y} \rangle = \langle \mathbf{y}, \mathbf{x} \rangle$

(b) $\langle \mathbf{x}, \mathbf{y} + \mathbf{z} \rangle = \langle \mathbf{x}, \mathbf{y} \rangle + \langle \mathbf{x}, \mathbf{z} \rangle$

(c) $\langle a\mathbf{x}, \mathbf{y} \rangle = \langle \mathbf{x}, a\mathbf{y} \rangle = a\langle \mathbf{x}, \mathbf{y} \rangle$ for any $a \in \mathbb{R}$.

(d) $\|\mathbf{x}\|^2 = \langle \mathbf{x}, \mathbf{x} \rangle$

(e) $\langle \mathbf{x}, \mathbf{y} \rangle = \frac{1}{4}\left\{\|\mathbf{x}+\mathbf{y}\|^2 + \|\mathbf{x}-\mathbf{y}\|^2\right\}$ *(Polarization Identity)*

(f) $\|\mathbf{x}+\mathbf{y}\|^2 + \|\mathbf{x}-\mathbf{y}\|^2 = 2\left\{\|\mathbf{x}\|^2 + \|\mathbf{y}\|^2\right\}$

9. *(The Vector Space M(m, n))* Let $M(2,2)$ denote the set of all 2×2 matrices. We can define vector addition and scalar multiplication by

$$\mathbf{x} + \mathbf{y} = \begin{bmatrix} x_1 & x_2 \\ x_3 & x_4 \end{bmatrix} + \begin{bmatrix} y_1 & y_2 \\ y_3 & y_4 \end{bmatrix} = \begin{bmatrix} x_1 + y_1 & x_2 + y_2 \\ x_3 + y_3 & x_4 + y_4 \end{bmatrix}$$

and

$$a\mathbf{x} = a\begin{bmatrix} x_1 & x_2 \\ x_3 & x_4 \end{bmatrix} = \begin{bmatrix} ax_1 & ax_2 \\ ax_3 & ax_4 \end{bmatrix}$$

for all $a \in \mathbb{R}, \mathbf{x}, \mathbf{y} \in M(2,2)$. Show that $M(2,2)$ is a vector space. In a similar fashion, we can generalize this result and show that the set of all $m \times n$ matrices, denoted by $M(m, n)$ is also a vector space.

10. Let P_2 denote the set of all polynomials of degree *less than or equal* to 2. A typical element $\mathbf{p}(x) \in P_2$ looks like

$$\mathbf{p}(x) = a_0 + a_1 x + a_2 x^2$$

where $a_0, a_1, a_2 \in \mathbb{R}$. After providing suitable definitions of vector addition and scalar multiplication, show that P_2 is a vector space.

11. Complete the proof of Theorem-1.

12. *(The Vector Space P_n)* Let P_n denote the set of all polynomials of degree *less than or equal* to n. Here, a typical element $\mathbf{p}(x) \in P_n$ has the form

$$\mathbf{p}(x) = a_0 + a_1 x + a_2 x^2 + \ldots + a_n x^n$$

where $a_0, a_1, \ldots, a_n \in \mathbb{R}$. Generalize the result of Exercise-10 by showing that P_n is a vector space.

10. Subspaces of Vector Spaces

Given a vector space V, in turns out that various *subsets* of V can form vector spaces in their own right, under the same operations of vector addition and scalar multiplication as V. Whenever this happens, we call the given subset a **subspace** of V.

To be a bit more precise, let W be a non-empty subset of a vector space V. We shall say that W is a **subspace** of V if W itself is a vector space under the *same operations* of vector addition and scalar multiplication as defined in V.

Now, if we are given a subset W of a vector space V, and we want to determine whether W is a subspace of V, then we could simply check whether W satisfies each of the vector space axioms. A less tedious method, however, is given in our first theorem.

Theorem 10.1: *(Test for a Subspace)*

If W is any *non-empty* subset of a vector space V, then W is a subspace of V if and only if W is *closed* under the *same* operations of vector addition and scalar multiplication as defined on W. That is, W is a subspace of V if and only if the following two conditions hold:
(a) If $\mathbf{u}, \mathbf{v} \in W$ then $\mathbf{u} + \mathbf{v} \in W$.
(b) If $\mathbf{u} \in W$ then $a\mathbf{u} \in W$ for all scalars a.

Proof: If W is a subspace of V, then since W is a vector space itself, both conditions (a) and (b) must hold, since these are the two closure axioms (1) and (6) that appear in the definition of a vector space.

On the other hand, assume that W is closed under vector addition and scalar multiplication so that (a) and (b) hold; then W satisfies the closure axioms (1) and (6) for a vector space. Also, if $\mathbf{u}, \mathbf{v}$, and $\mathbf{w}$ are vectors in W then these vectors are certainly in V so the vector space axioms (2), (3), (7), (8), (9), and (10) must also hold true for W.

So, it remains to show that W satisfies vector space axioms (4) and (5). Since $a\mathbf{u} \in W$ for all scalars a, then in particular $(0)\mathbf{u} = \mathbf{0} \in W$ so axiom (4) is satisfied, and $(-1)\mathbf{u} = -\mathbf{u} \in W$ so axiom (6) is holds as well, and the proof is complete. □.

Given any vector space V, it is clear that $V \subseteq V$, so every vector space is a subspace of itself. Also, since $\mathbf{0} \in V$, then the set $\{\mathbf{0}\}$ must also be a subspace of V, called the **zero subspace**. The subspaces V and $\{\mathbf{0}\}$ are often called **trivial subspaces** of V, while any other subspaces of V - if they exist - are said to be **proper subspaces** or **non-trivial subspaces** of V.

Example-1: Consider the vector space $\mathbb{R}^3$. Let W be the set of all vectors $\mathbf{y}$ of the form $\mathbf{y} = b\mathbf{x}$ where $b \neq 0$ is a fixed real number, and $\mathbf{x} \in \mathbb{R}^3$. Then W is a subspace of $\mathbb{R}^3$. To see this, we first note that W is not empty. Next, we assume that $\mathbf{y}_1 = b\mathbf{x}_1$ and $\mathbf{y}_2 = b\mathbf{x}_2$ are vectors in W, so that

$$\mathbf{y}_1 + \mathbf{y}_2 = b\mathbf{x}_1 + b\mathbf{x}_2 = b(\mathbf{x}_1 + \mathbf{x}_2) \in W$$

since $\mathbf{x}_1 + \mathbf{x}_2 \in \mathbb{R}^3$. Also if $\mathbf{y} \in W$ then $\mathbf{y} = b\mathbf{x}_1$ for some $\mathbf{x}_1 \in \mathbb{R}^3$. Consequently, for any real number a, we have

$$a\mathbf{y} = a(b\mathbf{x}_1) = b(a\mathbf{x}_1) \in W$$

since $a\mathbf{x}_1 \in \mathbb{R}^3$. Thus, by the previous theorem, W is a subspace of $\mathbb{R}^3$. Geometrically, W is the set of all points lying on a line which passes through the origin of $\mathbb{R}^3$. This situation is shown in Figure 10.1 below. Finally, if we set $b = 0$, then $W = \{\mathbf{0}\}$, the zero subspace.

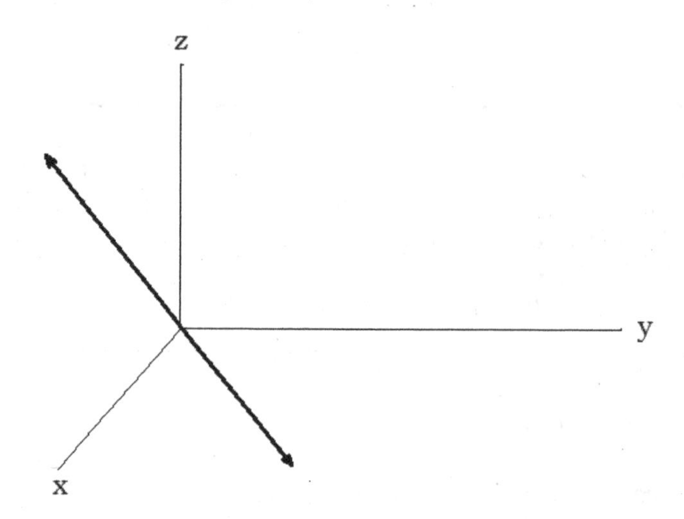

Figure 10.1: The subspace W is the line $\mathbf{y} = b\mathbf{x}$ in $\mathbb{R}^3$.

Example-2: Once again, consider the vector space $\mathbb{R}^3$. Let W be the set of all vectors $\mathbf{x} = [x_1, x_2, x_3]^T \in \mathbb{R}^3$ such that

$$\mathbf{c} \cdot \mathbf{x} = c_1 x_1 + c_2 x_2 + c_3 x_3 = 0$$

where $\mathbf{c} = [c_1, c_2, c_3]$ is any fixed non-zero row vector. Clearly, W is not empty, since $\mathbf{0} \in W$. Now, assume that $\mathbf{x}, \mathbf{y} \in W$ so that we have both $\mathbf{c} \cdot \mathbf{x} = 0$ and $\mathbf{c} \cdot \mathbf{y} = 0$, where $\mathbf{x}, \mathbf{y} \in \mathbb{R}^3$. Clearly, $\mathbf{x} + \mathbf{y} \in W$ since

$$\mathbf{c} \cdot (\mathbf{x} + \mathbf{y}) = (\mathbf{c} \cdot \mathbf{x}) + (\mathbf{c} \cdot \mathbf{y}) = 0$$

Similarly, if $\mathbf{x} \in W$, then $a\mathbf{x} \in W$ since

$$\mathbf{c} \cdot (a\mathbf{x}) = a(\mathbf{c} \cdot \mathbf{x}) = 0$$

So, by the previous theorem, we conclude that W is a subspace of $\mathbb{R}^3$. In terms of geometry, W represents a plane through the origin of $\mathbb{R}^3$. If we allow $\mathbf{c} = \mathbf{0}$, then W reduces to the zero subspace $\{\mathbf{0}\}$ and it then consists of a single point, i.e., the origin of $\mathbb{R}^3$. This state of affairs is shown in Figure 10.2 below.

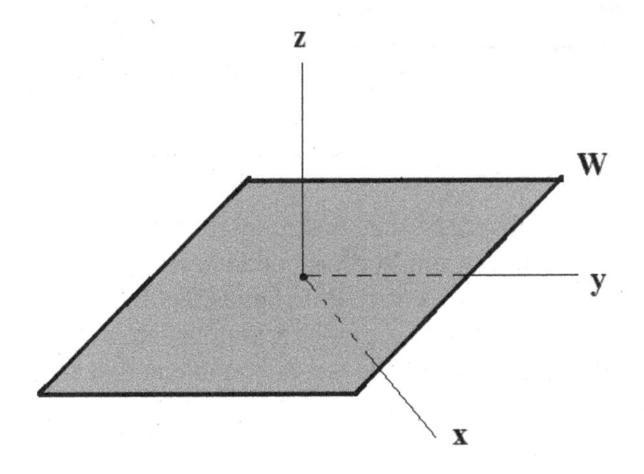

Figure 10.2: The subspace W is a plane through the origin in $\mathbb{R}^3$.

Example-3: Recall that $M(2,2)$, the set of all square matrices of order two is a vector space. Let W consist of all square matrices A of order two such that

$$A = \begin{bmatrix} a_{11} & a_{12} \\ a_{21} & a_{22} \end{bmatrix} \quad \text{where } Tr(A) = a_{11} + a_{22} = 0$$

Stated differently, if W is the set of all 2×2 matrices that are **trace free**. Then W is a subspace of $M(2,2)$. To see this, let $A, B \in W$ so that we have $Tr(A) = Tr(B) = 0$. Observe that

$$Tr(A + B) = Tr(A) + Tr(B) = 0$$

so $A + B \in W$. Also, for any real number c we have

$$Tr(cA) = c \cdot Tr(A) = 0$$

so $cA \in W$ whenever $A \in W$.

Example-4: Consider the vector space $\mathbb{R}^3$. Let W be the set of all vectors $\mathbf{x} = [x_1, x_2, x_3]^T \in \mathbb{R}^3$ that are *solutions* of the *homogeneous* system of linear equations $A\mathbf{x} = \mathbf{0}$, where A is a square matrix of order three. Then W is a subspace of $\mathbb{R}^3$ which is called the **solution space** of the system. To see this, we first note that W is not empty, since $\mathbf{0} \in W$. Next, assume that $\mathbf{x}, \mathbf{y} \in W$ so that $A\mathbf{x} = \mathbf{0}$ and $A\mathbf{y} = \mathbf{0}$. Then $\mathbf{x} + \mathbf{y} \in W$ since $A(\mathbf{x} + \mathbf{y}) = A\mathbf{x} + A\mathbf{y} = \mathbf{0}$. Also, for any real number c, we have $A(c\mathbf{x}) = c(A\mathbf{x}) = \mathbf{0}$ so $c\mathbf{x} \in W$ as well.

From the first two examples above, we see that the only subspaces of $\mathbb{R}^3$ are the zero subspace $\{\mathbf{0}\}$, all lines through the origin, all planes through the origin and the space $\mathbb{R}^3$ itself. Note that the *intersection* of any two of these subspaces is again a subspace of $\mathbb{R}^3$. This is no accident, as the next theorem shows.

Theorem 10.2: *(The Intersection of Subspaces)*

If W_1 and W_2 are subspaces of a vector space V, then $W_1 \cap W_2$ is a subspace of V as well.

Proof: Since W_1 and W_2 are subspaces, they both contain the zero vector $\mathbf{0}$, and so, the set $W_1 \cap W_2$ cannot be empty. Assume that $\mathbf{x},\mathbf{y} \in W_1 \cap W_2$ so that $\mathbf{x},\mathbf{y} \in W_1$ and $\mathbf{x},\mathbf{y} \in W_2$. But $\mathbf{x},\mathbf{y} \in W_1$ implies that both $\mathbf{x}+\mathbf{y} \in W_1$ and $c\mathbf{x} \in W_1$ for any scalar c. Similarly, $\mathbf{x},\mathbf{y} \in W_2$ implies that $\mathbf{x}+\mathbf{y} \in W_2$ and $c\mathbf{x} \in W_2$ for any scalar c. Consequently, we can conclude that both $\mathbf{x}+\mathbf{y} \in W_1 \cap W_2$ and $c\mathbf{x} \in W_1 \cap W_2$. $\square$.

Exercise Set 10

In Exercises 1-4, use Theorem-10.1 to verify that the given set W is a subspace of V.

1. $V = \mathbb{R}^3,\ W = \left\{[x_1,x_2,0]^T : x_1,x_2 \in \mathbb{R}\right\}$

2. $V = \mathbb{R}^2,\ W = \left\{[x_1,x_2,3x_1+2x_2]^T : x_1,x_2 \in \mathbb{R}\right\}$

3. $V = M(2,2)$, W is the set of all 2×2 matrices A of the form

$$A = \begin{bmatrix} a & 0 \\ 0 & b \end{bmatrix} \quad \text{where } a,b \in \mathbb{R}.$$

4. Let $\mathbf{x},\mathbf{y}$ be any two vectors in a vector space V. Let W denote the set of all vectors of the form $a\mathbf{x} + b\mathbf{y}$ where a, b are arbitrary scalars. Show that W is a subspace of V.

5. Is the *union* of any two subspaces of a vector space V also subspace of V?

6. Let A be a *fixed* square matrix in the vector space $M(2,2)$. Let W denote the set of all matrices $X \in M(2,2)$ such that $AX = 0$. Show that W is a subspace of $M(2,2)$.

11. Linear Dependence and Independence

Suppose that $\mathbf{v}_1, \mathbf{v}_2, \ldots, \mathbf{v}_m$ are vectors in a vector space V. If a vector $\mathbf{v} \in V$ can be written in the form

$$\mathbf{v} = c_1\mathbf{v}_1 + c_2\mathbf{v}_2 + \ldots + c_m\mathbf{v}_m$$

where $c_1, c_2, \ldots, c_m$ are scalars, then we shall say the $\mathbf{v}$ is a **linear combination** of the vectors $\mathbf{v}_1, \mathbf{v}_2, \ldots, \mathbf{v}_m$.

Example-1: Show that the vector $\mathbf{v} = [1, 17, 8]^T \in \mathbb{R}^3$ can be written as a linear combination of the vectors:

$$\mathbf{v}_1 = \begin{bmatrix} -1 \\ 2 \\ 3 \end{bmatrix}, \mathbf{v}_2 = \begin{bmatrix} 0 \\ 1 \\ 2 \end{bmatrix}, \text{ and } \mathbf{v}_3 = \begin{bmatrix} 1 \\ 4 \\ 0 \end{bmatrix}$$

Solution: We must find scalars $c_1, c_2,$ and c_3 such that

$$\mathbf{v} = c_1\mathbf{v}_1 + c_2\mathbf{v}_2 + c_3\mathbf{v}_3$$

Consequently, we can write

$$\begin{bmatrix} 1 \\ 17 \\ 8 \end{bmatrix} = c_1\begin{bmatrix} -1 \\ 2 \\ 3 \end{bmatrix} + c_2\begin{bmatrix} 0 \\ 1 \\ 2 \end{bmatrix} + c_3\begin{bmatrix} 1 \\ 4 \\ 0 \end{bmatrix}$$

This last equation, in turn, leads to the system of equations

$$\begin{aligned} -c_1 + c_3 &= 1 \\ 2c_1 + c_2 + 4c_3 &= 17 \\ 3c_1 + 2c_2 &= 8 \end{aligned}$$

whose solution is $c_1 = 2, c_2 = 1$, and $c_3 = 3$. Consequently,

$$\mathbf{v} = 2\begin{bmatrix} -1 \\ 2 \\ 3 \end{bmatrix} + \begin{bmatrix} 0 \\ 1 \\ 2 \end{bmatrix} + 3\begin{bmatrix} 1 \\ 4 \\ 0 \end{bmatrix}$$

as required.

Given any set of vectors in a vector space, it turns out that the set of all linear combinations of those vectors forms a vector subspace of the given space. Before we can demonstrate this idea, we need an important definition.

Definition: *(Spanning Set)*

Suppose that $S = \{\mathbf{v}_1, \mathbf{v}_2, \ldots, \mathbf{v}_m\}$ is a set of vectors in a vector space V. If *every* vector $\mathbf{v} \in V$ can be written as a linear combination of the vectors in S, then we call S a **spanning set** for V, or equivalently, we say that S **spans** V. Furthermore, we shall denote the set of all linear combinations of the vectors in S by *span*(S), i.e.,

$$span(S) \equiv span\{\mathbf{v}_1, \mathbf{v}_2, \ldots, \mathbf{v}_m\} = \left\{\mathbf{v} \in \mathbf{S} \ : \ \mathbf{v} = c_1\mathbf{v}_1 + c_2\mathbf{v}_2 + \ldots + c_m\mathbf{v}_m\right\} \tag{11.1}$$

where $c_1, c_2, \ldots, c_m$ are scalars.

Theorem 11.1: *(Span is a Subspace)*

If $S = \{\mathbf{v}_1, \mathbf{v}_2, \ldots, \mathbf{v}_m\}$ is a subset of the vector space V, then *span*(S) is a subspace of V. Furthermore, *span*(S) is the smallest subspace of V that contains the set S.

Proof: Suppose that the vectors $\mathbf{x}, \mathbf{y} \in span(S)$ so that we can write

$$\mathbf{x} = c_1\mathbf{v}_1 + c_2\mathbf{v}_2 + \ldots + c_m\mathbf{v}_m$$
$$\mathbf{y} = d_1\mathbf{v}_1 + d_2\mathbf{v}_2 + \ldots + d_m\mathbf{v}_m$$

where $c_1, c_2, \ldots, c_m$ and $d_1, d_2, \ldots, d_m$ are scalars. Now, observe that

$$\mathbf{x} + \mathbf{y} = (c_1 + d_1)\mathbf{v}_1 + (c_2 + d_2)\mathbf{v}_2 + \ldots + (c_m + d_m)\mathbf{v}_m$$

so $\mathbf{x} + \mathbf{y} \in span(S)$.. Also, for any scalar λ we have

$$\lambda\mathbf{x} = (\lambda c_1)\mathbf{v}_1 + (\lambda c_2)\mathbf{v}_2 + \ldots + (\lambda c_m)\mathbf{v}_m$$

so $\lambda\mathbf{x} \in span(S)$ as well. Since *span*(S) is closed under the operations of vector addition and scalar multiplication, then it is a subspace of V as claimed.

To establish the second half of the theorem, let W be any subspace of V that contains S. Since W is a subspace, it must be closed under vector addition and scalar multiplication; thus, if W contains S, then it must also contain all linear combinations of the vectors in S. In other words, $S \subseteq W$. $\square$.

Given a set of vectors $\{\mathbf{v}_1, \mathbf{v}_2, \ldots, \mathbf{v}_m\}$ in a vector space V, consider the equation

$$c_1\mathbf{v}_1 + c_2\mathbf{v}_2 + \ldots + c_m\mathbf{v}_m = \mathbf{0} \tag{11.2}$$

It is clear that (11.2) always has the **trivial solution**

$$c_1 = 0, \ c_2 = 0, \ldots, c_m = 0 \tag{11.3}$$

If this is the only solution to (11.2), then we say that S is a **linearly independent** set of vectors. On the other hand, if there exists at least one non-trivial solution to (11.2), then S is said to be a **linearly dependent** set. We formalize this concept in our next definition.

Definition: *(Linear Dependence and Linear Independence)*

Let $S = \{\mathbf{v}_1, \mathbf{v}_2, \ldots, \mathbf{v}_m\}$ be a set of vectors in a vector space V. If the equation

$$c_1\mathbf{v}_1 + c_2\mathbf{v}_2 + \ldots + c_m\mathbf{v}_m = \mathbf{0}$$

only has the trivial solution $c_1 = 0,\ c_2 = 0, \ldots, c_m = 0$, then S is called a **linearly independent** set; however, if there is at least one non-trivial solution then S is called a **linearly dependent** set.

Example-2: Determine whether the following vectors in $\mathbb{R}^3$ form a linearly dependent or linearly independent set:

$$\mathbf{v}_1 = \begin{bmatrix} 1 \\ -1 \\ 0 \end{bmatrix}, \quad \mathbf{v}_2 = \begin{bmatrix} 0 \\ 2 \\ 1 \end{bmatrix}, \quad \mathbf{v}_3 = \begin{bmatrix} 2 \\ 4 \\ 3 \end{bmatrix}$$

Solution: We assume that

$$c_1\mathbf{v}_1 + c_2\mathbf{v}_2 + c_3\mathbf{v}_3 = \mathbf{0}$$

Consequently, we must have

$$c_1\begin{bmatrix} 1 \\ -1 \\ 0 \end{bmatrix} + c_2\begin{bmatrix} 0 \\ 2 \\ 1 \end{bmatrix} + c_3\begin{bmatrix} 2 \\ 4 \\ 3 \end{bmatrix} = \begin{bmatrix} 0 \\ 0 \\ 0 \end{bmatrix}$$

which, in turn, leads us to consider the linear system:

$$\begin{aligned} c_1 + 0c_2 + 2c_3 &= 0 \\ -c_1 + 2c_2 + 4c_3 &= 0 \\ 0c_1 + c_2 + 3c_3 &= 0 \end{aligned}$$

This system has infinitely many *non-trivial* solutions of the form

$$c_1 = -2t, \quad c_2 = -3t, \quad c_3 = t$$

and so, the original set of vectors is linear dependent.

Example-3: Consider the vectors

$$\mathbf{e}_1 = \begin{bmatrix} 1 \\ 0 \\ 0 \end{bmatrix}, \quad \mathbf{e}_2 = \begin{bmatrix} 0 \\ 1 \\ 0 \end{bmatrix}, \quad \mathbf{e}_3 = \begin{bmatrix} 0 \\ 0 \\ 1 \end{bmatrix}$$

in $\mathbb{R}^3$. It's easy to see that these vectors are linearly independent, since the vector equation

$$c_1\mathbf{e}_1 + c_2\mathbf{e}_2 + c_3\mathbf{e}_3 = \mathbf{0}$$

implies that

$$c_1 \begin{bmatrix} 1 \\ 0 \\ 0 \end{bmatrix} + c_2 \begin{bmatrix} 0 \\ 1 \\ 0 \end{bmatrix} + c_3 \begin{bmatrix} 0 \\ 0 \\ 1 \end{bmatrix} = \begin{bmatrix} c_1 \\ c_2 \\ c_3 \end{bmatrix} = \begin{bmatrix} 0 \\ 0 \\ 0 \end{bmatrix}$$

Consequently, $c_1 = c_2 = c_3 = 0$, and the set of vectors is linearly independent.

Theorem 11.2: *(An Important Property of Linearly Dependent Sets)*

Let $S = \{\mathbf{v}_1, \mathbf{v}_2, \ldots, \mathbf{v}_m\}$ where $m \geq 2$ be a set of vectors in a vector space V. Then S is linearly dependent if and only if at least one vector in S can be written as a linear combination of the remaining vectors in S.

Proof: First, let's assume that S is a linearly dependent set. This means that the vector equation

$$c_1\mathbf{v}_1 + c_2\mathbf{v}_2 + c_3\mathbf{v}_3 + \ \ldots \ + c_m\mathbf{v}_m = \mathbf{0}$$

has at least one non-trivial solution, and consequently, at least one of the scalars $c_1, c_2, \ldots, c_m$ must be non-zero. Without loss of generality, let's suppose that $c_1 \neq 0$. Then we can solve the equation for $\mathbf{v}_1$ to obtain

$$\mathbf{v}_1 = -\frac{c_2}{c_1}\mathbf{v}_2 - \frac{c_3}{c_1}\mathbf{v}_3 - \ldots - \frac{c_m}{c_1}\mathbf{v}_m$$

So, we have expressed $\mathbf{v}_1$ *linear combination* of the remaining vectors $\mathbf{v}_2, \mathbf{v}_3, \ldots, \mathbf{v}_m$. On the other hand, suppose that a given vector in the set, say $\mathbf{v}_1$ can be written as a linear combination of the remaining vectors in S, so that

$$\mathbf{v}_1 = \lambda_2\mathbf{v}_2 + \lambda_3\mathbf{v}_3 + \ldots + \lambda_m\mathbf{v}_m$$

for some scalars $\lambda_2, \lambda_3, \ldots, \lambda_m$. But then the vector equation

$$(-1)\mathbf{v}_1 + \lambda_2\mathbf{v}_2 + \lambda_3\mathbf{v}_3 + \ldots + \lambda_m\mathbf{v}_m = \mathbf{0}$$

must hold, and at least one of the coefficients in this equation is different from zero. □.

A simple consequence of this theorem is that any two vectors $\mathbf{v}_1$ and $\mathbf{v}_2$ in vector space V are linearly dependent if and only if one vector is a scalar multiple of the other vector.

Theorem 11.3: *(Linear Dependence in $\mathbb{R}^n$)*

Let $S = \{\mathbf{v}_1, \mathbf{v}_2, \ldots, \mathbf{v}_m\}$ be any set of vectors in the vector space $\mathbb{R}^n$. If $m > n$, then S is a linearly dependent set.

Proof: Assume that

$$\mathbf{v}_1 = \begin{bmatrix} a_{11} \\ a_{21} \\ \cdots \\ a_{n1} \end{bmatrix}, \quad \mathbf{v}_2 = \begin{bmatrix} a_{12} \\ a_{22} \\ \cdots \\ a_{n2} \end{bmatrix}, \ldots, \mathbf{v}_m = \begin{bmatrix} a_{1m} \\ a_{2m} \\ \cdots \\ a_{nm} \end{bmatrix}$$

Then the vector equation

$$c_1\mathbf{v}_1 + c_2\mathbf{v}_2 + \ldots + c_m\mathbf{v}_m = \mathbf{0}$$

leads to a linear homogeneous system of n-equations in m-unknowns:

$$\begin{bmatrix} a_{11} & a_{12} & \cdots & a_{1m} \\ a_{21} & a_{22} & \cdots & a_{2m} \\ \cdots & \cdots & \cdots & \cdots \\ a_{n1} & a_{n2} & \cdots & a_{nm} \end{bmatrix} \begin{bmatrix} c_1 \\ c_2 \\ \cdots \\ c_m \end{bmatrix} = \begin{bmatrix} 0 \\ 0 \\ \cdots \\ 0 \end{bmatrix}$$

Since $m > n$, we have more unknowns than equations, and therefore, the system always has at least one non-trivial solution. Thus, S must be a linearly dependent set of vectors. □.

Exercise Set 11

In Exercises 1-3, determine whether the given set of vectors S spans $\mathbb{R}^3$.

1. $S = \left\{ \begin{bmatrix} 1 \\ 2 \\ 3 \end{bmatrix}, \begin{bmatrix} -1 \\ 2 \\ 1 \end{bmatrix}, \begin{bmatrix} 1 \\ 0 \\ 1 \end{bmatrix} \right\}$

2. $S = \left\{ \begin{bmatrix} 1 \\ 0 \\ 1 \end{bmatrix}, \begin{bmatrix} 1 \\ 2 \\ 0 \end{bmatrix}, \begin{bmatrix} 5 \\ 6 \\ 2 \end{bmatrix} \right\}$

3. $S = \left\{ \begin{bmatrix} 2 \\ 0 \\ 4 \end{bmatrix}, \begin{bmatrix} 1 \\ 1 \\ 2 \end{bmatrix}, \begin{bmatrix} 5 \\ 1 \\ 11 \end{bmatrix} \right\}$

In Exercises 4-6, determine whether the given set S is linearly dependent or linearly independent.

4. $S = \left\{ \begin{bmatrix} 2 \\ 8 \end{bmatrix}, \begin{bmatrix} -1 \\ -4 \end{bmatrix} \right\}$

5. $S = \left\{ \begin{bmatrix} 1 \\ 2 \\ 3 \end{bmatrix}, \begin{bmatrix} 1 \\ 2 \\ 1 \end{bmatrix}, \begin{bmatrix} 1 \\ 3 \\ 1 \end{bmatrix} \right\}$

6. $S = \left\{ \begin{bmatrix} 0 \\ 1 \\ 2 \end{bmatrix}, \begin{bmatrix} -1 \\ 2 \\ 1 \end{bmatrix}, \begin{bmatrix} 1 \\ 0 \\ 1 \end{bmatrix} \right\}$

In Exercises 7-9, determine whether the given set S of vectors in P_2 is linearly dependent or linearly independent.

7. $S = \{1 - x^2, 1 + x + x^2\}$

8. $S = \{2 + 3x, 1 + x^2, 2\}$

9. $S = \{1, 1 - x^2, 1 + x\}$

10. If $S = \{\mathbf{v}_1, \mathbf{v}_2, \ldots, \mathbf{v}_m\}$ is a linearly independent set of vectors in a vector space V, then show that any non-empty subset of S is also linearly independent.

11. Show that any set of vectors which contains the zero vector must be linearly dependent.

12. Prove that any two vectors $\mathbf{v}_1$ and $\mathbf{v}_2$ in vector space V are linearly dependent if and only if one vector is a scalar multiple of the other vector.

13. Let A be the $n \times n$ matrix whose columns consist of the n-vectors $\mathbf{v}_1, \mathbf{v}_2, \ldots, \mathbf{v}_n$ in $\mathbb{R}^n$. Show that these vectors are linearly dependent if and only if $\det(A) = 0$.

14. Show that $S = \{1, x, x^2, \ldots, x^n\}$ is a linearly independent set of vectors in the vector space P_n. [**Hint**: First assume there exist scalars $c_0, c_1, \ldots, c_n$ such that $c_0(1) + c_1 x + c_2 x^2 + \ldots + c_n x^n = 0$, and differentiate this relation n-times to obtain a linear system.]

15. Let A be any $m \times n$ matrix which is in row-echelon form. If the rows of A are viewed as vectors in $\mathbb{R}^n$, then show that the *non-zero* rows of A must be *linearly independent*.

12. Basis and Dimension

Intuitively, we know that $\mathbb{R}^3$ is a three dimensional space, and furthermore, the only possible non-trivial subspaces of $\mathbb{R}^3$ consist of one dimensional lines through the origin, and a two-dimensional planes through the origin. In this lesson, we shall formalize this type of thinking by introducing the important concepts of the *basis* and *dimension* of a vector space.

Definition: *(Basis and Components Relative to a Basis)*

A set of vectors $B = \{\mathbf{v}_1, \mathbf{v}_2, \ldots, \mathbf{v}_n\}$ in a vector space V is called a **basis** for V if and only if:
(a) B is a linearly independent set.
(b) B spans V; that is, for any vector $\mathbf{v} \in V$, there exist scalars $c_1, c_2, \ldots, c_n$ such that

$$\mathbf{v} = \sum_{k=1}^{n} c_k \mathbf{v}_k = c_1\mathbf{v}_1 + c_2\mathbf{v}_2 + \ldots + c_n\mathbf{v}_n \tag{12.1}$$

We call the scalars $c_1, c_2, \ldots, c_n$ in (12.1), the **components** of $\mathbf{v}$ *relative* to the basis B.

Theorem 12.1: *(Uniqueness of Vector Components Relative to a Basis)*

If a set of vectors $B = \{\mathbf{v}_1, \mathbf{v}_2, \ldots, \mathbf{v}_n\}$ is a basis for a vector space V, then for any vector $\mathbf{v} \in V$, the components of $\mathbf{v}$ relative to B are *unique*.

Proof: Assume that for any $\mathbf{v} \in V$ we have both

$$\mathbf{v} = c_1\mathbf{v}_1 + c_2\mathbf{v}_2 + \ldots + c_n\mathbf{v}_n$$
$$\mathbf{v} = c_1'\mathbf{v}_1 + c_2'\mathbf{v}_2 + \ldots + c_n'\mathbf{v}_n$$

Subtracting the second equation from the first one, we obtain

$$(c_1 - c_1')\mathbf{v}_1 + (c_2 - c_2')\mathbf{v}_2 + \ldots + (c_n - c_n')\mathbf{v}_n = \mathbf{0}$$

But, since the vectors $\mathbf{v}_1, \mathbf{v}_2, \ldots, \mathbf{v}_n$ comprising the basis must be linearly independent, the last equation implies

$$c_1 = c_1', \; c_2 = c_2', \ldots, c_n = c_n'$$

so the components of $\mathbf{v}$ relative to B are unique. □.

Example-1: *(Basis for P_n)* Show that the set $B = \{1, x, x^2, x^3\}$ is a basis for P_3.

Solution: It is obvious that B spans P_3 since any vector $\mathbf{p}(x) \in P_3$ can be written in the form

$$\mathbf{p}(x) = c_0(1) + c_1x + c_2x^2 + c_3x^3$$

So, we need only show that B is a linearly independent set. To this end, assume that there exist scalars $c_0, c_1, \ldots, c_3$ such that:

$$c_0(1) + c_1x + c_2x^2 + c_3x^3 = 0$$

Differentiating this relation three times, we obtain the linear homogeneous system

$$A\mathbf{c} \equiv \begin{bmatrix} 1 & x & x^2 & x^3 \\ 0 & 1 & 2x & 3x^2 \\ 0 & 0 & 2 & 6x \\ 0 & 0 & 0 & 6 \end{bmatrix} \begin{bmatrix} c_0 \\ c_1 \\ c_2 \\ c_3 \end{bmatrix} = \mathbf{0}$$

Since $\det(A) = 12 \neq 0$, this system only has the trivial solution, and we conclude that B is a linearly independent set. In a similar fashion, it's easy to show that the set $B = \{1, x, x^2, \ldots, x^n\}$ is a basis for the space P_n.

Example-2: *(Standard Basis for $\mathbb{R}^3$)* It's easy to see that the vectors

$$\mathbf{e}_1 = \begin{bmatrix} 1 \\ 0 \\ 0 \end{bmatrix}, \quad \mathbf{e}_2 = \begin{bmatrix} 0 \\ 1 \\ 0 \end{bmatrix}, \quad \mathbf{e}_3 = \begin{bmatrix} 0 \\ 0 \\ 1 \end{bmatrix}$$

form a basis for $\mathbb{R}^3$.

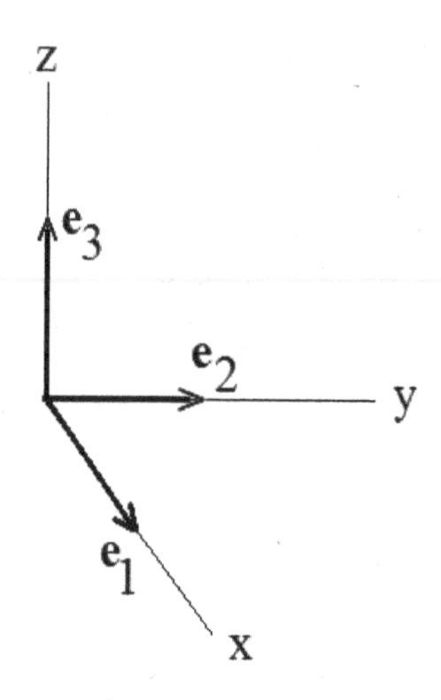

Figure 12.1: The standard basis vectors for $\mathbb{R}^3$.

We have already shown in the last lesson that the set $B = \{\mathbf{e}_1, \mathbf{e}_2, \mathbf{e}_3\}$ is linearly independent. Furthermore, given any vector $\mathbf{v} \in \mathbb{R}^3$ we can write

$$\mathbf{v} = \begin{bmatrix} v_1 \\ v_2 \\ v_3 \end{bmatrix} = v_1\mathbf{e}_1 + v_2\mathbf{e}_2 + v_3\mathbf{e}_3$$

so B spans $\mathbb{R}^3$ as well. We call B the **standard basis** for $\mathbb{R}^3$.

Theorem 12.2: *(Number of Basis Vectors is Unique)*

If

$$B = \{\mathbf{v}_1, \mathbf{v}_2, \ldots, \mathbf{v}_m\} \text{ and } B' = \{\mathbf{v}'_1, \mathbf{v}'_2, \ldots, \mathbf{v}'_n\}$$

are both *finite* basis sets for a vector space V, then $m = n$. In other words, any two bases for a vector space contain the *same number* of basis vectors.

Proof: First, we assume that $m < n$. Since B is a basis for V, we can express each vector in B' in terms of the basis vectors in B, to obtain

$$\begin{aligned} \mathbf{v}'_1 &= a_{11}\mathbf{v}_1 + a_{21}\mathbf{v}_2 + \ldots + a_{m1}\mathbf{v}_m \\ \mathbf{v}'_2 &= a_{12}\mathbf{v}_1 + a_{22}\mathbf{v}_2 + \ldots + a_{m2}\mathbf{v}_m \\ &\vdots \\ \mathbf{v}'_n &= a_{1n}\mathbf{v}_1 + a_{2n}\mathbf{v}_2 + \ldots + a_{mn}\mathbf{v}_m \end{aligned}$$

Written equivalently, in terms of components relative to B, we have

$$\mathbf{v}'_1 = \begin{bmatrix} a_{11} \\ a_{21} \\ \vdots \\ a_{m1} \end{bmatrix}, \quad \mathbf{v}'_2 = \begin{bmatrix} a_{12} \\ a_{22} \\ \vdots \\ a_{m2} \end{bmatrix}, \quad \mathbf{v}'_n = \begin{bmatrix} a_{1n} \\ a_{2n} \\ \vdots \\ a_{mn} \end{bmatrix} \qquad \text{(a)}$$

Now, since B' is a basis for V, we know that the vectors $\mathbf{v}'_1, \mathbf{v}'_2, \ldots, \mathbf{v}'_n$ must be linearly independent. Consequently, the vector equation

$$c_1\mathbf{v}'_1 + c_2\mathbf{v}'_2 + \ldots + c_n\mathbf{v}'_n = \mathbf{0} \qquad \text{(b)}$$

can only have the trivial solution $c_1 = c_2 = \ldots = c_n = 0$. However, if we substitute (a) into (b), we obtain

$$\begin{bmatrix} a_{11} \\ a_{21} \\ \vdots \\ a_{m1} \end{bmatrix} c_1 + \begin{bmatrix} a_{12} \\ a_{22} \\ \vdots \\ a_{m2} \end{bmatrix} c_2 + \quad + \begin{bmatrix} a_{1n} \\ a_{2n} \\ \vdots \\ a_{mn} \end{bmatrix} c_n = \begin{bmatrix} 0 \\ 0 \\ \vdots \\ 0 \end{bmatrix}$$

or equivalently, the linear homogeneous system

$$\begin{bmatrix} a_{11} & a_{12} & \cdots & a_{1n} \\ a_{21} & a_{22} & \cdots & a_{2n} \\ \vdots & \vdots & & \vdots \\ a_{m1} & a_{m2} & \cdots & a_{mn} \end{bmatrix} \begin{bmatrix} c_1 \\ c_2 \\ \vdots \\ c_n \end{bmatrix} = \begin{bmatrix} 0 \\ 0 \\ \vdots \\ 0 \end{bmatrix} \tag{c}$$

Now, since by hypothesis $m < n$, the system (c) has more unknowns than equations, and so this system must have at least one non–trivial solution. Clearly, this contradicts the fact that (b) can only have the trivial solution. Thus, we conclude that $m \geq n$.

If we now assume that $m > n$ and we express each vector in B in terms of the basis vectors in B', and follow a similar line of reasoning, we again arrive at a contradiction. We conclude that $m = n$. □.

This important theorem tells us that since any two bases for a given vector space must contain the same number of basis vectors, then this number is an *intrinsic property* of any vector space. If a *finite set* of n-vectors $\{\mathbf{v}_1, \mathbf{v}_2, \ldots, \mathbf{v}_n\}$ is a basis for a vector space V, then we say that V is a **finite dimensional** vector space of **dimension** n, and we write

$$\dim(V) = n \tag{12.2}$$

By convention, the zero vector space is assigned a dimension of zero.

Example-3: It's easy to see that

$$\dim(\mathbb{R}^n) = n \text{ and } \dim(P_n) = n + 1$$

If no *finite* basis exists for a vector space V, then we say that V is **infinite dimensional**. A simple example of an infinite dimensional vector space is the set $P(x)$ of all real polynomials of arbitrarily large degree. For the time being, however, we shall assume that we are working with finite dimensional spaces.

Theorem 12.3: *(A Convenient Test for Linear Dependence)*

Any set of $(n + 1)$-vectors in an n-dimensional vector space V is linearly dependent.

Proof: Assume that $\{\mathbf{v}_1, \mathbf{v}_2, \ldots, \mathbf{v}_{n+1}\}$ is any collection of $(n + 1)$-vectors in V, and assume that $B = \{\mathbf{e}_1, \mathbf{e}_2, \ldots, \mathbf{e}_n\}$ is a basis for V. Suppose further, that the components of the given vectors relative to B are

$$\mathbf{v}_1 = \begin{bmatrix} a_{11} \\ a_{21} \\ \vdots \\ a_{n1} \end{bmatrix}, \quad \mathbf{v}_2 = \begin{bmatrix} a_{12} \\ a_{22} \\ \vdots \\ a_{n2} \end{bmatrix}, \quad \mathbf{v}_{n+1} = \begin{bmatrix} a_{1(n+1)} \\ a_{2(n+1)} \\ \vdots \\ a_{n(n+1)} \end{bmatrix} \qquad \text{(e)}$$

Now, we show that the vector equation

$$c_1\mathbf{v}_1 + c_2\mathbf{v}_2 + \ldots + c_{n+1}\mathbf{v}_{n+1} = \mathbf{0} \qquad \text{(f)}$$

has at least one *non-trivial* solution.

If we substitute (e) into (f), we obtain the linear homogeneous system

$$\begin{bmatrix} a_{11} & a_{12} & \cdots & a_{1(n+1)} \\ a_{21} & a_{22} & \cdots & a_{2(n+1)} \\ \vdots & \vdots & & \vdots \\ a_{n1} & a_{n2} & \cdots & a_{n(n+1)} \end{bmatrix} \begin{bmatrix} c_1 \\ c_2 \\ \vdots \\ c_{n+1} \end{bmatrix} = \begin{bmatrix} 0 \\ 0 \\ \vdots \\ 0 \end{bmatrix} \qquad \text{(g)}$$

Since the linear system (f) has $(n + 1)$-unknowns but only n-equations, we know that it has at least one non-trivial solution. We conclude that the set $\{\mathbf{v}_1, \mathbf{v}_2, \ldots, \mathbf{v}_{n+1}\}$ is linearly dependent. □.

Corollary-12.1: *(Non-Uniqueness of a Basis)*

If V is a vector space where $\dim(V) = n$, then any set of n-linearly independent vectors is a basis for V.

Proof: Assume that $\{\mathbf{v}_1, \mathbf{v}_2, \ldots, \mathbf{v}_n\}$ is a linearly independent set of vectors in V, and let $\mathbf{v} \in V$ be arbitrary. In order to establish the theorem, it suffices to show that the vectors $\{\mathbf{v}_1, \mathbf{v}_2, \ldots, \mathbf{v}_n\}$ span V.

By the previous theorem, the set of vectors $\{\mathbf{v}, \mathbf{v}_1, \mathbf{v}_2, \ldots, \mathbf{v}_n\}$ must be linearly dependent. That is, there must exist scalars $c_0, c_1, \ldots, c_n$, not all zero, such that

$$c_0\mathbf{v} + c_1\mathbf{v}_1 + c_2\mathbf{v}_2 + \ldots + c_n\mathbf{v}_n = \mathbf{0}$$

Now, we must have $c_0 \neq 0$, since if $c_0 = 0$, then this would contradict the linear independence of the set $\{\mathbf{v}_1, \mathbf{v}_2, \ldots, \mathbf{v}_n\}$. Thus, we can solve this equation for $\mathbf{v}$ to obtain

$$\mathbf{v} = -\frac{c_1}{c_0}\mathbf{v}_1 - \frac{c_2}{c_0}\mathbf{v}_2 - \ldots - \frac{c_n}{c_0}\mathbf{v}_n$$

Since $\mathbf{v} \in V$ was arbitrary, we conclude that the linearly independent set of vectors $\{\mathbf{v}_1, \mathbf{v}_2, \ldots, \mathbf{v}_n\}$ spans V, and thus, is a basis for V. □.

Theorem 12.4: *(Basis Extension Theorem)*

Let $S_r = \{\mathbf{v}_1, \mathbf{v}_2, \ldots, \mathbf{v}_r\}$ be a linearly independent set of r-vectors in an n-dimensional vector space V, where $r < n$. Then, there exist vectors $\mathbf{v}_{r+1}, \mathbf{v}_{r+2}, \ldots, \mathbf{v}_n$ in V such that set of vectors

$$S_n = \{\mathbf{v}_1, \mathbf{v}_2, \ldots, \mathbf{v}_r, \mathbf{v}_{r+1}, \mathbf{v}_{r+2}, \ldots, \mathbf{v}_n\}$$

is a *basis* for V.

Proof: Observe that there must exist some vector $\mathbf{v}_{r+1} \in V$ such that $\mathbf{v}_{r+1} \notin span(S_r)$, for otberwise, S_r would span V, and we would have the $\dim(V) = r$, contrary to hypothesis. We now show that the set of vectors $S_{r+1} = \{\mathbf{v}_1, \mathbf{v}_2, \ldots, \mathbf{v}_r, \mathbf{v}_{r+1}\}$ is linearly independent.

To this end, suppose there exist scalars $\lambda_1, \lambda_2, \ldots, \lambda_{r+1}$ such that

$$\lambda_1\mathbf{v}_1 + \lambda_2\mathbf{v}_2 + \ldots + \lambda_r\mathbf{v}_r + \lambda_{r+1}\mathbf{v}_{r+1} = \mathbf{0}$$

If $\lambda_{r+1} \neq 0$, then $\mathbf{v}_{r+1}$ could be written as a linear combination of the vectors in S_r, which is not possible by our construction, so we must have $\lambda_{r+1} = 0$. Consequently, the last equation reduces to

$$\lambda_1\mathbf{v}_1 + \lambda_2\mathbf{v}_2 + \ldots + \lambda_r\mathbf{v}_r = \mathbf{0}$$

which, in turn, implies that $\lambda_1 = \lambda_2 = \ldots = \lambda_r = 0$ as well, owing to the linear independence of the set S_r. Thus S_{r+1} is a linearly independent set, as claimed. Now, if $r + 1 = n$, then we are done. If $r + 1 < n$, then in a similar fashion, we can find a vector $\mathbf{v}_{r+2} \in V$ such that $\mathbf{v}_{r+2} \notin span(S_{r+1})$ and which guarantees that the set of vectors $S_{r+2} = \{\mathbf{v}_1, \mathbf{v}_2, \ldots, \mathbf{v}_r, \mathbf{v}_{r+1}, \mathbf{v}_{r+2}\}$ is linearly independent. If $r + 2 = n$, then the set S_{r+2} is a basis for V; if not, we continue our procedure until we obtain the set S_n. □.

Exercise Set 12

---

1. Determine whether the following sets of vectors are bases for $\mathbb{R}^2$.

(a) $\left\{\begin{bmatrix} 2 \\ 4 \end{bmatrix}, \begin{bmatrix} -1 \\ 3 \end{bmatrix}\right\}$

(b) $\left\{\begin{bmatrix} -1 \\ 2 \end{bmatrix}, \begin{bmatrix} 0 \\ 1 \end{bmatrix}\right\}$

(c) $\left\{ \begin{bmatrix} 1 \\ 0 \end{bmatrix}, \begin{bmatrix} 2 \\ 5 \end{bmatrix}, \begin{bmatrix} 1 \\ 1 \end{bmatrix} \right\}$

2. Determine whether each of the following sets of vectors are bases for P_3.

(a) $\{3, 1+x, 2+x^2+x^3\}$
(b) $\{x^2-1, x, 1+x^3, 1-x^3\}$
(c) $\{3x-x^2, 2x+1, 1+x^3, x^3-x\}$

3. Show that the matrices

$$E_1 = \begin{bmatrix} 1 & 0 \\ 0 & 0 \end{bmatrix}, \ E_2 = \begin{bmatrix} 0 & 1 \\ 0 & 0 \end{bmatrix}, \ E_3 = \begin{bmatrix} 0 & 0 \\ 1 & 0 \end{bmatrix}, \ E_4 = \begin{bmatrix} 0 & 0 \\ 0 & 1 \end{bmatrix}$$

form a basis for the vector space $M(2,2)$ of all 2×2 matrices.

4. If $M(n,n)$ is the vector space of all $n \times n$ matrices, then show that $\dim(M(n,n)) = n^2$.

5. Prove that if the set $\{\mathbf{v}_1, \mathbf{v}_2, \ldots, \mathbf{v}_n\}$ is a basis for V, then the set of vectors $\{c_1\mathbf{v}_1, c_2\mathbf{v}_2, \ldots, c_n\mathbf{v}_n\}$, where $c_1, c_2, \ldots, c_n$ are all *non-zero* scalars, is also a basis for V.

6. If $\{\mathbf{v}_1, \mathbf{v}_2, \ldots, \mathbf{v}_n\}$ is a basis for V, and A is any *invertible* $n \times n$ matrix, then show that the set of vectors $\{A\mathbf{v}_1, A\mathbf{v}_2, \ldots, A\mathbf{v}_n\}$ is also a basis for V.

7. Show that if W is a subspace of a finite dimensional space V, then we must have $\dim(W) \leq \dim(V)$.

13. The Rank of a Matrix

In this lesson, we shall explore some of the vector spaces that can be constructed from the rows and columns of matrices. Our analysis will lead to a nice procedure for finding a basis for any vector space which is spanned by a finite set of vectors. Additionally, our analysis will lead to the notion of the *rank* of matrix; this concept, in turn, will provide us with an important tool for understanding the theory of general linear systems in the next lesson.

Definition: *(Row Space and Column Space)*

Given an $m \times n$ matrix

$$A = \begin{bmatrix} a_{11} & a_{12} & \dots & a_{1n} \\ a_{21} & a_{22} & \dots & a_{2n} \\ \dots & \dots & \dots & \dots \\ a_{m1} & a_{m2} & \dots & a_{mn} \end{bmatrix} \tag{13.1}$$

the vectors

$$\begin{aligned} \mathbf{r}_1 &= \begin{bmatrix} a_{11} & a_{12} & \cdots & a_{1n} \end{bmatrix} \\ \mathbf{r}_2 &= \begin{bmatrix} a_{21} & a_{22} & \cdots & a_{2n} \end{bmatrix} \\ &\vdots \\ \mathbf{r}_m &= \begin{bmatrix} a_{m1} & a_{m2} & \cdots & a_{mn} \end{bmatrix} \end{aligned} \tag{13.2}$$

are called the **row vectors** of A, while the vectors

$$\mathbf{c}_1 = \begin{bmatrix} a_{11} \\ a_{21} \\ \vdots \\ a_{m1} \end{bmatrix}, \quad \mathbf{c}_2 = \begin{bmatrix} a_{12} \\ a_{22} \\ \vdots \\ a_{m2} \end{bmatrix}, \quad \dots \quad \mathbf{c}_n = \begin{bmatrix} a_{1n} \\ a_{2n} \\ \vdots \\ a_{mn} \end{bmatrix} \tag{13.3}$$

are called the **column vectors** of A.

In the previous definition, observe that each row vector of A can be viewed as an element of the vector space $\mathbb{R}^n$, while each column vector of A is a vector in $\mathbb{R}^m$. By Theorem 11.1, the set of all linear combinations of the row vectors of A will form a subspace of $\mathbb{R}^n$, while the set of all linear combinations of the column vectors of A will be a subspace of $\mathbb{R}^m$. This observation motivates our next definition.

Definition: *(Row Space and Column Space)*

Given the $m \times n$ matrix A in (13.1), the **row space** of A, denoted by row(A), is the subspace of $\mathbb{R}^n$ which is spanned by the row vectors (13.2) of A,

$$\text{row}(A) = span\{\mathbf{r}_1, \mathbf{r}_2, \ldots, \mathbf{r}_m\} \quad (13.4$$

while the **column space** of A, denoted by col(A), is the subspace of $\mathbb{R}^m$ which is spanned by the column vectors (13.3) of A,

$$\text{col}(A) = span\{\mathbf{c}_1, \mathbf{c}_2, \ldots, \mathbf{c}_n\} \quad (13.5$$

Since given any matrix A, we know that row(A) and col(A) are vector spaces, then we should be able to find bases for these spaces, and hence determine the dimensions of these spaces as well. In order to accomplish this task, we first state and prove two useful lemmas.

Lemma 13.1: *(Significance of Row-Echelon Form)*

Let A be any $m \times n$ matrix. If we reduce A to *row-echelon form* to obtain the matrix B, then the *non-zero* row vectors of B are linearly independent.

Proof: Assume that after an appropriate sequence of elementary row operations, we reduce A to row-echelon form, to obtain the matrix B. Suppose that B has k non-zero row vectors $\mathbf{r}_1, \mathbf{r}_2, \ldots, \mathbf{r}_k$ while $\mathbf{r}_{k+1} = \mathbf{0}, \ldots, \mathbf{r}_m = \mathbf{0}$.

Assume that for $1 \le i \le k$, the *leading one* of the row vector $\mathbf{r}_i$ occurs in the column j_i. Then, by definition of row-echelon form, we must have

$$j_1 < j_2 < \ldots < j_k$$

Now $\mathbf{r}_1$ has its leading one in column j_1. Clearly $\mathbf{r}_1$ cannot be a linear combination of the remaining row vectors $\mathbf{r}_2, \ldots, \mathbf{r}_k$ since all these vectors must have zero as their j_1 component. Similarly, $\mathbf{r}_2$ has its leading one in column j_2 so $\mathbf{r}_2$ cannot be a linear combinaton of the row vectors $\mathbf{r}_3, \ldots, \mathbf{r}_k$ since all of these row vectors all must have zero as their j_2 component. Continuing in this fashion, we see that each row vector $\mathbf{r}_i$ cannot be a linear combination of the succeeding row vectors.$\mathbf{r}_{i+1}, \mathbf{r}_{i+2}, \ldots, \mathbf{r}_k$.

Now suppose there exist scalars $\lambda_1, \lambda_2, \ldots, \lambda_k$ such that

$$\lambda_1 \mathbf{r}_1 + \lambda_2 \mathbf{r}_2 + \lambda_3 \mathbf{r}_3 + \ldots + \lambda_k \mathbf{r}_k = \mathbf{0}$$

Since $\mathbf{r}_1$ is not a linear combination of the row vectors $\mathbf{r}_2, \ldots, \mathbf{r}_k$, then we must have $\lambda_1 = 0$ - since otherwise, we could solve for $\mathbf{r}_1$ in terms of the remaining row vectors. So,

$$\lambda_2\mathbf{r}_2 + \lambda_3\mathbf{r}_3 + \ldots + \lambda_k\mathbf{r}_k = \mathbf{0}$$

Similarly, since $\mathbf{r}_2$ is not a linear combination of the row vectors $\mathbf{r}_3, \ldots, \mathbf{r}_k$, then $\lambda_2 = 0$. Continuing in this fashion, we find $\lambda_3 = \ldots = \lambda_{k-1} = 0$, and we are finally left with the equation $\lambda_k\mathbf{r}_k = 0$, from which we conclude that $\lambda_k = 0$ (since $\mathbf{r}_k \neq 0$). Thus, the non-zero row vectors of B must be linearly independent. □.

Lemma 13.2

Let A be any $m \times n$ matrix. If we reduce A to *row-echelon form* to obtain the matrix B, then both matrices have the same row space, i.e.,

$$\text{row}(A) = \text{row}(B) \tag{13.6}$$

Proof: Since the rows of B are obtained by applying a sequence of elementary row operations to the rows of A, then each row of B must be a linear combination of the rows of A and consequently,

$$\text{row}(B) \subseteq \text{row}(A).$$

On the other hand, starting with B, and working backwards, we can perform the *reverse* sequence of *inverse* row operations on B to again recover the matrix A. Thus, each row of A is also a linear combination of the rows of B so

$$\text{row}(A) \subseteq \text{row}(B).$$

Thus, $\text{row}(A) = \text{row}(B)$ as claimed. □.

The previous lemma simply states that when we perform elementary row operations on any given matrix, then its row space is not affected. That is, the row space of the resulting matrix is always the same as the original matrix. We now can combine the two previous lemmas to create a useful theorem.

Theorem 13.3: *(Finding a Basis for the Row Space of a Matrix)*

Let A be any $m \times n$ matrix. If we reduce A to *row-echelon form* to obtain the matrix B, then the *non-zero* row vectors of B form a *basis* for the row space of A.

Proof: Follows directly from Lemma 13.1 and Lemma 13.2. □.

Example-1: Find a basis for the space S which is spanned by the vectors

$$\mathbf{v}_1 = \begin{bmatrix} 1 & 0 & 2 & 1 \end{bmatrix}, \mathbf{v}_2 = \begin{bmatrix} 1 & 3 & -1 & 2 \end{bmatrix}, \mathbf{v}_3 = \begin{bmatrix} 3 & 6 & 0 & 5 \end{bmatrix}$$

Solution: First, we form the matrix

$$A = \begin{bmatrix} 1 & 0 & 2 & 1 \\ 1 & 3 & -1 & 2 \\ 3 & 6 & 0 & 5 \end{bmatrix}$$

and reduce this matrix to row-echelon form to obtain

$$B = \begin{bmatrix} 1 & 0 & 2 & 1 \\ 0 & 1 & -1 & 1/3 \\ 0 & 0 & 0 & 0 \end{bmatrix}$$

By the previous theorem, the non-zero row vectors of B will serve as a basis for S. Consequently, the set of vectors

$$\left\{ \begin{bmatrix} 1 & 0 & 2 & 1 \end{bmatrix}, \begin{bmatrix} 0 & -1 & 1 & 1/3 \end{bmatrix} \right\}$$

is a basis for S.

Example-2: Given the matrix

$$A = \begin{bmatrix} 1 & 0 & -1 & 1 \\ 2 & 2 & 0 & 1 \\ 0 & -2 & -2 & 1 \end{bmatrix},$$

find:
(**a**) a basis for the row space of A
(**b**) a basis for the column space of A.

Solution:
(**a**) We first find a basis for the row space of A. Transforming this matrix to row-echelon form, we get

$$\begin{bmatrix} 1 & 0 & -1 & 1 \\ 0 & 1 & 1 & -1/2 \\ 0 & 0 & 0 & 0 \end{bmatrix}$$

By the previous theorem, the non-zero rows of this matrix form a basis for the row space of A. We conclude that the set

$$\left\{ \begin{bmatrix} 1 & 0 & -1 & 1 \end{bmatrix}, \begin{bmatrix} 0 & 1 & 1 & -1/2 \end{bmatrix} \right\}$$

is a basis for the row space of A and $\dim(\text{row}(A)) = 2$.

(**b**) Next, we find a basis for the column space of A. Observe that the column space of any matrix must be the same as the row space of its transpose. So, we take the transpose of A to obtain

$$A^T = \begin{bmatrix} 1 & 2 & 0 \\ 0 & 2 & -2 \\ -1 & 0 & -2 \\ 1 & 1 & 1 \end{bmatrix}$$

Next, we reduce A^T to row-echelon form to obtain

$$\begin{bmatrix} 1 & 2 & 0 \\ 0 & 1 & -1 \\ 0 & 0 & 0 \\ 0 & 0 & 0 \end{bmatrix}$$

Consequently, the row vectors $\begin{bmatrix} 1 & 2 & 0 \end{bmatrix}$ and $\begin{bmatrix} 0 & 1 & -1 \end{bmatrix}$ form a basis for the row space of A^T, and hence, their respective transposes will form a basis for the column space of A, i.e., the set of vectors

$$\left\{ \begin{bmatrix} 1 \\ 2 \\ 0 \end{bmatrix}, \begin{bmatrix} 0 \\ 1 \\ -1 \end{bmatrix} \right\}$$

is a basis for the column space of A and $\dim(\text{col}(A)) = 2$.

Observe that in the previous example, the row space and column space of A had the *same dimension*; this is no accident, as we will see in the next theorem.

Theorem 13.4: *(Row Space & Column Space Have the Same Dimension)*

Given any matrix A, its row space and column space always have the same dimension, i.e.,

$$\dim(\text{row}(A)) = \dim(\text{col}(A)) \tag{13.7}$$

Proof: Assume that $A = [a_{ij}]$ is an $m \times n$ where $\dim(\text{row}(A)) = k$. Suppose further, that the set of vectors $B = \{\mathbf{b}_1, \mathbf{b}_2, \ldots, \mathbf{b}_k\}$ is a basis for the row space of A where

$$\mathbf{b}_1 = \begin{bmatrix} b_{11} & b_{12} & \dots & b_{1n} \end{bmatrix}$$
$$\mathbf{b}_2 = \begin{bmatrix} b_{21} & b_{22} & \dots & b_{2n} \end{bmatrix}$$
$$\vdots$$
$$\mathbf{b}_k = \begin{bmatrix} b_{k1} & b_{k2} & \dots & b_{kn} \end{bmatrix}$$

Let $\mathbf{r}_1, \mathbf{r}_2, \dots, \mathbf{r}_m$ denote the m row vectors of A. Since B is a basis, there must exist scalars λ_{ip} such that

$$\mathbf{r}_i = \sum_{p=1}^{k} \lambda_{ip} \mathbf{b}_p \qquad (1 \le i \le m)$$

or equivalently,

$$\begin{aligned} \mathbf{r}_1 &= \lambda_{11}\mathbf{b}_1 + \lambda_{12}\mathbf{b}_2 + \dots + \lambda_{1k}\mathbf{b}_k \\ \mathbf{r}_2 &= \lambda_{21}\mathbf{b}_1 + \lambda_{22}\mathbf{b}_2 + \dots + \lambda_{2k}\mathbf{b}_k \\ &\vdots \\ \mathbf{r}_m &= \lambda_{m1}\mathbf{b}_1 + \lambda_{m2}\mathbf{b}_2 + \dots + \lambda_{mk}\mathbf{b}_k \end{aligned}$$

Since this is a set of *vector* equations, each equation must hold for each fixed component of the vectors on the left and right of each equation. So, if we equate the j-th component of each vector on the left and right of each equation, then we obtain set of scalar equations

$$\begin{aligned} a_{1j} &= \lambda_{11}b_{1j} + \lambda_{12}b_{2j} + \dots + \lambda_{1k}b_{kj} \\ a_{2j} &= \lambda_{21}b_{1j} + \lambda_{22}b_{2j} + \dots + \lambda_{2k}b_{kj} \\ &\vdots \\ a_{mj} &= \lambda_{m1}b_{1j} + \lambda_{m2}b_{2j} + \dots + \lambda_{mk}b_{kj} \end{aligned}$$

Now, we can rewrite the last set of equations in the form

$$\begin{bmatrix} a_{1j} \\ a_{2j} \\ \vdots \\ a_{mj} \end{bmatrix} = b_{1j} \begin{bmatrix} \lambda_{11} \\ \lambda_{21} \\ \vdots \\ \lambda_{m1} \end{bmatrix} + b_{2j} \begin{bmatrix} \lambda_{12} \\ \lambda_{22} \\ \vdots \\ \lambda_{m2} \end{bmatrix} + \dots + b_{kj} \begin{bmatrix} \lambda_{1k} \\ \lambda_{2k} \\ \vdots \\ \lambda_{mk} \end{bmatrix}$$

which clearly shows that the j-th column vector of A can be written as a linear combination of at most k-vectors. Since our choice for j is arbitrary, we must have

$$\dim(\mathrm{col}(A)) \le \dim(\mathrm{row}(A)) = k$$

Finally, if we consider the matrix A^T and perform the same analysis, we will find that

$$\dim(\text{col}(A^T)) \leq \dim(\text{row}(A^T))$$

which, in turn, implies that

$$\dim(\text{row}(A)) \leq \dim(\text{col}(A))$$

so the result follows. □.

Since the row space and column space of any given matrix must have the same dimension, this number is an important property of any matrix, and is called its **rank**.

Definition: *(Rank of a Matrix)*

The **rank** of a matrix A, denoted by $rank(A)$, is the dimension of the row space and column space of A, i.e.,

$$rank(A) = \dim(\text{row}(A)) = \dim(\text{col}(A)) \tag{13.8}$$

Exercise Set 13

---

1. Write down the row vectors and column vectors of the matrix

$$\begin{bmatrix} 1 & 0 & 1 \\ 2 & 1 & 3 \end{bmatrix}$$

In exercises 2-4, for each given matrix, find (a) a basis for its row space, (b) a basis for its column space, and (c) its rank.

2.

$$\begin{bmatrix} 1 & 3 & 2 \\ 3 & 0 & -1 \end{bmatrix}$$

3.

$$\begin{bmatrix} 1 & -1 & 2 \\ 4 & -1 & 3 \\ 3 & 0 & 1 \end{bmatrix}$$

4.

$$\begin{bmatrix} 2 & 1 & 1 & 0 \\ 0 & 1 & 1 & 2 \\ -1 & 3 & 1 & 0 \end{bmatrix}$$

5. Given the matrix

$$A = \begin{bmatrix} 1 & 2 & 1 \\ -1 & 3 & 2 \\ 0 & 1 & 2 \\ 1 & -2 & 1 \end{bmatrix},$$

show by direct calculation that $\dim(\text{row}(A)) = \dim(\text{col}(A)) = 3$, thus verifying Theorem 13.4.

6. Find a basis for the subspace of $\mathbb{R}^4$ which is spanned by each set of vectors

(a)

$$\left\{ \begin{bmatrix} 1 \\ 2 \\ 0 \\ 3 \end{bmatrix}, \begin{bmatrix} -1 \\ 3 \\ 0 \\ 1 \end{bmatrix}, \begin{bmatrix} -2 \\ 5 \\ 2 \\ 1 \end{bmatrix} \right\}$$

(b)

$$\left\{ \begin{bmatrix} 1 \\ -2 \\ 0 \\ 1 \end{bmatrix}, \begin{bmatrix} 0 \\ 2 \\ 1 \\ 1 \end{bmatrix}, \begin{bmatrix} 3 \\ 5 \\ 2 \\ -1 \end{bmatrix}, \begin{bmatrix} 2 \\ 2 \\ 1 \\ 4 \end{bmatrix} \right\}$$

7. If A is an $m \times n$ matrix with $m \neq n$, then show that either the row vectors of A or the column vectors of A must be linearly dependent.

8. Show that if A is an invertible $n \times n$ matrix, then $rank(A) = n$.

14. Linear Systems Revisited

In our previous work, we introduced linear systems, and developed several algorithms for their solution; however, little was said about their underlying theory. Now that we have developed sufficiently many theoretical tools and concepts, we can turn our attention towards exploring the basic theory of such systems.

Recall that an $m \times n$ linear system $A\mathbf{x} = \mathbf{b}$ is a set of m equations in n unknowns which can be written in matrix form as

$$\begin{bmatrix} a_{11} & a_{12} & \cdots & a_{1n} \\ a_{21} & a_{22} & \cdots & a_{2n} \\ \vdots & \vdots & & \vdots \\ a_{m1} & a_{m2} & \cdots & a_{mn} \end{bmatrix} \begin{bmatrix} x_1 \\ x_2 \\ \vdots \\ x_n \end{bmatrix} = \begin{bmatrix} b_1 \\ b_2 \\ \vdots \\ b_n \end{bmatrix} \tag{14.1}$$

where we make the identifications

$$A = \begin{bmatrix} a_{11} & a_{12} & \cdots & a_{1n} \\ a_{21} & a_{22} & \cdots & a_{2n} \\ \vdots & \vdots & & \vdots \\ a_{m1} & a_{m2} & \cdots & a_{mn} \end{bmatrix}, \quad \mathbf{x} = \begin{bmatrix} x_1 \\ x_2 \\ \vdots \\ x_n \end{bmatrix}, \quad \mathbf{b} = \begin{bmatrix} b_1 \\ b_2 \\ \vdots \\ b_n \end{bmatrix} \tag{14.2}$$

Recall further that we assume the matrix A and the column vector $\mathbf{b}$ are given, and our problem is to find all **solution vectors x** that satisfy (14.1).

Theorem 14.1: *(Existence of Solutions)*

Let $A\mathbf{x} = \mathbf{b}$ be the $m \times n$ linear system given by (14.1). Then this system has *at least one solution* if and only if the matrix A and the augmented matrix $[A|\mathbf{b}]$ have the same rank, i.e.,

$$rank(A) = rank([A|\mathbf{b}]) \tag{14.3}$$

Proof: First, assume that (14.1) has a solution. Observe that we can rewrite (14.1) in the form

$$x_1\mathbf{c}_1 + x_2\mathbf{c}_2 + \ldots + x_n\mathbf{c}_n = \mathbf{b} \tag{14.4}$$

where $\mathbf{c}_1, \mathbf{c}_2, \ldots, \mathbf{c}_n$ are the column vectors of A. But from (14.4), we see that $\mathbf{b}$ is a linear combination of the column vectors of A, so $\mathbf{b} \in \text{col}(A)$, and consequently,

$$rank(A) = rank([A|\mathbf{b}])$$

On the other hand, if $rank(A) = rank([A|\mathbf{b}])$, then clearly, $\mathbf{b} \in \text{col}(A)$. So there must exist scalars, say $x_1, x_2, \ldots, x_n$ such that

$$x_1\mathbf{c}_1 + x_2\mathbf{c}_2 + \ldots + x_n\mathbf{c}_n = \mathbf{b}$$

Therefore, $\mathbf{x} = [x_1, x_2, \ldots, x_n]^T$ is a solution of (14.1). $\square$.

The next theorem is of particular importance to the theory of linear systems since it provides us with a sufficient condition which guarantees that any given linear system will have a *unique solution.*

Theorem 14.2: *(Sufficient Condition for Uniqueness of a Solution)*

Let $A\mathbf{x} = \mathbf{b}$ be the $m \times n$ linear system given by (14.1). If

$$rank(A) = rank([A|\mathbf{b}]) = n$$

then the system has a *unique solution.*

Proof: By the previous theorem, we know that (14.1) must have at least one solution. Assume that $\mathbf{x}$ and $\mathbf{x}'$ are solutions of (14.1). By (14.3), we must have

$$x_1\mathbf{c}_1 + x_2\mathbf{c}_2 + \ldots + x_n\mathbf{c}_n = \mathbf{b} \quad \text{and} \quad x'_1\mathbf{c}_1 + x'_2\mathbf{c}_2 + \ldots + x'_n\mathbf{c}_n = \mathbf{b}$$

Upon subtracting these equations, we get

$$(x_1 - x'_1)\mathbf{c}_1 + (x_2 - x'_2)\mathbf{c}_2 + \ldots + (x_n - x'_n)\mathbf{c}_n = \mathbf{0}$$

But $Rank(A) = n$ so the column vectors $\mathbf{c}_1, \mathbf{c}_2, \ldots \mathbf{c}_n$ are linearly independent, and consequently,

$$(x_1 - x'_1) = (x_2 - x'_2) = \ \ldots \ = (x_n - x'_n) = 0$$

We conclude that $\mathbf{x} = \mathbf{x}'$ and the solution is unique. $\square$.

Theorem 14.3: *(Sufficient Condition for Infinitely Many Solutions)*

Let $A\mathbf{x} = \mathbf{b}$ be the $m \times n$ linear system given by (14.1). If

$$rank(A) = rank([A|\mathbf{b}]) = r$$

where $r < n$, then the system has *infinitely many solutions.*

Proof: Assume that $rank(A) = r < n$. Then A must have r linearly independent column vectors. Without loss of generality, assume that the first r column vectors of A, say $\mathbf{c}_1, \mathbf{c}_2, \ldots, \mathbf{c}_r$ are linearly independent. Consequently, we can write the remaining column vectors $\mathbf{c}_{r+1},\ldots,\mathbf{c}_n$, in terms of these linearly independent vectors as follows

$$\mathbf{c}_{r+1} = \sum_{k=1}^{r} \lambda_k^{(r+1)} \mathbf{c}_k,\ \mathbf{c}_{r+2} = \sum_{k=1}^{r} \lambda_k^{(r+2)} \mathbf{c}_k,\ldots,\mathbf{c}_n = \sum_{k=1}^{r} \lambda_k^{(n)} \mathbf{c}_k. \qquad \text{(a)}$$

where the λ's are *uniquely determined* scalars, and the superscripts $(r+1),(r+2),\ldots,(n)$ are not exponents, but simply indices. Since the linear system must have at least one solution $\mathbf{x}$ then we can write

$$x_1\mathbf{c}_1 + x_2\mathbf{c}_2 + \ldots + x_r\mathbf{c}_r + x_{r+1}\mathbf{c}_{r+1} + \ldots + x_n\mathbf{c}_n = \mathbf{b} \qquad \text{(b)}$$

Substituting (a) into (b) we obtain

$$\mathbf{c}_1 x_1' + \mathbf{c}_2 x_2' + \ldots + \mathbf{c}_r x_r' = \mathbf{b} \qquad \text{(c)}$$

where

$$x_1' = x_1 + h_1,\ x_2' = x_2 + h_2,\ldots,\ x_r' = x_r + h \qquad \text{(d)}$$

and the quantities $h_1,h_2,\ldots,h_r$ are given by

$$\begin{aligned} h_1 &= \lambda_1^{(r+1)} x_{r+1} + \lambda_1^{(r+2)} x_{r+2} + \ldots + \lambda_1^{(n)} x_n \\ h_2 &= \lambda_2^{(r+1)} x_{r+1} + \lambda_2^{(r+2)} x_{r+2} + \ldots + \lambda_2^{(n)} x_n \\ &\vdots \\ h_r &= \lambda_r^{(r+1)} x_{r+1} + \lambda_r^{(r+2)} x_{r+2} + \ldots + \lambda_r^{(n)} x_n \end{aligned} \qquad \text{(e)}$$

Since the column vectors $\mathbf{c}_1,\mathbf{c}_2,\ldots,\mathbf{c}_r$ are linearly independent, then the quantities $x_1',x_2',\ldots,x_r'$ that appear in (c) must be *unique*, and are completely determined. Now, we can arbitrarily choose the $(n-r)$ quantities $x_{r+1},\ldots,x_n$, then calculate the quantities $h_1,h_2,\ldots,h_r$ in (e), and finally solve for the quantities $x_1,x_2,\ldots,x_r$ in (d) to obtain

$$x_1 = x_1' - h_1, x_2 = x_2' - h_2, \ldots, x_r = x_r' - h_r$$

We are then guaranteed that $\mathbf{x} = [x_1,x_2,\ldots,x_n]^T$ is a solution of (14.1). Clearly, we can repeat this procedure as often as we please, thus generating infinitely many solutions. □.

Definition: *(Solution Space)*

The set of all vectors $\mathbf{x} \in \mathbb{R}^n$ that are solutions of the $m \times n$ linear homogeneous system $A\mathbf{x} = \mathbf{0}$ is called the **solution space** (or **null space**) of A.

It turns out that the solution space of a homogeneous system is particularly significant since it forms a vector subspace of $\mathbb{R}^n$. This is the content of the next theorem.

Theorem 14.4: *(Solution Space is a Subspace of $\mathbb{R}^n$)*

If $A\mathbf{x} = \mathbf{0}$ is an $m \times n$ linear homogeneous system, then its solution space is a vector subspace of $\mathbb{R}^n$.

Proof: Clearly, the solution space of A, i.e., the set of all solutions to the $m \times n$ linear system $A\mathbf{x} = \mathbf{0}$ must be a subset of $\mathbb{R}^n$. Furthermore, the solution space is not empty since $\mathbf{x} = \mathbf{0}$ is a solution. If $\mathbf{x}_1$ and $\mathbf{x}_2$ are solutions of $A\mathbf{x} = \mathbf{0}$, then observe that

$$A(\mathbf{x}_1 + \mathbf{x}_2) = A\mathbf{x}_1 + A\mathbf{x}_2 = \mathbf{0} + \mathbf{0} = \mathbf{0}$$

and for any scalar λ,

$$A(\lambda\mathbf{x}) = \lambda(A\mathbf{x}) = \lambda(\mathbf{0}) = \mathbf{0}$$

Thus, $(\mathbf{x}_1 + \mathbf{x}_2)$ and $\lambda\mathbf{x}$ are also solutions of $A\mathbf{x} = \mathbf{0}$ so the solution space is closed under the operations of vector addition and scalar multiplication; it is therefore, a subspace of $\mathbb{R}^n$. $\square$.

The reader should be careful to note that the set of solutions to the *nonhomogeneous* system $A\mathbf{x} = \mathbf{b}$ (where $\mathbf{b} \neq \mathbf{0}$) *does not* form a subspace of $\mathbb{R}^n$, since this set of vectors is not closed under the operations of vector addition and scalar multiplication.

In contrast, since the solution space of the $m \times n$ linear homogeneous system $A\mathbf{x} = \mathbf{0}$ is a vector space in its own right, it makes sense to talk about its dimension. The dimension of the solution space of A is sometimes called the **nullity** of A, which we denote by *nullity*(A). An expression for the nullity of A is given in the next theorem.

Theorem 14.5: *(Solutions of a Homogeneous System)*

Let $A\mathbf{x} = \mathbf{0}$ be an $m \times n$ linear homogeneous system where

$$rank(A) = r < n.$$

Then this system has infinitely many solutions, and the dimension of the solution space of A is $n - r$, i.e.,

$$nullity(A) = n - r \tag{14.5}$$

Proof: Clearly, $rank(A) = rank([A|\mathbf{0}] = r < n$. So by Theorem 14.3, the system $A\mathbf{x} = \mathbf{0}$ must have infinitely many solutions. It remains to show that the dimension of the solution space is $n - r$. Proceeding as in the proof of Theorem 14.3, we know that A must have r linearly independent column vectors. Without loss of generality, we again assume that the first r column vectors of A, say

$\mathbf{c}_1, \mathbf{c}_2, \ldots, \mathbf{c}_r$ are linearly independent. Consequently, we can again write the remaining column vectors $\mathbf{c}_{r+1}, \ldots, \mathbf{c}_n$ as

$$\mathbf{c}_{r+1} = \sum_{k=1}^{r} \lambda_k^{(r+1)} \mathbf{c}_k, \quad \mathbf{c}_{r+2} = \sum_{k=1}^{r} \lambda_k^{(r+2)} \mathbf{c}_k, \ldots, \mathbf{c}_n = \sum_{k=1}^{r} \lambda_k^{(n)} \mathbf{c}_k. \qquad \text{(A)}$$

where the λ's are *uniquely determined* scalars. Since the linear system must have at least one solution $\mathbf{x}$, then we can write

$$x_1\mathbf{c}_1 + x_2\mathbf{c}_2 + \ldots + x_r\mathbf{c}_r + x_{r+1}\mathbf{c}_{r+1} + \ldots + x_n\mathbf{c}_n = \mathbf{0} \qquad \text{(B)}$$

Substituting (A) into (B), we obtain

$$\mathbf{c}_1(x_1 + h_1) + \mathbf{c}_2(x_2 + h_2) + \ldots + \mathbf{c}_r(x_r + h_r) = \mathbf{0} \qquad \text{(C)}$$

where the quantities $h_1, h_2, \ldots, h_r$ are again given by

$$\begin{aligned} h_1 &= \lambda_1^{(r+1)} x_{r+1} + \lambda_1^{(r+2)} x_{r+2} + \ldots + \lambda_1^{(n)} x_n \\ h_2 &= \lambda_2^{(r+1)} x_{r+1} + \lambda_2^{(r+2)} x_{r+2} + \ldots + \lambda_2^{(n)} x_n \\ &\vdots \\ h_r &= \lambda_r^{(r+1)} x_{r+1} + \lambda_r^{(r+2)} x_{r+2} + \ldots + \lambda_r^{(n)} x_n \end{aligned} \qquad \text{(D)}$$

Since the column vectors $\mathbf{c}_1, \mathbf{c}_2, \ldots, \mathbf{c}_r$ are linearly independent, then the quantities in the parentheses of (C) must all be zero, and we then obtain the equations

$$\begin{aligned} x_1 &= -\left(\lambda_1^{(r+1)} x_{r+1} + \lambda_1^{(r+2)} x_{r+2} + \ldots + \lambda_1^{(n)} x_n\right) \\ x_2 &= -\left(\lambda_2^{(r+1)} x_{r+1} + \lambda_2^{(r+2)} x_{r+2} + \ldots + \lambda_2^{(n)} x_n\right) \\ &\vdots \\ x_r &= -\left(\lambda_r^{(r+1)} x_{r+1} + \lambda_r^{(r+2)} x_{r+2} + \ldots + \lambda_r^{(n)} x_n\right) \end{aligned} \qquad \text{(E)}$$

Now, let $x_{r+1} = t_1, x_{r+2} = t_2, \ldots, x_n = t_{n-r}$ where $t_1, t_2, \ldots, t_{n-r}$ are arbitrary parameters. Then it is easy to see that the vector

$$\mathbf{x} = \begin{bmatrix} x_1 \\ x_2 \\ \vdots \\ x_r \\ x_{r+1} \\ \vdots \\ x_n \end{bmatrix} = \begin{bmatrix} -\left(\lambda_1^{(r+1)} t_1 + \lambda_1^{(r+2)} t_2 + \ldots + \lambda_1^{(n)} t_{n-r}\right) \\ -\left(\lambda_2^{(r+1)} t_1 + \lambda_2^{(r+2)} t_2 + \ldots + \lambda_2^{(n)} t_{n-r}\right) \\ \vdots \\ -\left(\lambda_r^{(r+1)} t_1 + \lambda_r^{(r+2)} t_2 + \ldots + \lambda_r^{(n)} t_{n-r}\right) \\ t_1 \\ \vdots \\ t_{n-r} \end{bmatrix} \qquad \text{(F)}$$

is a solution of the given linear system. Furthermore, we can rewrite the last equation in the form

$$\mathbf{x} = t_1 \begin{bmatrix} -\lambda_1^{(r+1)} \\ -\lambda_2^{(r+1)} \\ \vdots \\ -\lambda_r^{(r+1)} \\ 1 \\ 0 \\ 0 \\ \vdots \\ 0 \end{bmatrix} + t_2 \begin{bmatrix} -\lambda_1^{(r+2)} \\ -\lambda_2^{(r+2)} \\ \vdots \\ -\lambda_r^{(r+2)} \\ 0 \\ 1 \\ 0 \\ \vdots \\ 0 \end{bmatrix} +\ldots+ t_{n-r} \begin{bmatrix} -\lambda_1^{(n)} \\ -\lambda_2^{(n)} \\ \vdots \\ -\lambda_r^{(n)} \\ 0 \\ 0 \\ 0 \\ \vdots \\ 1 \end{bmatrix} \tag{G}$$

We see that the $(n-r)$ column vectors in (G) span the solution space of A We leave it to the reader to show that these $(n-r)$ column vectors are linearly independent. Thus, they form a basis for the solution space of A, and the dimension of this solution space is $(n-r)$. □.

The previous theorem tells us that if we are given a linear homogeneous system $A\mathbf{x} = \mathbf{0}$ where the rank of A is less than the number of unknowns, then the solution space will contain infinitely many solutions. Furthermore, the nullity of A will be equal to the difference between the number of unknowns and the rank of A.

Stated differently, under the same assumptions as in Theorem 14.5, we have shown that there must exist $(n-r)$ basis vectors, say $\{\mathbf{x}_1, \mathbf{x}_2, \ldots, \mathbf{x}_{n-r}\}$ for the solution space of $A\mathbf{x} = \mathbf{0}$ such that *any* solution of this system can be written in the form

$$\mathbf{x} = \lambda_1\mathbf{x}_1 + \lambda_2\mathbf{x}_2 +\ldots+\lambda_{n-r}\mathbf{x}_{n-r} \tag{14.6}$$

We shall call (14.6) the **general solution** of the homogeneous system.

Theorem 14.6: *(General Solution of a Non-Homogeneous System)*

Assume that the non-homogeneous linear system $A\mathbf{x} = \mathbf{b}$ (where $\mathbf{b} \neq \mathbf{0}$) has at least one particular solution, $\mathbf{x}_p$. Then *all* solutions of this system can be written in the form

$$\mathbf{x} = \mathbf{x}_p + \mathbf{x}_h \tag{14.7}$$

where $\mathbf{x}_h$ is the *general solution* of the corresponding homogeneous linear system $A\mathbf{x} = \mathbf{0}$.

Proof: Let $\mathbf{x}$ be *any* solution of the non-homogeneous system, and let $\mathbf{x}_p$ be any fixed solution of that system. We must have both

$$A\mathbf{x} = \mathbf{b} \text{ and } A\mathbf{x}_p = \mathbf{b}$$

Subtracting the last two equations, we get

$$A\mathbf{x} - A\mathbf{x}_p = A(\mathbf{x} - \mathbf{x}_p) = \mathbf{0}$$

and consequently, $\mathbf{x} - \mathbf{x}_p$ is a solution of the corresponding homogeneous system, say $\mathbf{x}_h$. But then $\mathbf{x} = \mathbf{x}_p + \mathbf{x}_h$ as claimed. □.

Since any solution of the linear system $A\mathbf{x} = \mathbf{b}$ can be obtained from (14.7), we call this the **general solution** of the non-homogeneous system. Thus, the previous theorem tells us that *the general solution of a non-homogeneous system can always be written as the sum of a particular (fixed) solution of the given system and the general solution of the corresponding homogeneous system.*

Example-1: Discuss the general solution of the 3×4 linear system $A\mathbf{x} = \mathbf{b}$ given by

$$\begin{bmatrix} 2 & 3 & 1 & 1 \\ 1 & -1 & -1 & -1 \\ 1 & 1 & 1 & 2 \end{bmatrix} \begin{bmatrix} x_1 \\ x_2 \\ x_3 \\ x_4 \end{bmatrix} = \begin{bmatrix} 2 \\ 1 \\ 0 \end{bmatrix}$$

Solution: Let's solve this system and see what happens. We reduce the augmented matrix to reduced-row echelon form to get

$$\left[\begin{array}{cccc|c} 2 & 3 & 1 & 1 & 2 \\ 1 & -1 & -1 & -1 & 1 \\ 1 & 1 & 1 & 2 & 0 \end{array}\right] \Rightarrow \left[\begin{array}{cccc|c} 1 & 0 & 0 & \frac{1}{2} & \frac{1}{2} \\ 0 & 1 & 0 & -\frac{3}{4} & \frac{3}{4} \\ 0 & 0 & 1 & \frac{9}{4} & -\frac{5}{4} \end{array}\right]$$

The linear system which corresponds to the last matrix is

$$x_1 + \frac{1}{2}x_4 = \frac{1}{2}, \quad x_2 - \frac{3}{4}x_4 = \frac{3}{4}, \quad x_3 + \frac{9}{4}x_4 = -\frac{5}{4}$$

If we let $x_4 = s$, where s is an arbitrary parameter, then the solution vector $\mathbf{x}$ is given by

$$\mathbf{x} = \begin{bmatrix} x_1 \\ x_2 \\ x_3 \\ x_4 \end{bmatrix} = \begin{bmatrix} \frac{1}{2} \\ \frac{3}{4} \\ -\frac{5}{4} \\ 0 \end{bmatrix} + s \begin{bmatrix} -\frac{1}{2} \\ \frac{3}{4} \\ -\frac{9}{4} \\ 1 \end{bmatrix} = \mathbf{x}_p + \mathbf{x}_h$$

Observe that here $n = 4$ (since the system contains 4 unknowns), and *rank*(*A*) = 3 (since there are 3 non-zero rows in the row-reduced matrix), so according to Theorem 14.5, the dimension of the solution space of the *homogeneous system* is 4 – 3 = 1. Thus, the general solution $\mathbf{x}_h$ of the homogeneous system must be

$$\mathbf{x}_h = s\begin{bmatrix} -\frac{1}{2} \\ \frac{3}{4} \\ -\frac{9}{4} \\ 1 \end{bmatrix}$$

while a particular solution $\mathbf{x}_p$ of the *non-homogeneous system* is

$$\mathbf{x}_p = \begin{bmatrix} \frac{1}{2} \\ \frac{3}{4} \\ -\frac{5}{4} \\ 0 \end{bmatrix}$$

Consequently, the general solution of the non-homogeneous system is given by $\mathbf{x} = \mathbf{x}_p + \mathbf{x}_h$, as claimed in the Theorem 14.6.

Exercise Set 14

In exercises 1-4, find the rank of each matrix by reducing each matrix to row-echelon form.

1.

$$\begin{bmatrix} 1 & 2 \\ 3 & 1 \end{bmatrix}$$

2.

$$\begin{bmatrix} -1 & 1 \\ 1 & 2 \\ 3 & 0 \end{bmatrix}$$

3.

$$\begin{bmatrix} 2 & 0 & 1 & -1 \\ 6 & 1 & 0 & 3 \\ 2 & 1 & 0 & -1 \end{bmatrix}$$

4.

$$\begin{bmatrix} 1 & -1 & 2 & 1 \\ 3 & 1 & 1 & 2 \\ 0 & 1 & 2 & 1 \\ 3 & 1 & 2 & -1 \end{bmatrix}$$

In Exercises 5-8, find a basis for the solution space of each linear homogeneous system, and write down its general solution.

5.

$$\begin{bmatrix} 2 & 1 \\ -1 & 4 \end{bmatrix}\begin{bmatrix} x \\ y \end{bmatrix} = \begin{bmatrix} 0 \\ 0 \end{bmatrix}$$

6.

$$\begin{bmatrix} 1 & 0 & -1 \\ 2 & 1 & 3 \\ -1 & -1 & -4 \end{bmatrix}\begin{bmatrix} x \\ y \\ z \end{bmatrix} = \begin{bmatrix} 0 \\ 0 \\ 0 \end{bmatrix}$$

7.

$$\begin{bmatrix} 1 & 2 & 1 & 0 \\ 3 & 1 & 2 & -1 \\ 3 & 0 & 2 & 1 \end{bmatrix}\begin{bmatrix} x \\ y \\ z \\ w \end{bmatrix} = \begin{bmatrix} 0 \\ 0 \\ 0 \\ 0 \end{bmatrix}$$

8.

$$\begin{bmatrix} 1 & -1 & 1 \\ 0 & 2 & 1 \end{bmatrix}\begin{bmatrix} x \\ y \\ z \end{bmatrix} = \begin{bmatrix} 0 \\ 0 \end{bmatrix}$$

In exercises 9-13, if the non-homogeneous system is consistent, then find its general solution, and write in the form $\mathbf{x} = \mathbf{x}_p + \mathbf{x}_h$ *given by (14.7).*

9.

$$\begin{bmatrix} 1 & 1 \\ -3 & -4 \end{bmatrix} \begin{bmatrix} x \\ y \end{bmatrix} = \begin{bmatrix} 2 \\ 1 \end{bmatrix}$$

10.

$$\begin{bmatrix} 1 & 0 & -1 \\ 2 & 1 & 3 \\ -1 & 1 & -4 \end{bmatrix} \begin{bmatrix} x \\ y \\ z \end{bmatrix} = \begin{bmatrix} 1 \\ 3 \\ 0 \end{bmatrix}$$

11.

$$\begin{bmatrix} 1 & 2 & 1 & 0 \\ 3 & 1 & 2 & -1 \\ 3 & 0 & 2 & 1 \end{bmatrix} \begin{bmatrix} x \\ y \\ z \\ w \end{bmatrix} = \begin{bmatrix} 3 \\ 4 \\ 3 \\ 0 \end{bmatrix}$$

12.

$$\begin{bmatrix} 1 & -1 & 1 \\ 0 & 2 & 1 \end{bmatrix} \begin{bmatrix} x \\ y \\ z \end{bmatrix} = \begin{bmatrix} 1 \\ 0 \end{bmatrix}$$

13.

$$\begin{bmatrix} 1 & -1 & 1 & 1 \\ 2 & 1 & 1 & 1 \\ 1 & 2 & 0 & 0 \\ 3 & 0 & 2 & 2 \end{bmatrix} \begin{bmatrix} x \\ y \\ z \\ w \end{bmatrix} = \begin{bmatrix} 1 \\ 3 \\ 2 \\ 4 \end{bmatrix}$$

Unit IV: More About Vector Spaces

Amalie Emmy Noether (1882-1935)

Emmy Noether was born in Erlangen, Germany to a talented Jewish family. She attended the University of Erlangen, and completed her doctoral work in 1907. Subsequently, she held positions at the Mathematical Institute of Erlangen, then at the University of Gottingen, where she worked with David Hilbert and Felix Klein. In 1932, due to Nazi oppression, she was forced to leave Gottingen, and was granted a professorship at Bryn Mawr College in Pennsylvania, a position she held until her death.

During her productive career, she made major contributions to algebra, and the theory of algebraic invariants where she studied mathematical expressions that remain invariant under certain groups of transformations. Later on, Noether investigated continuous symmetries, and demonstrated that each conservation law in Physics can be associated with a differentiable symmetry of a physical system. This important work, today known as Noether's Theorem, continues to have extreme importance in Theoretical Physics. Noether is generally recognized as one of the greatest mathematicians of the twentieth century.

15. Sums and Direct Sums of Subspaces

We now turn our attention to alternative means of building subspaces of a vector space V. We have already seen that if S and T are subspaces of a vector space V, then $S \cap T$ is again a subspace of V. We can also build subspaces by the operation of addition. Let's see how this works.

Definition: *(Sum of Subspaces)*

Let S and T be subspaces of a vector space V. Then the **sum** of S and T, denoted by $S + T$, is the set of vectors given by

$$S + T = \{\mathbf{v} : \mathbf{v} = \mathbf{s} + \mathbf{t}, \text{ where } \mathbf{s} \in S \text{ and } \mathbf{t} \in T\} \tag{15.1}$$

Now, it turns outs that if we are given any two subspaces of a vector space V, then their sum will be a subspace of V. This gives us a new way to build subspaces, and is the content of the first theorem.

Theorem 15.1: *($S + T$ is a subspace of V)*

If S and T are subspaces of a vector space V, then $S + T$ is also a subspace of V.

Proof: Let $\mathbf{v}_1, \mathbf{v}_2 \in S + T$. By definition, there exist vectors $\mathbf{s}_1, \mathbf{s}_2 \in S$ and $\mathbf{t}_1, \mathbf{t}_2 \in T$ such that $\mathbf{v}_1 = \mathbf{s}_1 + \mathbf{t}_1$ and $\mathbf{v}_2 = \mathbf{s}_2 + \mathbf{t}_2$. Since S and T are subspaces of V, they are closed under the operations of vector addition and scalar multiplication. Consequently,

$$\mathbf{v}_1 + \mathbf{v}_2 = (\mathbf{s}_1 + \mathbf{s}_2) + (\mathbf{t}_1 + \mathbf{t}_2) \in S + T$$

since $\mathbf{s}_1 + \mathbf{s}_2 \in S$ and $\mathbf{t}_1 + \mathbf{t}_2 \in T$. Similarly, for any scalar λ, we have

$$\lambda \mathbf{v}_1 = \lambda(\mathbf{s}_1 + \mathbf{t}_1) = \lambda \mathbf{s}_1 + \lambda \mathbf{t}_1 \in S + T$$

since $\lambda \mathbf{s}_1 \in S$ and $\lambda \mathbf{t}_1 \in T$. This completes the proof. □.

Example-1: Let $V = \mathbb{R}^2$ and consider the subspaces of $\mathbb{R}^2$ given by

$$S = \left\{ \begin{bmatrix} a \\ 2a \end{bmatrix} : a \in \mathbb{R} \right\} \text{ and } T = \left\{ \begin{bmatrix} 0 \\ b \end{bmatrix} : b \in \mathbb{R} \right\}$$

Here, S consists of all vectors along the line $y = 2x$ in the xy-plane, while T consists of all vectors lying along the y-axis. Clearly, any plane vector

$\mathbf{v} \in \mathbb{R}^2$ can be written as the sum of a vector in S and a vector in T since

$$\mathbf{v} = \begin{bmatrix} v_1 \\ v_2 \end{bmatrix} = \begin{bmatrix} v_1 \\ 2v_1 \end{bmatrix} + \begin{bmatrix} 0 \\ v_2 - 2v_1 \end{bmatrix}$$

Consequently, $\mathbb{R}^2 = S + T$.

Theorem 15.2: *(Dimension of the Sum of Subspaces)*

If S and T are finite dimensional subspaces of a vector space V then

$$\dim(S+T) = \dim(S) + \dim(T) - \dim(S \cap T) \tag{15.2}$$

Proof: Clearly, $S \cap T$ is a subspace of S and T. Now, assume that $\{\boldsymbol{\varepsilon}_1, \boldsymbol{\varepsilon}_2, \ldots, \boldsymbol{\varepsilon}_r\}$ is a basis for $S \cap T$ so $\dim(S \cap T) = r$. Also, let $\dim(S) = m$ and $\dim(T) = n$. By the Basis Extension Theorem (Theorem 12.4), we can extend the basis for $S \cap T$ to obtain a basis B for S and a basis B' for T, such that

$$B = \{\boldsymbol{\varepsilon}_1, \boldsymbol{\varepsilon}_2, \ldots, \boldsymbol{\varepsilon}_r, \mathbf{s}_1, \mathbf{s}_2, \ldots, \mathbf{s}_{m-r}\}$$
$$B' = \{\boldsymbol{\varepsilon}_1, \boldsymbol{\varepsilon}_2, \ldots, \boldsymbol{\varepsilon}_r, \mathbf{t}_1, \mathbf{t}_2, \ldots, \mathbf{t}_{n-r}\}$$

where the vectors $\mathbf{s}_1, \mathbf{s}_2, \ldots, \mathbf{s}_{m-r} \in S$ and the vectors $\mathbf{t}_1, \mathbf{t}_2, \ldots, \mathbf{t}_{n-r} \in T$. This state of affairs is shown in Figure 15.1 below.

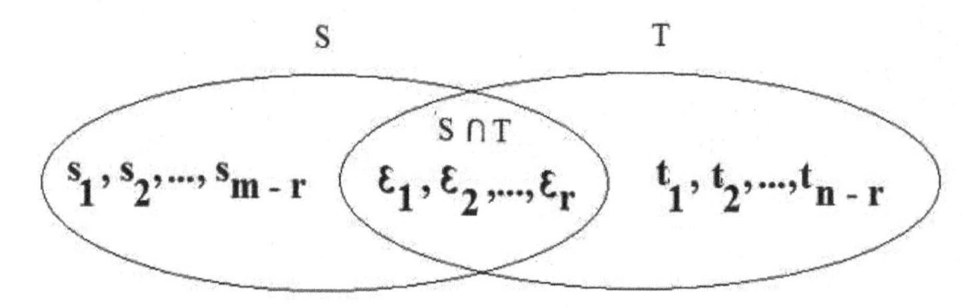

Figure 15.1: General setting for the proof of Theorem 15.2.

We now show that

$$B \cup B' = \{\boldsymbol{\varepsilon}_1, \boldsymbol{\varepsilon}_2, \ldots, \boldsymbol{\varepsilon}_r, \mathbf{s}_1, \mathbf{s}_2, \ldots, \mathbf{s}_{m-r}, \mathbf{t}_1, \mathbf{t}_2, \ldots, \mathbf{t}_{n-r}\}$$

is a basis for $S+T$. It is obvious that $B \cup B'$ spans $S+T$, so we only have to show that $B \cup B'$ is a linearly independent set. To this end, suppose there exist scalars $a_1, a_2, \ldots, a_r$; $b_1, b_2, \ldots, b_{m-r}$, and $c_1, c_2, \ldots c_{n-r}$ such that

$$(a_1\boldsymbol{\varepsilon}_1 + \ldots + a_r\boldsymbol{\varepsilon}_r + b_1\mathbf{s}_1 + \ldots + b_{m-r}\mathbf{s}_{m-r}) + c_1\mathbf{t}_1 + \ldots + c_{n-r}\mathbf{t}_{n-r} = \mathbf{0} \tag{a}$$

Let $\mathbf{s}$ denote the quantity in the parentheses in (a); it is clear that $\mathbf{s} \in S$ We can solve this equation for $\mathbf{s}$ to obtain

$$\mathbf{s} = -(c_1\mathbf{t}_1 + \ldots + c_{n-r}\mathbf{t}_{n-r}) \tag{b}$$

so that $\mathbf{s} \in T$ as well, and consequently, $\mathbf{s} \in S \cap T$. Thus, there must exist scalars, say $a'_1, a'_2, \ldots, a'_r$, such that

$$\mathbf{s} = a'_1\boldsymbol{\varepsilon}_1 + a'_2\boldsymbol{\varepsilon}_2 + \ldots + a'_r\boldsymbol{\varepsilon}_r \tag{c}$$

Combining (b) and (c), we find that

$$a'_1\boldsymbol{\varepsilon}_1 + a'_2\boldsymbol{\varepsilon}_2 + \ldots + a'_r\boldsymbol{\varepsilon}_r + (c_1\mathbf{t}_1 + c_2\mathbf{t}_2 + \ldots + c_{n-r}\mathbf{t}_{n-r}) = \mathbf{0} \tag{d}$$

which implies $a'_1 = a'_2 = \ldots = a'_r = 0$ and $c_1 = c_2 = \ldots = c_{n-r} = 0$ since B' is a basis for T. Substituting the result $c_1 = c_2 = \ldots = c_{n-r} = 0$ into (a) gives

$$a_1\boldsymbol{\varepsilon}_1 + \ldots + a_r\boldsymbol{\varepsilon}_r + b_1\mathbf{s}_1 + \ldots + b_{m-r}\mathbf{s}_{m-r} = \mathbf{0}$$

from which it follows that $a_1 = a_2 = \ldots = a_r = 0$ and $b_1 = b_2 = \ldots = b_{m-r} = 0$, since B. is a basis for S. Thus, $B \cup B'$ is a linearly independent set that spans $S \cup T$ and so, it is a basis for $S + T$. Finally, observe that since $B \cup B'$ contains $m + n - r$ linearly independent vectors, then

$$\dim(S + T) = m + n - r = \dim(S) + \dim(T) - \dim(S \cap T)$$

as claimed. □.

Definition: *(Direct Sum of Subspaces)*

Let S and T be subspaces of a vector space V. We say that V is the **direct sum** of S and T, and write

$$V = S \oplus T \tag{15.3}$$

if for every $\mathbf{v} \in V$ there exist vectors $\mathbf{s} \in S$ and $\mathbf{t} \in T$ such that

$$\mathbf{v} = \mathbf{s} + \mathbf{t} \tag{15.4}$$

where the decomposition is *unique*.

Theorem 15.3: *(Necessary & Sufficient Condition for a Direct Sum)*

Let S and T be subspaces of a vector space V. Then $V = S \oplus T$ if and only if
(a) $V = S + T$ and
(b) $S \cap T = \{\mathbf{0}\}$.

Proof: First, assume that $V = S \oplus T$; it then follows immediately that $V = S + T$. To establish (b), assume that $\mathbf{v} \in S \cap T$. Since we have both

$$\mathbf{v} = \mathbf{0} + \mathbf{v} \text{ and } \mathbf{v} = \mathbf{v} + \mathbf{0}$$

and the decomposition of any vector must be *unique*, then $\mathbf{v} = \mathbf{0}$, and consequently $S \cap T = \{\mathbf{0}\}$.

Now suppose that both properties (a) and (b) are satisfied. Given an arbitrary $\mathbf{v} \in V$ there must exist vectors $\mathbf{s} \in S$ and $\mathbf{t} \in T$ such that $\mathbf{v} = \mathbf{s} + \mathbf{t}$. We must show that this decomposition is unique. To this end, suppose that there also exist vectors $\mathbf{s}' \in S$ and $\mathbf{t}' \in T$ such that

$$\mathbf{v} = \mathbf{s} + \mathbf{t} = \mathbf{s}' + \mathbf{t}'$$

Then $\mathbf{s} - \mathbf{s}' = \mathbf{t}' - \mathbf{t}$. Consequently, $\mathbf{s} - \mathbf{s}' \in S \cap T$ and $\mathbf{t}' - \mathbf{t} \in S \cap T$. But by property (b), the only element in common to both S and T is the zero vector, so

$$\mathbf{s} - \mathbf{s}' = \mathbf{t}' - \mathbf{t} = \mathbf{0}$$

and the decomposition is unique. □.

Example-2: Let $V = \mathbb{R}^3$ and consider the subspaces of $\mathbb{R}^3$ given by

$$S = \left\{ \begin{bmatrix} a \\ b \\ 0 \end{bmatrix} : a, b \in \mathbb{R} \right\} \text{ and } T = \left\{ \begin{bmatrix} c \\ c \\ c \end{bmatrix} : c \in \mathbb{R} \right\}$$

Here S represents the set of all vectors in the xy-plane, and T is a line through the origin. Clearly, $V = S + T$ since if we are given an arbitrary vector $\mathbf{v} \in V$, we can write

$$\mathbf{v} = \begin{bmatrix} v_1 \\ v_2 \\ v_3 \end{bmatrix} = \begin{bmatrix} v_1 - v_3 \\ v_2 - v_3 \\ 0 \end{bmatrix} + \begin{bmatrix} v_3 \\ v_3 \\ v_3 \end{bmatrix} = \mathbf{s} + \mathbf{t}$$

where

$$\mathbf{s} = \begin{bmatrix} v_1 - v_3 \\ v_2 - v_3 \\ 0 \end{bmatrix} \in S \text{ and } \mathbf{t} = \begin{bmatrix} v_3 \\ v_3 \\ v_3 \end{bmatrix} \in T.$$

Also, $S \cap T = \{\mathbf{0}\}$, so by Theorem 15.3, we conclude that $\mathbb{R}^3 = S \oplus T$.

Corollary 15.1: *(Dimension of $S \oplus T$)*

If V is a vector space and $V = S \oplus T$, then

$$\dim(V) = \dim(S \oplus T) = \dim(S) + \dim(T) \tag{15.5}$$

Proof: If $V = S \oplus T$, then by Theorem 15.3, $V = S + T$ and $S \cap T = \{\mathbf{0}\}$ Consequently, $\dim(S \cap T) = 0$, and

$$\dim(S \oplus T) = \dim(V) = \dim(S + T) = \dim(S) + \dim(T)$$

as claimed. □.

Now, if S and T are subspaces of a vector space V, and $V = S \oplus T$, then S and T are called **complementary subspaces** of V. Equivalently, we say that S is the **complement** of T, and T is the **complement** of S. The next theorem shows that any given subspace of a finite dimensional vector space always has a complement.

Theorem 15.4: *(Existence of a Complementary Subspace)*

If V is any n-dimensional vector space, and T is an r-dimensional subspace of V, then there exists an $(n - r)$-dimensional subspace S of V such that $V = S \oplus T$.

Proof: If $T = \{\mathbf{0}\}$, or if $T = V$, then there is nothing to show. So, assume that the set $B = \{\mathbf{t}_1, \mathbf{t}_2, \ldots, \mathbf{t}_r\}$ is a basis for T, where $1 \le r < n$. By the Basis Extension Theorem, we know there exists a set B' of $(n - r)$ vectors, say $B' = \{\mathbf{s}_1, \mathbf{s}_2, \ldots, \mathbf{s}_{n-r}\}$, such that

$$B' \cup B = \{\mathbf{s}_1, \mathbf{s}_2, \ldots, \mathbf{s}_{n-r}, \mathbf{t}_1, \mathbf{t}_2, \ldots, \mathbf{t}_r\}$$

is a basis for V.

Now, let $S = span\{\mathbf{s}_1, \mathbf{s}_2, \ldots, \mathbf{s}_{n-r}\}$ so S is a subspace of V. We shall show that $V = S \oplus T$. Given any $\mathbf{v} \in V$, there exist scalars $\lambda_1, \lambda_2, \ldots, \lambda_n$ such that

$$\mathbf{v} = \lambda_1\mathbf{s}_1 + \lambda_2\mathbf{s}_2 + \ldots, \lambda_{n-r}\mathbf{s}_{n-r} + \lambda_{(n-r)+1}\mathbf{t}_1 + \lambda_{(n-r)+2}\mathbf{t}_2 + \ldots + \lambda_n\mathbf{t}_r$$

or equivalently, $\mathbf{v} = \mathbf{s} + \mathbf{t}$, where

$$\mathbf{s} = \lambda_1\mathbf{s}_1 + \lambda_2\mathbf{s}_2 + \ldots, \lambda_{n-r}\mathbf{s}_{n-r} \in S$$

$$\mathbf{t} = \lambda_{(n-r)+1}\mathbf{t}_1 + \lambda_{(n-r)+2}\mathbf{t}_2 + \ldots + \lambda_n\mathbf{t}_r \in T$$

Thus, $V = S + T$.

We now show that $S \cap T = \{\mathbf{0}\}$. To this end, assume that an arbitrary vector $\mathbf{v} \in S \cap T$, so $\mathbf{v} \in S$ and $\mathbf{v} \in T$. Now, $\mathbf{v} \in S$ implies there exist scalars $a_1, a_2, \ldots a_{n-r}$ such that

$$\mathbf{v} = a_1\mathbf{s}_1 + a_2\mathbf{s}_2 + \ldots + a_{n-r}\mathbf{s}_{n-r} \qquad \text{(A)}$$

Similarly, $\mathbf{v} \in T$ implies there exist scalars $b_1, b_2, \ldots, b_r$ such that

$$\mathbf{v} = b_1\mathbf{t}_1 + b_2\mathbf{t}_2 + \ldots + b_r\mathbf{t}_r \tag{B}$$

Subtracting (B) from (A), we obtain

$$a_1\mathbf{s}_1 + a_2\mathbf{s}_2 + \ldots + a_{n-r}\mathbf{s}_{n-r} - b_1\mathbf{t}_1 - b_2\mathbf{t}_2 - \ldots - b_r\mathbf{t}_r = \mathbf{0}$$

which implies $a_1 = a_2 = \cdots = a_{n-r} = 0$, and $b_1 = b_2 = \cdots b_r = 0$ since $B' \cup B$ is a basis for V. Thus, $S \cap T = \{\mathbf{0}\}$. Recapping, we have shown that $V = S + T$ and $S \cap T = \{\mathbf{0}\}$, so the result follows immediately from Theorem 15.3. □.

Exercise Set 15

---

1. Given that S and T are the subspaces of $\mathbb{R}^3$ given by

$$S = \left\{ \begin{bmatrix} a \\ b \\ 0 \end{bmatrix} : a, b \in \mathbb{R} \right\} \text{ and } T = \left\{ \begin{bmatrix} 0 \\ 0 \\ c \end{bmatrix} : c \in R \right\}$$

(a) Show that $\mathbb{R}^3 = S \oplus T$

(b) Describe the subspaces S and T geometrically.

2. Let P_4 denote the vector space of all polynomials with degree less than or equal to 4. Let S be the subspace of P_4 given by

$$S = \{\mathbf{p}(x) \in P_4 : \mathbf{p}(0) = \mathbf{0}\}$$

(a) Find the complementary subspace T of S.

(b) Find a basis for T.

(c) Verify that $P_4 = S \oplus T$.

3. Let $V = M(2,2)$, the space of all 2×2 matrices. Consider the subspaces of $M(2,2)$ given by

$$S = \left\{ A \in M(2,2) : A = \begin{bmatrix} a_{11} & a_{12} \\ 0 & a_{22} \end{bmatrix} \right\}$$

$$T = \left\{ B \in M(2,2) : B = \begin{bmatrix} b_{11} & 0 \\ b_{21} & b_{22} \end{bmatrix} \right\}$$

(a) Describe the elements of the subspaces S and T.

(b) What is the subspace $S \cap T$?

(c) Show that $V = S + T$ but $V \neq S \oplus T$.

(d) Verify that $\dim(S + T) = \dim(S) + \dim(T) - \dim(S \cap T)$.

4. Let $V = M(2,2)$, the vector space of all 2×2 matrices. Consider the sets of matrices

$$S = \{A \in M(2,2) : A^T = A\}$$
$$T = \{B \in M(2,2) : B^T = -B\}$$

(a) Show that S and T are subspaces of V.
(b) Describe the elements of the sets S and T.
(c) Show that $V = S \oplus T$.
(d) What are the dimensions of the subspaces S and T?
(e) Verify that $\dim(S \oplus T) = \dim(S) + \dim(T)$.

5. Let $V = P_n$ the space of polynomial functions of degree less than or equal to n. Consider the sets of polynomials

$$E_n = \{\mathbf{p}(x) \in P_n : \mathbf{p}(-x) = \mathbf{p}(x) \text{ for all } x \in \mathbb{R}\}$$
$$O_n = \{\mathbf{p}(x) \in P_n : \mathbf{p}(-x) = -\mathbf{p}(x) \text{ for all } x \in \mathbb{R}\}$$

(a) Describe the elements of the sets E_n and O_n.
(b) Show that E_n is a subspace of V.
(c) Show that O_n is a subspace of V.
(d) What are the dimensions of the subspaces E_n and O_n?
(e) Show that $P_n = E_n \oplus O_n$.

6. Let $V = C[-\infty, \infty]$ be the space of real valued functions that are continuous everywhere on $\mathbb{R}$. Consider the sets of functions

$$E = \{\mathbf{f}(x) \in C[-\infty, \infty] : \mathbf{f}(-x) = \mathbf{f}(x) \text{ for all } x \in \mathbb{R}\}$$
$$O = \{\mathbf{f}(x) \in C[-\infty, \infty] : \mathbf{f}(-x) = -\mathbf{f}(x) \text{ for all } x \in \mathbb{R}\}$$

The elements of E and O are called **even functions** and **odd functions**, respectively.
(a) Show that E and O are subspaces of $C[-\infty, \infty]$.
(b) Show that any function $\mathbf{f}(x) \in C[-\infty, \infty]$ can be written in the form:

$$\mathbf{f}(x) = \mathbf{f}_E(x) + \mathbf{f}_O(x)$$

where

$$\mathbf{f}_E(x) = \frac{f(x) + f(-x)}{2} \quad \text{and} \quad \mathbf{f}_O(x) = \frac{f(x) - f(-x)}{2}$$

such that $\mathbf{f}_E(x) \in E$ and $\mathbf{f}_O(x) \in O$. Consequently, any function can be written as the sum of its "even part" $\mathbf{f}_E(x)$ and its "odd part" $\mathbf{f}_O(x)$.
(c) Show that $C[-\infty, \infty] = E \oplus O$.

16. Quotient Spaces

In this lesson, we continue to examine various ways of building vector spaces. Given a vector space V and a subspace W of V, we will learn how to construct a new vector space, called the **quotient of V by W,** and it is denoted by V/W.

Definition: *(Coset of a Subspace)*

Let W be a subspace of a vector space V. For each vector $\mathbf{v} \in V$, we define the set of vectors $\mathbf{v} + W$ by

$$\mathbf{v} + W = \{\mathbf{v} + \mathbf{w} : \mathbf{w} \in W\} \tag{16.1}$$

We shall call the set $\mathbf{v} + W$ is called a **coset** of W in V.

We denote the set of all cosets W in V by V/W, which is called the **quotient of V by W.** We should be careful to note that each element $\mathbf{v} + W$ of V/W is not a vector in the ordinary sense, but is an entire *set* of vectors. A simple example will help to clarify this concept.

Example-1: Let $V = \mathbb{R}^2$ and consider the subspace W given by

$$W = \left\{ \begin{bmatrix} 0 \\ y \end{bmatrix} : y \in \mathbb{R} \right\}$$

Describe the cosets of W in V.

Solution: It's easy to see that W consists of all vectors lying along the y-axis of $\mathbb{R}^2$. Let $\mathbf{x}_0$ be a fixed vector in V, and assume that $\mathbf{x} \in \mathbf{x}_0 + W$.
Then $\mathbf{x} - \mathbf{x}_0 \in W$ so that

$$\mathbf{x} - \mathbf{x}_0 = t \begin{bmatrix} 0 \\ 1 \end{bmatrix} \quad \text{where } t \in \mathbb{R}$$

Consequently, as shown in Figure 16.1, $\mathbf{x}_0 + W$ consists of all vectors lying along the vertical line which passes through the tip of the vector $\mathbf{x}_0$. Since $\mathbf{x}_0$ was arbitrary, we see that the cosets of W are simply all vertical lines in the plane.

Next, we state and prove two lemmas that will be useful in exploring the properties of cosets.

Lemma 16.1

Let W be a subspace of a vector space V. Then for any vectors $\mathbf{u}, \mathbf{v} \in V$, we have $\mathbf{u} \in \mathbf{v} + W$ if and only if $(\mathbf{u} - \mathbf{v}) \in W$.

Proof: First, assume that $\mathbf{u} \in \mathbf{v} + W$. By definition, there exists a vector $\mathbf{w}_0 \in W$ such that $\mathbf{u} = \mathbf{v} + \mathbf{w}_0$. But then $(\mathbf{u} - \mathbf{v}) = \mathbf{w}_0 \in W$. On the other hand, assume that $(\mathbf{u} - \mathbf{v}) \in W$. Then there exists a vector $\mathbf{w}_0 \in W$ such that $(\mathbf{u} - \mathbf{v}) = \mathbf{w}_0$. Consequently, $\mathbf{u} = \mathbf{v} + \mathbf{w}_0 \in \mathbf{v} + W$. □.

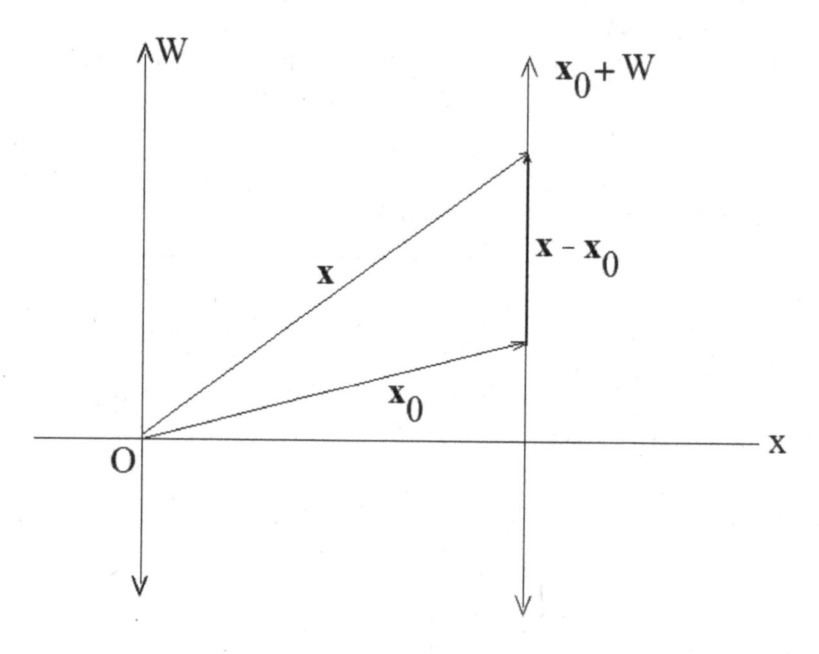

Figure 16.1: A typical coset of W in $\mathbb{R}^2$.

Lemma 16.2: *(Condition for Equality of Cosets)*

Let W be a subspace of a vector space V, and let V/W be the set of all cosets of W in V. Then for any cosets $\mathbf{u} + W, \mathbf{v} + W \in V/W$:

$$\mathbf{u} + W = \mathbf{v} + W \quad \text{if and only if} \quad (\mathbf{u} - \mathbf{v}) \in W. \tag{16.2}$$

Proof: First, assume that $\mathbf{u} + W = \mathbf{v} + W$. Then there exist vectors $\mathbf{w}, \mathbf{w}' \in W$ such that $\mathbf{u} + \mathbf{w} = \mathbf{v} + \mathbf{w}'$, hence, $(\mathbf{u} - \mathbf{v}) = (\mathbf{w}' - \mathbf{w}) \in W$. Now suppose that $(\mathbf{u} - \mathbf{v}) \in W$, and let $\mathbf{x} \in \mathbf{u} + W$. Observe that

$$\mathbf{x} - \mathbf{v} = (\mathbf{x} - \mathbf{u}) + (\mathbf{u} - \mathbf{v}) \in W$$

so $\mathbf{x} \in \mathbf{v} + W$ as well, and $\mathbf{u} + W \subseteq \mathbf{v} + W$.

On the other hand, if $\mathbf{x} \in \mathbf{v} + W$ then

$$\mathbf{x} - \mathbf{u} = (\mathbf{x} - \mathbf{v}) - (\mathbf{u} - \mathbf{v}) \in W$$

so $\mathbf{x} \in \mathbf{u} + W$ and $\mathbf{v} + W \subseteq \mathbf{u} + W$. Consequently,

$$\mathbf{u} + W = \mathbf{v} + W$$

as required. □.

Theorem 16.3: *(Cosets of V/W Form a Partition of V)*

Let W be a subspace of a vector space V, and let V/W denote the set of all cosets of W in V. Then the cosets of V/W form a partition of V; that is,
(**a**) Each vector $\mathbf{v} \in V$ belongs to some coset in V/W, so that

$$V = \bigcup_{\mathbf{v} \in V} (\mathbf{v} + W) \tag{16.3}$$

(**b**) Any two cosets $\mathbf{u} + W$, $\mathbf{v} + W$ in V/W are either equal or disjoint; that is,

$$\text{either} \quad \mathbf{u} + W = \mathbf{v} + W \quad \text{or} \quad (\mathbf{u} + W) \cap (\mathbf{v} + W) = \varnothing \tag{16.4}$$

Proof: We first establish (16.3). Assume that $\mathbf{x} \in V$. Clearly $\mathbf{x} \in \mathbf{x} + W$ since $\mathbf{x} - \mathbf{x} = \mathbf{0} \in W$, and therefore,

$$\mathbf{x} \in \bigcup_{\mathbf{v} \in V} (\mathbf{v} + W)$$

Thus, V is a subset of the union of all cosets in V/W:

$$V \subseteq \bigcup_{\mathbf{v} \in V} (\mathbf{v} + W)$$

Now suppose that $\mathbf{x}$ belongs to the union of all cosets in V/W. Then, there exists some vector $\mathbf{v}_0 \in V$ such that $\mathbf{x} \in \mathbf{v}_0 + W$. Then by the previous lemma, $\mathbf{x} - \mathbf{v}_0 \in W \subseteq V$, and consequently, $\mathbf{x} \in V$. Thus,

$$\bigcup_{\mathbf{v} \in V} (\mathbf{v} + W) \subseteq V$$

as well, so (16.3) follows immediately.

Now assume that the cosets $\mathbf{u} + W$ and $\mathbf{v} + W$ are not disjoint, i.e.,

$$(\mathbf{u} + W) \cap (\mathbf{v} + W) \neq \varnothing$$

So, there exists a vector $\mathbf{y} \in V$ such that $\mathbf{y} \in \mathbf{u} + W$ and $\mathbf{y} \in \mathbf{v} + W$. In other words, there exist vectors $\mathbf{w}, \mathbf{w}' \in W$ such that

$$\mathbf{y} = \mathbf{u} + \mathbf{w} \quad \text{and} \quad \mathbf{y} = \mathbf{v} + \mathbf{w}'$$

Subtracting, we find $\mathbf{u} - \mathbf{v} = \mathbf{w}' - \mathbf{w}$, so $\mathbf{u} - \mathbf{v} \in W$. Consequently, by

(16.2), it follows immediately that $(\mathbf{u} + W) = (\mathbf{v} + W)$. Thus, we have shown that

$$(\mathbf{u} + W) \cap (\mathbf{v} + W) \neq \varnothing \implies \mathbf{u} + W = \mathbf{v} + W$$

Now, the contrapositive of this statement must also hold true, i.e.,

$$\mathbf{u} + W \neq \mathbf{v} + W \implies (\mathbf{u} + W) \cap (\mathbf{v} + W) = \varnothing$$

Thus, any two cosets in V/W are either identical or disjoint. $\square$.

If V is a vector space, and W is a subspace of V, it turns out that the set V/W, equipped with the operations of coset addition and the multiplication of cosets by scalars, forms a vector space in its own right; it is called a **quotient space.** This is the content of the next theorem.

Theorem 16.4: *(V/W is a Vector Space)*

Let W be a subspace of a vector space V, and let V/W denote the set of all cosets of W in V. Then V/W is a vector space, under the operations of addition, and scalar multiplication given by

$$(\mathbf{u} + W) + (\mathbf{v} + W) = (\mathbf{u} + \mathbf{v}) + W \tag{16.5}$$

$$\lambda(\mathbf{u} + W) = \lambda\mathbf{u} + W \quad \text{for all scalars } \lambda \tag{16.6}$$

Proof: We must first show that the above operations are well-defined. That is, if $\mathbf{u}' + W = \mathbf{u} + W$ and $\mathbf{v}' + W = \mathbf{v} + W$, we must show that

$$(\mathbf{u}' + \mathbf{v}') + W = (\mathbf{u} + \mathbf{v}) + W$$

$$\lambda(\mathbf{u}' + W) = \lambda\mathbf{u} + W$$

Now $\mathbf{u}' + W = \mathbf{u} + W$ implies $\mathbf{u}' - \mathbf{u} \in W$ and $\mathbf{v}' + W = \mathbf{v} + W$ implies that $\mathbf{v}' - \mathbf{v} \in W$. Observe that

$$(\mathbf{u}' + \mathbf{v}') - (\mathbf{u} + \mathbf{v}) = (\mathbf{u}' - \mathbf{u}) + (\mathbf{v}' - \mathbf{v}) \in W$$

Consequently, by (16.2), we have

$$(\mathbf{u}' + \mathbf{v}') + W = (\mathbf{u} + \mathbf{v}) + W.$$

Similarly, $\mathbf{u}' - \mathbf{u} \in W$ implies that $\lambda(\mathbf{u}' - \mathbf{u}) = \lambda\mathbf{u}' - \lambda\mathbf{u} \in W$ so by (16.2), we also have

$$\lambda(\mathbf{u}' + W) = \lambda\mathbf{u} + W$$

as well. Thus, the operations in V/W are *well defined.*

We leave it to the reader show that the operation of coset addition is commutative and associative, that $\mathbf{0} + W$ is the identity element for coset addition, $-\mathbf{u} + W$ is the negative (or additive inverse) of any given element $\mathbf{u} + W$, etc., and that all of the remaining vector space axioms hold good for the quotient space V/W. □.

Theorem 16.5: *(The Dimension of V/W)*

If W is a subspace of the finite dimensional vector space V, then

$$\dim(V/W) = \dim(V) - \dim(W) \tag{16.7}$$

Proof: Let $\dim(V) = n, \dim(W) = r$, and let $\{\mathbf{w}_1, \mathbf{w}_2, \ldots, \mathbf{w}_r\}$ be a basis for W. By the Basis Extension Theorem, there exist vectors $\mathbf{v}_1, \mathbf{v}_2, \ldots, \mathbf{v}_{n-r}$ such that the set $B = \{\mathbf{v}_1, \mathbf{v}_2, \ldots, \mathbf{v}_{n-r}, \mathbf{w}_1, \mathbf{w}_2, \ldots, \mathbf{w}_r\}$ is a basis for V. We show that the set of cosets

$$B' = \{\mathbf{v}_1 + W, \mathbf{v}_2 + W, \ldots, \mathbf{v}_{n-r} + W\}$$

is a basis for the quotient space V/W.

We first show that B' spans the quotient space V/W. Let $\mathbf{v}_0 + W \in V/W$ be arbitrary. Recall that, by definition,

$$\mathbf{v}_0 + W = \{\mathbf{v}_0 + \mathbf{w} : \mathbf{w} \in W\}$$

Since $\mathbf{v}_0 \in V$, then there exist scalars $\lambda_1, \lambda_2, \ldots \lambda_{n-r}$ and $\mu_1, \mu_2, \ldots, \mu_r$ such that

$$\mathbf{v}_0 = \lambda_1\mathbf{v}_1 + \lambda_2\mathbf{v}_2 + \ldots + \lambda_{n-r}\mathbf{v}_{n-r} + \mu_1\mathbf{w}_1 + \mu_2\mathbf{w}_2 + \ldots + \mu_r\mathbf{w}_r$$

At the same time, since $\{\mathbf{w}_1, \mathbf{w}_2, \ldots, \mathbf{w}_r\}$ is a basis for W then for *any* $\mathbf{w} \in W$, there exist r scalars, say $\nu_1, \nu_2, \ldots, \nu_r$ such that

$$\mathbf{w} = \nu_1\mathbf{w}_1 + \nu_2\mathbf{w}_2 + \ldots + \nu_r\mathbf{w}_r$$

Thus,

$$\begin{aligned}
\mathbf{v}_0 + W &= (\lambda_1\mathbf{v}_1 + \ldots + \lambda_{n-r}\mathbf{v}_{n-r}) + (\mu_1 + \nu_1)\mathbf{w}_1 + \ldots + (\mu_r + \nu_r)\mathbf{w}_r \\
&= (\lambda_1\mathbf{v}_1 + \lambda_2\mathbf{v}_2 + \ldots + \lambda_{n-r}\mathbf{v}_{n-r}) + W \\
&= \lambda_1(\mathbf{v}_1 + W) + \lambda_2(\mathbf{v}_2 + W) + \ldots + \lambda_{n-r}(\mathbf{v}_{n-r} + W)
\end{aligned}$$

where the last step follows from (16.5).

Now suppose there exist scalars $c_1, c_2, \ldots, c_{n-r}$ such that

$$c_1(\mathbf{v}_1 + W) + c_2(\mathbf{v}_2 + W) + \ldots + c_{n-r}(\mathbf{v}_{n-r} + W) = \mathbf{0} + W$$

then

$$\left(c_1\mathbf{v}_1 + c_2\mathbf{v}_2 + \ldots + c_{n-r}\mathbf{v}_{n-r}\right) + W = \mathbf{0} + W$$

So, by Lemma 16.2, $c_1\mathbf{v}_1 + c_2\mathbf{v}_2 + \ldots + c_{n-r}\mathbf{v}_{n-r} \in W$. But then there exist r scalars $d_1, d_2, .. d_r$ such that

$$c_1\mathbf{v}_1 + c_2\mathbf{v}_2 + \ldots + c_{n-r}\mathbf{v}_{n-r} = d_1\mathbf{w}_1 + d_2\mathbf{w}_2 + \ldots + d_r\mathbf{w}_r$$

Rearranging terms in the last equation, we get

$$c_1\mathbf{v}_1 + c_2\mathbf{v}_2 + \ldots + c_{n-r}\mathbf{v}_{n-r} - d_1\mathbf{w}_1 - d_2\mathbf{w}_2 - \ldots - d_r\mathbf{w}_r = \mathbf{0}$$

But since B is a basis for V then $c_1 = c_2 = \ldots = c_{n-r} = d_1 = d_2 = \ldots = d_r = 0$. Thus, the $(n - r)$ cosets not only span V/W but they are linearly independent; so they form a basis for V/W. Finally, observe that

$$\dim(V/W) = n - r = \dim(V) - \dim(W)$$

This completes the proof. □.

The attentive reader will note that there is a very useful algorithm contained in the proof of the last theorem; it is an algorithm which enables us to find a basis for the quotient space V/W whenever we are given a basis for W. We state this as a corollary.

Corollary 16.6: *(Finding a Basis for V/W)*

Let W be a subspace of an n-dimensional vector space V, and let the set of vectors $\{\mathbf{w}_1, \mathbf{w}_2, \ldots, \mathbf{w}_r\}$ be a basis for W. If the set of n-vectors

$$\{\mathbf{v}_1, \mathbf{v}_2, \ldots, \mathbf{v}_{n-r}, \mathbf{w}_1, \mathbf{w}_2, \ldots, \mathbf{w}_r\}$$

is a basis for V, then the set of $(n - r)$-cosets

$$\{\mathbf{v}_1 + W, \mathbf{v}_2 + W, \ldots, \mathbf{v}_{n-r} + W\}$$

is a basis for the quotient space V/W.

Proof: See the proof of the previous theorem. □.

Example-2: Let $V = \mathbb{R}^3$, and consider the subspace W of $\mathbb{R}^3$ given by

$$W = \left\{[x, y, z]^T \in \mathbb{R}^3 : 2x + y + z = 0\right\}$$

Find a basis for the quotient space V/W, and describe its elements.

Solution: Let's first find a basis for W. For any $\mathbf{w} \in W$ we can write

$$\mathbf{w} = \begin{bmatrix} r \\ s \\ -2r - s \end{bmatrix} = r\begin{bmatrix} 1 \\ 0 \\ -2 \end{bmatrix} + s\begin{bmatrix} 0 \\ 1 \\ -1 \end{bmatrix}$$

where $r, s \in \mathbb{R}$ are parameters. Clearly, the vectors $\mathbf{w}_1 = [1, 0, -2]^T$ and $\mathbf{w}_2 = [0, 1, -1]^T$ span W and they are linearly independent. Thus, they form a basis for W.

Since $V = \mathbb{R}^3$ is three dimensional, we need to find an additional vector, say $\mathbf{v}$, such that the set of vectors $\{\mathbf{v}, \mathbf{w}_1, \mathbf{w}_2\}$ is linearly independent, and thus, it will form a basis for V. To this end, assume that $\mathbf{v} = [a, b, c]^T$ and form the matrix:

$$\begin{bmatrix} 1 & 0 & -2 \\ 0 & 1 & 1 \\ a & b & c \end{bmatrix}$$

We know that if we can reduce this matrix to *row-echelon form* where none of its rows are zero, then the resulting set of vectors will be linearly independent. In doing so, we obtain the matrix

$$\begin{bmatrix} 1 & 0 & -2 \\ 0 & 1 & 1 \\ 0 & 0 & 2a - b + c \end{bmatrix}$$

This matrix will certainly be in row-echelon form (with no zero rows) as long as $2a - b + c = 1$. Although many choices are possible, let's take $a = 1$, $b = 1$, and $c = 0$, so $\mathbf{v} = [1, 1, 0]^T$ and we are guaranteed that the set $\{\mathbf{v}, \mathbf{w}_1, \mathbf{w}_2\}$ is a basis for V.

By the previous lemma, a basis B' for the quotient space V/W is given by the single coset

$$B' = \left\{ \begin{bmatrix} 1 \\ 1 \\ 0 \end{bmatrix} + W \right\}$$

Now, if $\mathbf{x} = [x, y, z]^T \in V/W$, then

$$\mathbf{x} = \begin{bmatrix} x \\ y \\ z \end{bmatrix} = t \begin{bmatrix} 1 \\ 1 \\ 0 \end{bmatrix} + W \Rightarrow \begin{bmatrix} x - t \\ y - t \\ z \end{bmatrix} \in W$$

so that $2(x - t) + (y - t) + z = 0$, or $2x + y + z = 3t = c$ where c is an arbitrary constant. In other words, each coset consists of those vectors in V whose tips lie in the plane $2x + y + z = c$ which is parallel to the plane represented by W.

Exercise Set 16

1. Let $V = \mathbb{R}^2$ and let W be the subspace of $\mathbb{R}^2$ given by

$$W = \left\{ \begin{bmatrix} a \\ 0 \end{bmatrix} : a \neq 0 \right\}$$

(a) Describe a typical coset of V/W.
(b) Find a basis for the coset space V/W.
(c) What is the dimension of V/W ?
(d) Verify that equation (16.7) is satisfied.

2. Let $V = P_4$, the space of all real polynomials of degree ≤ 4, and let W be the subspace of P_4 given by

$$W = \{\mathbf{p}(x) \in P_4 : \mathbf{p}(x) \text{ is divisible by } x^2\}$$

(a) Write down a typical element of W.
(b) Find a basis for W.
(c) Extend the basis for W to obtain a basis for V.
(d) Find a basis for the quotient space V/W.
(e) Describe a typical element of the quotient space V/W.

3. Let $V = \mathbb{R}^3$ and let W be the subspace of $\mathbb{R}^3$ given by

$$W = \left\{ \begin{bmatrix} a \\ a \\ a \end{bmatrix} \in \mathbb{R}^3 : a \neq 0 \right\}$$

(a) Describe the subspace W.
(b) Find a basis for W.
(c) Extend the basis for W to obtain a basis for V.
(d) Find a basis for the quotient space V/W.
(e) Describe a typical element of the quotient space V/W.
(f) As a partial check on your work, verify that equation (16.7) is satisfied.

4. Let $V = P_2$, the space of all real polynomials of degree ≤ 3, and consider the subspace of P_3 given by

$$W = \{\mathbf{p}(x) \in P_2 : \mathbf{p}(x) \text{ is divisible by } x\}$$

(a) Find a basis for W.
(b) Show that the set

$$B = \{1 + W, x^3 + W\}$$

is a basis for the quotient space V/W.
(c) Verify that equation (16.7) is satisfied.

5. Complete the proof of Theorem 16.4.

6. Use Lemma 16.2 to show that the equality of cosets is an **equivalence relation,** i.e., it satisfies each of the following properties:

(a) *Reflexive property*: for any $\mathbf{u} + W \in V/W$,

$$\mathbf{u} + W = \mathbf{u} + W$$

(b) *Symmetric property:* for any $\mathbf{u} + W, \mathbf{v} + W \in V/W$,

$$\mathbf{u} + W = \mathbf{v} + W \implies \mathbf{v} + W = \mathbf{u} + W$$

(c) *Transitive property:* for any $\mathbf{u} + W, \mathbf{v} + W, \mathbf{w} + W \in V/W$,

$$\mathbf{u} + W = \mathbf{v} + W \text{ and } \mathbf{v} + W = \mathbf{w} + W \implies \mathbf{u} + W = \mathbf{w} + W$$

17. Change of Basis

From Theorem 12.1, we know that if we are given a basis B for an n-dimensional vector space V, say $B = \{\mathbf{b}_1, \mathbf{b}_2, \ldots, \mathbf{b}_n\}$, then any vector $\mathbf{x} \in V$ can be written *uniquely* as

$$\mathbf{x} = x_1\mathbf{v}_1 + x_2\mathbf{v}_2 + \ldots + x_n\mathbf{v}_n \tag{17.1}$$

where the scalars $x_1, x_2, \ldots, x_n$ are called the **components** (or the **coordinates**) of $\mathbf{x}$ relative to the basis B. The **coordinate vector** of $\mathbf{x}$ relative to the basis B is *defined* to be the vector in $\mathbb{R}^n$ given by

$$[\mathbf{x}]_B = \begin{bmatrix} x_1 \\ x_2 \\ \vdots \\ x_n \end{bmatrix} \tag{17.2}$$

The procedure for finding the coordinate vector of $\mathbf{x}$ relative to a given basis B usually requires us to solve a system of linear equations. Let's see how this actually works.

Example-1: Given that the set

$$B = \left\{ \begin{bmatrix} 2 \\ 1 \end{bmatrix}, \begin{bmatrix} -1 \\ 1 \end{bmatrix} \right\}$$

is a basis for $\mathbb{R}^2$, find the coordinate vector of $\mathbf{x} = [3, 4]^T$.

Solution: First, we write $\mathbf{x}$ in the form

$$\mathbf{x} = \begin{bmatrix} 3 \\ 4 \end{bmatrix} = x_1 \begin{bmatrix} 2 \\ 1 \end{bmatrix} + x_2 \begin{bmatrix} -1 \\ 1 \end{bmatrix}$$

and obtain the system of equations

$$\begin{aligned} 2x_1 - x_2 &= 3 \\ x_1 + x_2 &= 4 \end{aligned}$$

We find $x_1 = \frac{7}{3}$, and $x_2 = \frac{5}{3}$ so that

$$\mathbf{x} = \begin{bmatrix} 3 \\ 4 \end{bmatrix} = \frac{7}{3} \begin{bmatrix} 2 \\ 1 \end{bmatrix} + \frac{5}{3} \begin{bmatrix} -1 \\ 1 \end{bmatrix}$$

and we conclude that the coordinate vector of $\mathbf{x}$ relative to B is

$$[\mathbf{x}]_B = \begin{bmatrix} \frac{7}{3} \\ \frac{5}{3} \end{bmatrix}$$

From the previous example, we see that if we were to use another basis B' for $\mathbb{R}^2$ then we would expect the coordinate vector for $\mathbf{x}$ relative to B' to be different.

Now any vector space will usually admit several bases, and we can use any basis we wish to do our work. In some physical problems, however, one basis may be easier to work with than another. So a question which naturally arises is how do the components of a given vector change if we change bases?

In order to answer this question, let $B = \{\mathbf{b}_1, \mathbf{b}_2, \ldots, \mathbf{b}_n\}$ be a basis for a vector space V and $B' = \{\mathbf{b}'_1, \mathbf{b}'_2, \ldots, \mathbf{b}'_n\}$ be a new basis for V. For any vector $\mathbf{x} \in V$, we can write

$$\mathbf{x} = x_1\mathbf{b}_1 + x_2\mathbf{b}_2 + \ldots + x_n\mathbf{b}_n \tag{17.3}$$

and

$$\mathbf{x} = x'_1\mathbf{b}'_1 + x'_2\mathbf{b}'_2 + \ldots + x'_n\mathbf{b}'_n \tag{17.4}$$

Now, we can express each of the old basis vectors in terms of the new basis vectors as

$$\mathbf{b}_k = \sum_{j=1}^{n} p_{jk}\mathbf{b}'_j \tag{17.5}$$

where the p's are just scalars. The last equation can be written more explicitly as:

$$\begin{aligned} \mathbf{b}_1 &= p_{11}\mathbf{b}'_1 + p_{21}\mathbf{b}'_2 + \ldots + p_{n1}\mathbf{b}'_n \\ \mathbf{b}_2 &= p_{12}\mathbf{b}'_1 + p_{22}\mathbf{b}'_2 + \ldots + p_{n2}\mathbf{b}'_n \\ &\vdots \\ \mathbf{b}_n &= p_{1n}\mathbf{b}'_1 + p_{2n}\mathbf{b}'_2 + \ldots + p_{nn}\mathbf{b}'_n \end{aligned} \tag{17.6}$$

Substituting (17.6) into (17.3), we obtain

$$\begin{aligned} \mathbf{x} = {} & x_1\left(p_{11}\mathbf{b}'_1 + p_{21}\mathbf{b}'_2 + \ldots + p_{n1}\mathbf{b}'_n\right) + x_2\left(p_{12}\mathbf{b}'_1 + p_{22}\mathbf{b}'_2 + \ldots + p_{n2}\mathbf{b}'_n\right) \\ & + \ldots + x_n\left(p_{1n}\mathbf{b}'_1 + p_{2n}\mathbf{b}'_2 + \ldots + p_{nn}\mathbf{b}'_n\right) \end{aligned}$$

Comparing this result with (17.4), we see that

$$x_1' = p_{11}x_1 + p_{12}x_2 + \ldots + p_{1n}x_n$$
$$x_2' = p_{21}x_1 + p_{22}x_2 + \ldots + p_{2n}x_n$$
$$\vdots$$
$$x_n' = p_{n1}x_1 + p_{n2}x_2 + \ldots + p_{nn}x_n$$

In matrix form, the last set of equations becomes

$$[\mathbf{x}]_{B'} = P[\mathbf{x}]_B$$

where the matrix P is given by

$$P = \begin{bmatrix} p_{11} & p_{12} & \cdots & p_{1n} \\ p_{21} & p_{22} & \cdots & p_{2n} \\ \vdots & \vdots & & \vdots \\ p_{n1} & p_{n2} & \cdots & p_{nn} \end{bmatrix} \tag{17.7}$$

It's important to note from (17.6), that the *columns* of the matrix P are simply the *coordinate vectors* of the old basis vectors relative to the new basis vectors. We call P a **transition matrix**. We have proved the following useful theorem:

Theorem 17.1: *(Change of Basis)*

Let $B = \{\mathbf{b}_1, \mathbf{b}_2, \ldots, \mathbf{b}_n\}$ be a basis for a vector space V, and let $\mathbf{x}$ be an arbitrary vector in V. If we change the basis B for V to a new basis $B' = \{\mathbf{b}_1', \mathbf{b}_2', \ldots, \mathbf{b}_n'\}$, then there exists an $n \times n$ transition matrix P such that

$$[\mathbf{x}]_{B'} = P[\mathbf{x}]_B \tag{17.8}$$

which relates the new coordinate vector $[\mathbf{x}]_{B'}$ for $\mathbf{x}$ to the old coordinate vector $[\mathbf{x}]_B$ for $\mathbf{x}$. The columns of the transition matrix P are simply the *coordinate vectors* of the old basis vectors relative to the new basis vectors, i.e.,

$$P = \left[\ [\mathbf{b}_1]_{B'} \mid [\mathbf{b}_2]_{B'} \mid \ \cdots \ \mid [\mathbf{b}_n]_{B'}\ \right] \tag{17.9}$$

Example-2: Given the bases $B = \{\mathbf{b}_1, \mathbf{b}_2\}$ and $B' = \{\mathbf{b}_1', \mathbf{b}_2'\}$ for $\mathbb{R}^2$ where

$$\mathbf{b}_1 = \begin{bmatrix} 1 \\ 1 \end{bmatrix}, \ \mathbf{b}_2 = \begin{bmatrix} 1 \\ -1 \end{bmatrix}$$

and

$$\mathbf{b}_1' = \begin{bmatrix} 0 \\ -1 \end{bmatrix}, \quad \mathbf{b}_2' = \begin{bmatrix} 1 \\ 0 \end{bmatrix}$$

(**a**) Find the transition matrix P from B to B'.
(**b**) Given that a vector $\mathbf{x} \in \mathbb{R}^2$ has the coordinate vector

$$[\mathbf{x}]_B = \begin{bmatrix} 1 \\ 2 \end{bmatrix}$$

find its coordinate vector $[\mathbf{x}]_{B'}$ relative to the basis B'.

Solution:
(**a**) We must write each of the old basis vectors in terms of the new ones. By inspection, we find that

$$\mathbf{b}_1 = -\mathbf{b}_1' + \mathbf{b}_2' \implies [\mathbf{b}_1]_{B'} = \begin{bmatrix} -1 \\ 1 \end{bmatrix}$$

$$\mathbf{b}_2 = \mathbf{b}_1' + \mathbf{b}_2' \implies [\mathbf{b}_2]_{B'} = \begin{bmatrix} 1 \\ 1 \end{bmatrix}$$

Consequently, the transition matrix is

$$P = \begin{bmatrix} [\mathbf{b}_1]_{B'} & [\mathbf{b}_2]_{B'} \end{bmatrix} = \begin{bmatrix} -1 & 1 \\ 1 & 1 \end{bmatrix}$$

(**b**) Using (17.8), we find that

$$[\mathbf{x}]_{B'} = \begin{bmatrix} -1 & 1 \\ 1 & 1 \end{bmatrix} \begin{bmatrix} 1 \\ 2 \end{bmatrix} = \begin{bmatrix} 1 \\ 3 \end{bmatrix}$$

Consequently, in terms of the new basis vectors, $\mathbf{x} = \mathbf{b}_1' + 3\mathbf{b}_2'$.

Equation (17.8) is called a **coordinate transformation**, since it relates the new coordinates of any given vector to its old coordinates. In many applications, however, it is necessary to be able to "transform back" to the old coordinates after we are finished working in the new coordinates. This type of procedure involves finding an **inverse coordinate transformation** of the form

$$[\mathbf{x}]_B = Q[\mathbf{x}]_{B'}$$

where Q is some $n \times n$ matrix to be determined. The next theorem shows us how we can always find this matrix Q.

Theorem 17.2: *(Transition Matrices are Invertible)*

Let P be a transition matrix from a basis B to a new basis B' for a finite dimensional vector space V. Then P is *invertible*, and for any vector $\mathbf{x} \in V$, the *inverse coordinate transformation* is given by:

$$[\mathbf{x}]_B = P^{-1}[\mathbf{x}]_{B'} \tag{17.10}$$

Proof: Let $B = \{\mathbf{b}_1, \mathbf{b}_2, \ldots, \mathbf{b}_n\}$ and $B' = \{\mathbf{b}_1', \mathbf{b}_2', \ldots, \mathbf{b}_n'\}$ be bases for a vector space V. We can express each of the old basis vectors as linear combinations of the new basis vectors:

$$\mathbf{b}_k = \sum_{j=1}^{n} p_{jk}\mathbf{b}_j'$$

where the p's are scalars. Conversely, we can express each of the new basis vectors in terms of the old ones as

$$\mathbf{b}_j' = \sum_{m=1}^{n} q_{mj}\mathbf{b}_m$$

where the q's are scalars.

Substituting the second equation into the first one, we obtain

$$\begin{aligned}\mathbf{b}_k &= \sum_{j=1}^{n} p_{jk}\mathbf{b}_j' = \sum_{j=1}^{n} p_{jk}\left(\sum_{m=1}^{n} q_{mj}\mathbf{b}_m\right) \\ &= \sum_{m=1}^{n}\left(\sum_{j=1}^{n} q_{mj}p_{jk}\right)\mathbf{b}_m\end{aligned}$$

But observe that

$$\mathbf{b}_k = \sum_{m=1}^{n} \delta_{mk}\mathbf{b}_m$$

where δ_{mk} is the Kronecker delta of order n. Combining the last two equations gives

$$\sum_{m=1}^{n}\left[\delta_{mk} - \left(\sum_{j=1}^{n} q_{mj}p_{jk}\right)\right]\mathbf{b}_m = \mathbf{0}$$

But since the $\mathbf{b}$'s are linearly independent, we have

$$\sum_{j=1}^{n} q_{mj}p_{jk} = \delta_{mk} \ ,$$

which in the language of matrices, is simply $QP = I$, where $Q = [q_{ij}]$.

Consequently, $Q = P^{-1}$, and P is invertible.

Finally, let's multiply both sides of (17.8) by P^{-1}:

$$\begin{aligned} [\mathbf{x}]_{B'} &= P[\mathbf{x}]_B \\ &\Rightarrow P^{-1}[\mathbf{x}]_{B'} = P^{-1}P[\mathbf{x}]_B = I[\mathbf{x}]_B \\ &\Rightarrow P^{-1}[\mathbf{x}]_{B'} = [\mathbf{x}]_B \end{aligned}$$

which is (17.10). □.

Exercise Set 17

---

1. Given the basis $B = \{\mathbf{b}_1, \mathbf{b}_2\}$ for $\mathbb{R}^2$, and the vector $\mathbf{x} \in \mathbb{R}^2$, find the coordinate vector $[\mathbf{x}]_B$ of $\mathbf{x}$ relative to the basis B:

(a)

$$\mathbf{b}_1 = \begin{bmatrix} -1 \\ 1 \end{bmatrix}, \quad \mathbf{b}_2 = \begin{bmatrix} 0 \\ 1 \end{bmatrix}; \text{ and } \mathbf{x} = \begin{bmatrix} 2 \\ 3 \end{bmatrix}$$

(b)

$$\mathbf{b}_1 = \begin{bmatrix} 2 \\ 4 \end{bmatrix}, \quad \mathbf{b}_2 = \begin{bmatrix} 1 \\ 1 \end{bmatrix}; \text{ and } \mathbf{x} = \begin{bmatrix} -8 \\ 4 \end{bmatrix}$$

(c)

$$\mathbf{b}_1 = \begin{bmatrix} 1 \\ 2 \end{bmatrix}, \quad \mathbf{b}_2 = \begin{bmatrix} -1 \\ 1 \end{bmatrix}; \text{ and } \mathbf{x} = \begin{bmatrix} 1 \\ 2 \end{bmatrix}$$

2. Let B and B' both be bases for P_2 where

$$B = \{1, 2x, 1 + x^2\} \quad \text{and} \quad B' = \{x, 3 + x^2, x + x^2\}$$

Given two vectors $\mathbf{p}(x) = x^2 + 5x + 6$, and $\mathbf{q}(x) = x^2 - 1$ in P_2:

(a) Find the coordinate vectors $[\mathbf{p}(x)]_B$ and $[\mathbf{q}(x)]_B$.
(b) Find the transition matrix P from B to B'.
(c) Use P to find the coordinate vectors $[\mathbf{p}(x)]_{B'}$; and $[\mathbf{q}(x)]_{B'}$.
(d) Find the transition matrix P^{-1} from B' to B.
(e) Starting with the coordinate vectors $[\mathbf{p}(x)]_{B'}$; and $[\mathbf{q}(x)]_{B'}$, that were found in **(c)**, use P^{-1} to obtain the coordinate vectors $[\mathbf{p}(x)]_B$

and $[\mathbf{q}(x)]_B$. [The answer should agree with that found (**a**).]

3. Given the bases $S = \{\mathbf{s}_1, \mathbf{s}_2\}$ and $T = \{\mathbf{t}_1, \mathbf{t}_2\}$ for $\mathbb{R}^2$, where

$$\mathbf{s}_1 = \begin{bmatrix} 1 \\ 1 \end{bmatrix}, \mathbf{s}_2 = \begin{bmatrix} 1 \\ 0 \end{bmatrix}, \mathbf{t}_1 = \begin{bmatrix} 2 \\ 1 \end{bmatrix}, \mathbf{t}_2 = \begin{bmatrix} -1 \\ 2 \end{bmatrix}$$

(**a**) Find the transition matrix P from S to T.
(**b**) Find the transition matrix P^{-1} from T to S.
(**c**) Given the vectors $\mathbf{x}, \mathbf{y} \in \mathbb{R}^2$,

$$\mathbf{x} = \begin{bmatrix} 2 \\ 6 \end{bmatrix} \text{ and } \mathbf{y} = \begin{bmatrix} -4 \\ 2 \end{bmatrix}$$

find the coordinate vectors $[\mathbf{x}]_S$ and $[\mathbf{y}]_S$.
(**d**) Use the transition matrix P and the results of (**c**) to find the coordinate vectors $[\mathbf{x}]_T$ and $[\mathbf{y}]_T$.

4. Given the bases $S = \{\mathbf{s}_1, \mathbf{s}_2, \mathbf{s}_3\}$ and $T = \{\mathbf{t}_1, \mathbf{t}_2, \mathbf{t}_3\}$ for $\mathbb{R}^3$, where

$$\mathbf{s}_1 = \begin{bmatrix} 1 \\ 1 \\ 1 \end{bmatrix}, \mathbf{s}_2 = \begin{bmatrix} 0 \\ -1 \\ 0 \end{bmatrix}, \mathbf{s}_3 = \begin{bmatrix} 0 \\ 1 \\ 1 \end{bmatrix}$$

$$\mathbf{t}_1 = \begin{bmatrix} 1 \\ 1 \\ 0 \end{bmatrix}, \mathbf{t}_2 = \begin{bmatrix} 0 \\ 2 \\ 1 \end{bmatrix}, \mathbf{t}_3 = \begin{bmatrix} 0 \\ 0 \\ 2 \end{bmatrix}$$

(**a**) Find the transition matrix P from S to T.
(**b**) Find the transition matrix P^{-1} from T to S.
(**c**) Given the vectors $\mathbf{x}, \mathbf{y} \in \mathbb{R}^3$,

$$\mathbf{x} = \begin{bmatrix} -1 \\ 2 \\ 1 \end{bmatrix} \text{ and } \mathbf{y} = \begin{bmatrix} 1 \\ 0 \\ 1 \end{bmatrix}$$

find the coordinate vectors $[\mathbf{x}]_S$ and $[\mathbf{y}]_S$.
(**d**) Use the transition matrix P and the results of (**c**) to find the coordinate vectors $[\mathbf{x}]_T$ and $[\mathbf{y}]_T$.

5. Show that if B is a basis for an n-dimensional space V, then for any vectors $\mathbf{x}, \mathbf{y} \in V$ and for any scalar λ we have both:

$$[\mathbf{x}+\mathbf{y}]_B = [\mathbf{x}]_B + [\mathbf{y}]_B$$
$$[\lambda\mathbf{x}]_B = \lambda[\mathbf{x}]_B$$

6. Let B be a basis for an n-dimensional space V. Show that if $\{\mathbf{x}_1, \mathbf{x}_2, \ldots, \mathbf{x}_r\}$ is any set of r-linearly independent vectors in V, then the set of coordinate vectors $\{[\mathbf{x}_1]_B, [\mathbf{x}_2]_B, \ldots, [\mathbf{x}_r]_B\}$ forms a linearly independent set in $\mathbb{R}^n$.

7. Let B be a basis for an n-dimensional space V. Given any two vectors $\mathbf{x}, \mathbf{y} \in V$, show that $\mathbf{x} = \mathbf{y}$ if and only if $[\mathbf{x}]_B = [\mathbf{y}]_B$

18. Euclidean Spaces

We have seen that the concept of an abstract vector space is extremely useful, and this concept unexpectedly surfaces in some of the most unlikely places in mathematics. We can make this concept even more useful by providing some *additional structure* to a vector space. One way of doing this is to introduce an *inner product* into a vector space.

As we will see in the discussion that follows, this new concept will, in turn, enable us to introduce the familiar concepts of length, angle, and distance into a vector space. We begin with a definition.

Definition: *(Real Inner Product)*

Given a real vector space V, a (real) **inner product** is a rule that assigns a unique real number $\langle \mathbf{u}, \mathbf{v} \rangle$ to each ordered pair of vectors $\mathbf{u}, \mathbf{v} \in V$ such that the following properties are satisfied for all vectors $\mathbf{u}, \mathbf{v}, \mathbf{w} \in V$ and for all real numbers λ:

(a) $\langle \mathbf{u}, \mathbf{v} \rangle = \langle \mathbf{v}, \mathbf{u} \rangle$ (symmetric property)
(b) $\langle \mathbf{u} + \mathbf{v}, \mathbf{w} \rangle = \langle \mathbf{u}, \mathbf{w} \rangle + \langle \mathbf{v}, \mathbf{w} \rangle$ (linearity property)
(c) $\langle \lambda\mathbf{u}, \mathbf{v} \rangle = \lambda\langle \mathbf{u}, \mathbf{v} \rangle$ (homogeneity property)
(d) $\langle \mathbf{u}, \mathbf{u} \rangle \geq 0$ (positive definiteness property)
(e) $\langle \mathbf{u}, \mathbf{u} \rangle = 0$ if and only if $\mathbf{u} = \mathbf{0}$ (non-degeneracy property)

A real vector space which is equipped with a real inner product is called a **Euclidean space.** So, a Euclidean space really consists of two things: a real vector space V, and a real inner product which has been defined on V.

Example-1: Given the vectors $\mathbf{x} = [x_1, x_2, \ldots, x_n]^T$ and $\mathbf{y} = [y_1, y_2, \ldots, y_n]^T$ in $\mathbb{R}^n$, we define their **Euclidean inner product** as the real number:

$$\langle \mathbf{x}, \mathbf{y} \rangle = x_1y_1 + x_2y_2 + \ \ldots \ + x_ny_n \tag{18.1}$$

It's easy to show that (18.1) satisfies all of the properties of an inner product. So $\mathbb{R}^n$, equipped with this inner product, is a Euclidean space.

Example-2: Let P_n denote the space of all polynomials of degree less than or equal to n. If $\mathbf{p} = p(x)$ and $\mathbf{q} = q(x)$ are any two vectors in P_n, then the expression

$$\langle \mathbf{p}, \mathbf{q} \rangle = \int_a^b p(x)q(x)dx \tag{18.2}$$

where $a < b$, defines a real inner product on P_n. Let's verify that (18.2) satisfies all of the properties of an inner product.

(**a**) symmetric property:

$$\langle \mathbf{p}, \mathbf{q} \rangle = \int_a^b p(x)q(x)dx = \int_a^b q(x)p(x)dx = \langle \mathbf{q}, \mathbf{p} \rangle$$

(**b**) linearity property:

$$\begin{aligned}\langle \mathbf{p} + \mathbf{q}, \mathbf{r} \rangle &= \int_a^b (p(x) + q(x))r(x)dx \\ &= \int_a^b p(x)r(x)dx + \int_a^b q(x)r(x)dx \\ &= \langle \mathbf{p}, \mathbf{r} \rangle + \langle \mathbf{q}, \mathbf{r} \rangle\end{aligned}$$

(**c**) homogeneity property: For any real number λ,

$$\langle \lambda \mathbf{p}, \mathbf{q} \rangle = \int_a^b \lambda p(x)q(x)dx = \lambda \int_a^b p(x)q(x)dx = \lambda \langle \mathbf{p}, \mathbf{q} \rangle$$

(**d**) positive definiteness property: We must have

$$\langle \mathbf{p}, \mathbf{p} \rangle = \int_a^b p^2(x)dx \geq 0$$

since $p^2(x) \geq 0$ for any real number x, and the integral on the right represents the area under the graph of a *non-negative* function.
(**e**) non-degeneracy property: Clearly,

$$\langle \mathbf{p}, \mathbf{p} \rangle = \int_a^b p^2(x)dx = 0 \ \text{ if and only if } \ \mathbf{p} = \mathbf{0}$$

since once again, we note that $p^2(x) \geq 0$ and the integral on the right is just the area under the graph of a non-negative, continuous function.

Example-3: Let $M(m,n)$ be the space of $m \times n$ matrices. For any $A, B \in M(m,n)$ we can define the inner product

$$\langle A, B \rangle = Tr(B^T A) = \sum_{i=1}^{m} \sum_{j=1}^{n} a_{ij} b_{ij} \tag{18.3}$$

The reader should verify that the last expression on right-hand side of (18.3) is valid. Now, let's use properties (3.4a)-(3.4d) of the transpose operation to show that (18.3) satisfies all of the properties of an inner product.

(**a**) symmetric property:

$$\langle A, B \rangle = Tr(B^T A) = Tr\{(A^T B)^T\} = Tr(A^T B) = \langle B, A \rangle$$

(**b**) linearity property:

$$\langle A+B,C\rangle = Tr\{C^T(A+B)\} = Tr(C^TA)+Tr(C^TB)$$
$$= \langle A,C\rangle + \langle B,C\rangle$$

(c) homogeneity property: For any real number λ,

$$\begin{aligned}\langle \lambda A,B\rangle &= Tr\{B^T(\lambda A)\} \\ &= Tr\{\lambda(B^TA))\} \\ &= \lambda Tr(B^TA) \\ &= \lambda\langle A,B\rangle\end{aligned}$$

(d) positive definiteness property: We have

$$\langle A,A\rangle = Tr(A^TA) = \sum_{i=1}^{m}\sum_{j=1}^{n} a_{ij}^2 \geq 0$$

since each $a_{ij}^2 \geq 0$.
(e) non-degeneracy property: Clearly,

$$\langle A,A\rangle = 0 \text{ if and only if } A = 0$$

since from the expression in **(d)**, $\langle A,A\rangle = 0$ if and only if $a_{ij} = 0$ for all i and j where $1 \leq i \leq m$, and $1 \leq j \leq n$.

Theorem 18.1: *(Some Useful Properties of an Inner Product)*

A real inner product on a real vector space V satisfies the following additional properties for any vectors $\mathbf{u},\mathbf{v},\mathbf{w} \in V$ and for any scalar λ:
(f) $\langle \mathbf{u},\mathbf{v}+\mathbf{w}\rangle = \langle \mathbf{u},\mathbf{v}\rangle + \langle \mathbf{u},\mathbf{w}\rangle$
(g) $\langle \mathbf{u},\lambda\mathbf{v}\rangle = \lambda\langle \mathbf{u},\mathbf{v}\rangle$
(h) $\langle \mathbf{v},\mathbf{0}\rangle = \langle \mathbf{0},\mathbf{v}\rangle = 0$.

Proof: In order to establish **(f)**, observe that

$$\begin{aligned}\langle \mathbf{u},\mathbf{v}+\mathbf{w}\rangle &= \langle \mathbf{v}+\mathbf{w},\mathbf{u}\rangle && \text{(by symmetric property)} \\ &= \langle \mathbf{v},\mathbf{u}\rangle + \langle \mathbf{w},\mathbf{u}\rangle && \text{(by linearity property)} \\ &= \langle \mathbf{u},\mathbf{v}\rangle + \langle \mathbf{u},\mathbf{w}\rangle && \text{(by symmetric property)}\end{aligned}$$

The proofs of **(g)** and **(h)** are easy also, and are left as exercises. □.

Theorem 18.2: *(Cauchy-Schwarz Inequality)*

If V is an Euclidean space, then for any vectors $\mathbf{u},\mathbf{v} \in V$:

$$\langle \mathbf{u},\mathbf{v}\rangle^2 \leq \langle \mathbf{u},\mathbf{u}\rangle\langle \mathbf{v},\mathbf{v}\rangle \tag{18.4}$$

Proof: If $\mathbf{u} = \mathbf{0}$ then there is nothing to show, since we obtain $0 \le 0$, which is certainly true. So, let's assume that $\mathbf{u} \neq \mathbf{0}$. For any real number λ, it's clear that

$$\langle \lambda\mathbf{u} + \mathbf{v}, \lambda\mathbf{u} + \mathbf{v} \rangle \ge 0$$

Now, if we expand the left-hand side, we get

$$\langle \mathbf{u}, \mathbf{u} \rangle \lambda^2 + 2\langle \mathbf{u}, \mathbf{v} \rangle \lambda + \langle \mathbf{v}, \mathbf{v} \rangle \ge 0$$

But the last expression is just a quadratic inequality of the form $ax^2 + bx + c \ge 0$ where $a = \langle \mathbf{u}, \mathbf{u} \rangle$, $b = 2\langle \mathbf{u}, \mathbf{v} \rangle$, and $c = \langle \mathbf{v}, \mathbf{v} \rangle$. Thus, the corresponding quadratic equation $ax^2 + bx + c = 0$ must have either complex conjugate roots, or a repeated real root. Consequently, the discriminant $\Delta = b^2 - 4ac$ must be less than or equal to zero. Thus,

$$\Delta = b^2 - 4ac = 4\left[\langle \mathbf{u}, \mathbf{v} \rangle^2 - \langle \mathbf{u}, \mathbf{u} \rangle \langle \mathbf{v}, \mathbf{v} \rangle\right] \le 0$$

which, in turn, implies

$$\langle \mathbf{u}, \mathbf{v} \rangle^2 \le \langle \mathbf{u}, \mathbf{u} \rangle \langle \mathbf{v}, \mathbf{v} \rangle$$

as claimed. □.

Definition: *(Norm of a Vector & Distance Between Vectors)*

Given a Euclidean space V, the **norm** (or **length**) of any vector $\mathbf{v} \in V$ is given by

$$\|\mathbf{v}\| = \sqrt{\langle \mathbf{v}, \mathbf{v} \rangle} \tag{18.5}$$

and the **distance** $d(\mathbf{u}, \mathbf{v})$ between any two vectors $\mathbf{u}, \mathbf{v} \in V$ is defined by

$$d(\mathbf{u}, \mathbf{v}) = \|\mathbf{u} - \mathbf{v}\| \tag{18.6}$$

Observe that if $\mathbf{v}$ is a *non-zero* vector in a Euclidean space V, then the vector defined by

$$\mathbf{e}_v = \frac{1}{\|\mathbf{v}\|}\mathbf{v} \tag{18.7}$$

has a length of one, and is called a **unit vector**. Also, note that the vector $\mathbf{e}_v$ has the *same direction* as the given vector $\mathbf{v}$. Whenever we perform this procedure, we say that the vector $\mathbf{v}$ has been **normalized**, and the process is called **normalization**. A familiar example of unit vectors are the standard basis vectors $\mathbf{e}_1, \mathbf{e}_2, \mathbf{e}_3$ in the space $\mathbb{R}^3$, equipped with the Euclidean inner product (18.1).

We now turn our attention to exploring the basic properties of the norm.

Before doing so, however, we note in passing that the Cauchy-Scwartz inequality can be rewritten in the language of norms as

$$|\langle \mathbf{u}, \mathbf{v} \rangle| \leq \|\mathbf{u}\| \|\mathbf{v}\| \tag{18.8}$$

This is a useful form of the inequality, which is used in the proof of the following theorem.

Theorem 18.3: *(Properties of the Norm)*

Let V be a Euclidean space. Then for any vectors $\mathbf{u}, \mathbf{v} \in V$ and for any real number λ, the following properties are satisfied:

(a) $\|\mathbf{u}\| \geq 0$
(b) $\|\mathbf{u}\| = 0$ if and only if $\mathbf{u} = \mathbf{0}$
(c) $\|\lambda \mathbf{u}\| = |\lambda| \|\mathbf{u}\|$
(d) $\|\mathbf{u} + \mathbf{v}\| \leq \|\mathbf{u}\| + \|\mathbf{v}\|$ (Triangle Inequality)

Proof: We prove property (d), and leave the proofs of the remaining properties as exercises. To establish (d), observe that

$$\begin{aligned}
\|\mathbf{u} + \mathbf{v}\|^2 &= \langle \mathbf{u} + \mathbf{v}, \mathbf{u} + \mathbf{v} \rangle \\
&= \|\mathbf{u}\|^2 + 2\langle \mathbf{u}, \mathbf{v} \rangle + \|\mathbf{v}\|^2 \\
&\leq \|\mathbf{u}\|^2 + 2|\langle \mathbf{u}, \mathbf{v} \rangle| + \|\mathbf{v}\|^2 \\
&\leq \langle \mathbf{u}, \mathbf{u} \rangle + 2\|\mathbf{u}\| \|\mathbf{v}\| + \langle \mathbf{v}, \mathbf{v} \rangle \quad [\text{ by } (18.8)\ ] \\
&\leq (\|\mathbf{u}\| + \|\mathbf{v}\|)^2
\end{aligned}$$

So, upon taking square roots, property (**d**) follows immediately. □.

In the spaces $\mathbb{R}^2$ and $\mathbb{R}^3$, the Triangle Inequality can be interpreted in terms of distance. From Figure 18.1, we see that the triangle inequality states that the shortest distance between the any two points A and B is a straight line.

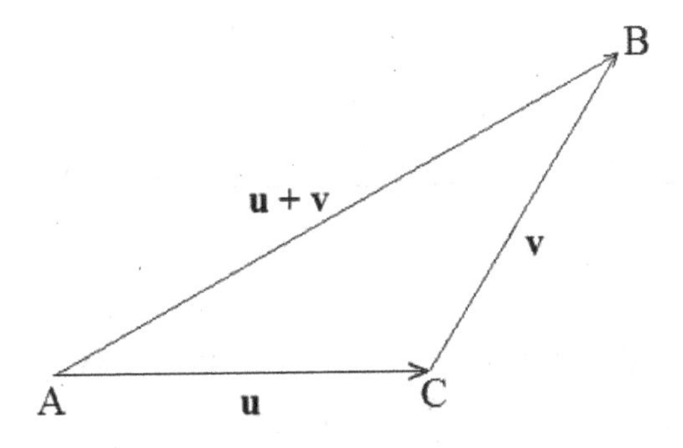

Figure 18.1: The Triangle Inequality

The Cauchy-Schwarz inequality allows us to define the angle between any two vectors in a Euclidean space V. To see this, observe that (18.7) guarantees that

$$-1 \leq \frac{\langle \mathbf{u}, \mathbf{v} \rangle}{\|\mathbf{u}\| \|\mathbf{v}\|} \leq 1$$

for any two vectors $\mathbf{u}, \mathbf{v} \in V$. As a direct result, there exists a unique real number $\angle(\mathbf{u}, \mathbf{v})$, called the **angle between u and v**, such that

$$\cos \angle(\mathbf{u}, \mathbf{v}) = \frac{\langle \mathbf{u}, \mathbf{v} \rangle}{\|\mathbf{u}\| \|\mathbf{v}\|} \tag{18.9}$$

and where $0 \leq \angle(\mathbf{u}, \mathbf{v}) \leq \pi$. Observe that (18.9) agrees with the usual formula for the angle between two vectors in the spaces $\mathbb{R}^2$ and $\mathbb{R}^3$.

Definition: *(Orthogonal Vectors & Orthogonal Complement)*

Given a Euclidean space V, any vectors $\mathbf{u}, \mathbf{v} \in V$ are said to be **orthogonal** and we write $\mathbf{u} \perp \mathbf{v}$ if $\langle \mathbf{u}, \mathbf{v} \rangle = 0$, i.e.,

$$\mathbf{u} \perp \mathbf{v} \text{ if and only if } \langle \mathbf{u}, \mathbf{v} \rangle = 0 \tag{18.10}$$

Furthermore, given any non-empty *subset* W of V, we define the **orthogonal complement** of W, denoted by $W^{\perp}$, by

$$W^{\perp} = \{\mathbf{v} \in V : \langle \mathbf{v}, \mathbf{w} \rangle = 0 \text{ for all } \mathbf{w} \in W\} \tag{18.11}$$

From (18.9), we see that any two *non-zero* vectors $\mathbf{u}$ and $\mathbf{v}$ in a Euclidean space are orthogonal if and only if they are *perpendicular* to each other. Also, the orthogonal complement of any subset W of a Euclidean space V simply consists of those vectors in V that are orthogonal to *each* vector in W.

Theorem 18.4: *(Generalized Pythagorean Theorem)*

Let V be an inner product space, and let $\mathbf{u}$ and $\mathbf{v}$ be any *orthogonal* vectors in V. Then

$$\|\mathbf{u} + \mathbf{v}\|^2 = \|\mathbf{u}\|^2 + \|\mathbf{v}\|^2 \tag{18.12}$$

Proof: Observe that

$$\begin{aligned} \|\mathbf{u} + \mathbf{v}\|^2 &= \langle \mathbf{u} + \mathbf{v}, \mathbf{u} + \mathbf{v} \rangle = \langle \mathbf{u}, \mathbf{u} \rangle + 2\langle \mathbf{u}, \mathbf{v} \rangle + \langle \mathbf{v}, \mathbf{v} \rangle \\ &= \langle \mathbf{u}, \mathbf{u} \rangle + \langle \mathbf{v}, \mathbf{v} \rangle \quad [\text{since } \langle \mathbf{u}, \mathbf{v} \rangle = 0\,] \\ &= \|\mathbf{u}\|^2 + \|\mathbf{v}\|^2 \end{aligned}$$

as claimed. □.

It's interesting to note that Theorem 18.4 can be generalized even further. That is, it can be shown to hold for any number of vectors in a Euclidean space as long as those vectors are *pairwise orthogonal*. The interested reader is referred to Exercise 16 below.

Exercise Set 18

In Exercises 1-9, using the space $\mathbb{R}^3$, *equipped with the Euclidean inner product*

$$\langle \mathbf{x}, \mathbf{y} \rangle = x_1 y_1 + x_2 y_2 + x_3 y_3$$

and the vectors

$$\mathbf{x} = \begin{bmatrix} 1 \\ 1 \\ 0 \end{bmatrix}, \quad \mathbf{y} = \begin{bmatrix} -2 \\ 1 \\ 1 \end{bmatrix}, \quad \mathbf{z} = \begin{bmatrix} 0 \\ 0 \\ 1 \end{bmatrix},$$

find each of the following expressions:

1. $\langle \mathbf{x}, \mathbf{y} \rangle$

2. $\langle \mathbf{x}, \mathbf{z} \rangle$

3. $\langle \mathbf{x}, \mathbf{y} + \mathbf{z} \rangle$

4. $\langle \mathbf{x}, 2\mathbf{y} + 3\mathbf{z} \rangle$

5. $\|\mathbf{x} + \mathbf{y}\|$

6. $\angle(\mathbf{x}, \mathbf{y})$

7. $\angle(\mathbf{y}, \mathbf{z})$

8. $d(\mathbf{x}, \mathbf{y})$

9. $d(\mathbf{y}, \mathbf{z})$

10. Let P_4 be the space of all polynomials of degree ≤ 4, and assume that it is equipped with the inner product

$$\langle \mathbf{p}, \mathbf{q} \rangle = \int_{-1}^{+1} p(x)q(x)dx$$

(a) Show that $\mathbf{p} = x^2$ and $\mathbf{q} = x$ are orthogonal.
(b) Given that $\mathbf{p} = x^2$, find $\|\mathbf{p}\|$.
(c) If $\mathbf{q} = x$ then find a *unit vector* $\mathbf{e}_q$ which has the same direction as $\mathbf{q}$.

11. Given that W is a non-empty subset of a Euclidean space V, show that the orthogonal complement of W is a subspace of V.

12. Given a Euclidean space V and arbitrary vectors $\mathbf{u}, \mathbf{v}, \mathbf{w} \in W$, show that

$$d(\mathbf{u}, \mathbf{w}) \leq d(\mathbf{u}, \mathbf{v}) + d(\mathbf{v}, \mathbf{w})$$

13. Given the space $C[-\pi, \pi]$ equipped with the inner product

$$\langle \mathbf{f}, \mathbf{g} \rangle = \int_{-\pi}^{+\pi} f(t)g(t)dt \ ,$$

consider the set of n-functions

$$\mathbf{f}_1 = \frac{1}{\sqrt{\pi}} \cos t, \ \mathbf{f}_2 = \frac{1}{\sqrt{\pi}} \cos 2t, \ldots, \mathbf{f}_n = \frac{1}{\sqrt{\pi}} \cos nt.$$

Show that

$$\langle \mathbf{f}_k, \mathbf{f}_m \rangle = \delta_{km}$$

where δ is the Kronecker delta of order n.

14. Show that equality holds in (18.8) if and only if the vectors $\mathbf{u}$ and $\mathbf{v}$ are linearly dependent.

15. Given the space $C[a, b]$ equipped with the inner product

$$\langle \mathbf{f}, \mathbf{g} \rangle = \int_a^b f(t)g(t)dt$$

show that for any $\mathbf{f}, \mathbf{g} \in C[a, b]$, we have

$$\left\{ \int_a^b f(t)g(t)dt \right\}^2 \leq \left\{ \int_a^b f^2(t)dt \right\} \left\{ \int_a^b g^2(t)dt \right\}$$

16. Let V be a Euclidean space, and let the vectors $\mathbf{v}_1, \mathbf{v}_2, \ldots, \mathbf{v}_n \in V$ be *pairwise orthogonal*, i.e., $\langle \mathbf{v}_k, \mathbf{v}_m \rangle = 0$ whenever $k \neq m$. Show that

$$\left\| \sum_{k=1}^{n} \mathbf{v}_k \right\|^2 = \sum_{k=1}^{n} \|\mathbf{v}_k\|^2$$

Note that this is just a generalization of Theorem 18.4.

19. Orthonormal Bases

We have already seen that vector spaces admit many different bases. In performing practical calculations, we can use any basis we please, and it's advantageous to choose a convenient basis which simplifies our work. As the reader will see, one of the nicest bases to work with is an orthonormal basis.

Definition: *(Orthonormal Set)*

A set of vectors $\{\boldsymbol{\varepsilon}_1, \boldsymbol{\varepsilon}_2, \ldots, \boldsymbol{\varepsilon}_r\}$ in a Euclidean space V is called an **orthonormal set** if

$$\langle \boldsymbol{\varepsilon}_i, \boldsymbol{\varepsilon}_j \rangle = \delta_{ij} = \begin{cases} 0, & \text{if } i \neq j \\ 1, & \text{if } i = j \end{cases} \tag{19.1}$$

for all $1 \leq i,j \leq r$.

Example-1: It's easy to verify that the standard basis vectors

$$\mathbf{e}_1 = \begin{bmatrix} 1 \\ 0 \\ 0 \end{bmatrix}, \quad \mathbf{e}_2 = \begin{bmatrix} 0 \\ 1 \\ 0 \end{bmatrix}, \quad \mathbf{e}_3 = \begin{bmatrix} 0 \\ 0 \\ 1 \end{bmatrix}$$

form an orthonormal set in $\mathbb{R}^3$, equipped with the Euclidean inner product.

Theorem 19.1: *(Orthonormal Sets are Linearly Independent)*

Let $S = \{\boldsymbol{\varepsilon}_1, \boldsymbol{\varepsilon}_2, \ldots, \boldsymbol{\varepsilon}_r\}$ be an orthonormal set of vectors in a Euclidean space V. Then the set S is linearly independent.

Proof: Assume there exist real numbers $\lambda_1, \lambda_2, \ldots, \lambda_r$ such that

$$\sum_{m=1}^{r} \lambda_m \boldsymbol{\varepsilon}_m = \lambda_1 \boldsymbol{\varepsilon}_1 + \lambda_2 \boldsymbol{\varepsilon}_2 + \ldots + \lambda_r \boldsymbol{\varepsilon}_r = \mathbf{0}$$

Let's take the inner product of both sides of this equation with a typical element $\boldsymbol{\varepsilon}_k$ of the set S, where k is *fixed*, and $1 \leq k \leq r$. We obtain

$$\left\langle \boldsymbol{\varepsilon}_k, \sum_{m=1}^{r} \lambda_m \boldsymbol{\varepsilon}_m \right\rangle = \langle \boldsymbol{\varepsilon}_k, \mathbf{0} \rangle = 0$$

Expanding the left-hand side, we get

$$\left\langle \boldsymbol{\varepsilon}_k, \sum_{m=1}^{r} \lambda_m \boldsymbol{\varepsilon}_m \right\rangle = \sum_{m=1}^{r} \lambda_m \langle \boldsymbol{\varepsilon}_k, \boldsymbol{\varepsilon}_m \rangle = \sum_{m=1}^{r} \lambda_m \delta_{km} = \lambda_k = 0$$

Since our choice of $\boldsymbol{\varepsilon}_k$ was completely arbitrary, we must have all of the λ's equal to zero. Thus, we conclude that S is a linearly independent set. □.

If V is an n-dimensional Euclidean space, we already know that any set of n-linearly independent vectors forms a basis for V. Thus, by the previous theorem, any set of n-orthonormal vectors will also form a basis for V; we call such a basis an **orthonormal basis.**

Theorem 19.2: *(Orthonormal Basis)*

Let $B = \{\boldsymbol{\varepsilon}_1, \boldsymbol{\varepsilon}_2, \ldots, \boldsymbol{\varepsilon}_n\}$ be an orthonormal set of vectors in an n-dimensional Euclidean space V. Then for any vector $\mathbf{v} \in V$,

$$\mathbf{v} = \sum_{k=1}^{n} v_k \boldsymbol{\varepsilon}_k \tag{19.2}$$

where the n-components v_k of $\mathbf{v}$ are given by

$$v_k = \langle \mathbf{v}, \boldsymbol{\varepsilon}_k \rangle \qquad (\text{for } 1 \le k \le n) \tag{19.3}$$

Proof: Since B is a basis for V, then B spans V and there must exist real numbers $v_1, v_2, \ldots, v_n$ such that

$$\mathbf{v} = v_1 \boldsymbol{\varepsilon}_1 + v_2 \boldsymbol{\varepsilon}_2 + \ldots + v_n \boldsymbol{\varepsilon}_n = \sum_{m=1}^{n} v_m \boldsymbol{\varepsilon}_m$$

which is (19.2). To establish (19.3), observe that for $1 \le k \le n$,

$$\begin{aligned} \langle \mathbf{v}, \boldsymbol{\varepsilon}_k \rangle &= \left\langle \sum_{m=1}^{n} v_m \boldsymbol{\varepsilon}_m, \boldsymbol{\varepsilon}_k \right\rangle = \sum_{m=1}^{n} v_m \langle \boldsymbol{\varepsilon}_m, \boldsymbol{\varepsilon}_k \rangle \\ &= \sum_{m=1}^{n} v_m \delta_{mk} \quad [\text{ by } (19.1)\ ] \\ &= v_k \end{aligned}$$

which is (19.3). □.

The expression given in (19.2) is called an **orthonormal expansion** of the vector $\mathbf{v}$ in terms of the orthonormal basis B. Observe how easily the expansion coefficients are obtained by (19.3); this stands in stark contrast to the usual situation, where we are dealing with an ordinary basis, and we must solve a linear system of equations to obtain the expansion coefficients.

Lemma 19.3: *(Projection Lemma)*

Let $\{\boldsymbol{\varepsilon}_1, \boldsymbol{\varepsilon}_2, \ldots, \boldsymbol{\varepsilon}_r\}$ be an orthonormal set of vectors in an n-dimensional Euclidean space V, and let $S = span\ \{\boldsymbol{\varepsilon}_1, \boldsymbol{\varepsilon}_2, \ldots, \boldsymbol{\varepsilon}_r\}$ so S is a subspace of V. Then for any $\mathbf{v} \in V$, there exist vectors $\mathbf{s} \in S$ and $\mathbf{s}_\perp \in S^\perp$ such that

$$\mathbf{v} = \mathbf{s} + \mathbf{s}_\perp \tag{19.4a}$$

where

$$\mathbf{s} = \sum_{k=1}^{r} \langle \mathbf{v}, \boldsymbol{\varepsilon}_k \rangle \boldsymbol{\varepsilon}_k \tag{19.4b}$$

$$\mathbf{s}_\perp = \mathbf{v} - \mathbf{s} \tag{19.4c}$$

Proof: From (19.4b), the vector $\mathbf{s} \in S$ since it is a linear combination of vectors in S. So, it suffices to show that the vector $\mathbf{s}_\perp$ belongs to the orthogonal complement, $S^\perp$. We first show that for any *fixed* vector $\boldsymbol{\varepsilon}_m$,

$$\langle \mathbf{v} - \mathbf{s}, \boldsymbol{\varepsilon}_m \rangle = 0. \tag{a}$$

To this end, observe that

$$\begin{aligned}
\langle \mathbf{v} - \mathbf{s}, \boldsymbol{\varepsilon}_m \rangle &= \left\langle \mathbf{v} - \sum_{k=1}^{r} \langle \mathbf{v}, \boldsymbol{\varepsilon}_k \rangle \boldsymbol{\varepsilon}_k,\ \boldsymbol{\varepsilon}_m \right\rangle = \langle \mathbf{v}, \boldsymbol{\varepsilon}_m \rangle - \left\langle \sum_{k=1}^{r} \langle \mathbf{v}, \boldsymbol{\varepsilon}_k \rangle \boldsymbol{\varepsilon}_k,\ \boldsymbol{\varepsilon}_m \right\rangle \\
&= \langle \mathbf{v}, \boldsymbol{\varepsilon}_m \rangle - \sum_{k=1}^{r} \langle \mathbf{v}, \boldsymbol{\varepsilon}_k \rangle \langle \boldsymbol{\varepsilon}_k, \boldsymbol{\varepsilon}_m \rangle \\
&= \langle \mathbf{v}, \boldsymbol{\varepsilon}_m \rangle - \sum_{k=1}^{r} \langle \mathbf{v}, \boldsymbol{\varepsilon}_k \rangle \delta_{km} \\
&= \langle \mathbf{v}, \boldsymbol{\varepsilon}_m \rangle - \langle \mathbf{v}, \boldsymbol{\varepsilon}_m \rangle = 0
\end{aligned}$$

so relation (a) is established. Now, for any vector $\mathbf{s}' \in S$, we have

$$\mathbf{s}' = \sum_{m=1}^{r} \langle \mathbf{s}', \boldsymbol{\varepsilon}_m \rangle \boldsymbol{\varepsilon}_m$$

Consequently,

$$\begin{aligned}
\langle \mathbf{v} - \mathbf{s}, \mathbf{s}' \rangle &= \left\langle \mathbf{v} - \mathbf{s}, \sum_{m=1}^{r} \langle \mathbf{s}', \boldsymbol{\varepsilon}_m \rangle \boldsymbol{\varepsilon}_m \right\rangle = \sum_{m=1}^{r} \langle \mathbf{s}', \boldsymbol{\varepsilon}_m \rangle \langle \mathbf{v} - \mathbf{s}, \boldsymbol{\varepsilon}_m \rangle \\
&= 0 \quad [\text{ by virtue of (a) }]
\end{aligned}$$

We have shown that $\langle \mathbf{v} - \mathbf{s}, \mathbf{s}' \rangle = 0$ for any $\mathbf{s}' \in S$, so $\mathbf{s}_\perp = \mathbf{v} - \mathbf{s} \in S^\perp$ as claimed. This completes the proof. $\square$.

The vector $\mathbf{s}$ which appears in (19.4b) is called the **orthogonal projection**

of v onto S, and we denote this by $proj_S\mathbf{v}$; that is,

$$\mathbf{s} = proj_S\mathbf{v} = \sum_{k=1}^{r}\langle \mathbf{v}, \boldsymbol{\varepsilon}_k\rangle \boldsymbol{\varepsilon}_k \tag{19.5}$$

The vector $\mathbf{s}_\perp$, defined in (19.4c), is called the **vector component of v orthogonal to** S.

The previous lemma guarantees that if we are given any subspace S of a Euclidean space V, then we can always write any given vector in V as the sum of a vector in S and a vector in the orthogonal complement $S^\perp$ of S; thus, (19.4a) is sometimes called an **orthogonal decomposition**.

For example, consider the space $\mathbb{R}^3$, equipped with the Euclidean inner product. We know that any plane S through the origin is a subspace of $\mathbb{R}^3$. So, any vector $\mathbf{v} \in \mathbb{R}^3$ can be written as $\mathbf{v} = \mathbf{s} + \mathbf{s}_\perp$ where $\mathbf{s} \in S$ and the vector $\mathbf{s}_\perp$ lives in the orthogonal complement $S^\perp$ to S. Here, observe that the orthogonal complement $S^\perp$ happens to be a line through the origin that is normal to S. This state of affairs is shown in Figure 19.1 below.

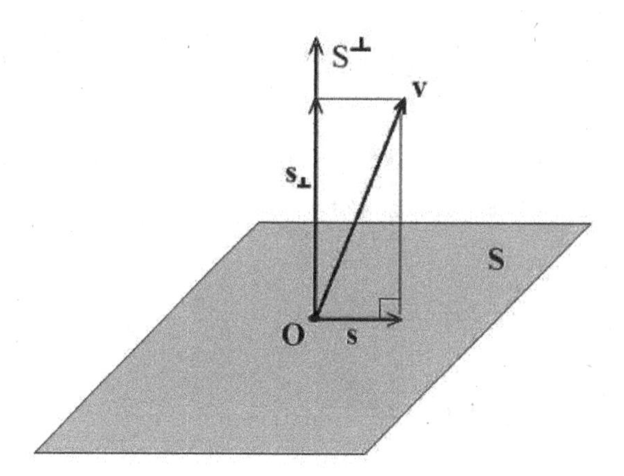

Figure 19.1: An orthogonal decomposition of $\mathbb{R}^3$.

Theorem 19.4: *(Gram-Schmidt Orthogonalization Procedure)*

Let $B = \{\mathbf{v}_1, \mathbf{v}_2, \ldots, \mathbf{v}_n\}$ be a basis for an n-dimensional Euclidean space V. Then there exists an *orthonormal basis* $B' = \{\boldsymbol{\varepsilon}_1, \boldsymbol{\varepsilon}_2, \ldots, \boldsymbol{\varepsilon}_n\}$ for V. Furthermore, the vectors in this orthonormal basis are given by

$$\boldsymbol{\varepsilon}_m = \frac{\mathbf{v}_m - \sum_{k=1}^{m-1}\langle \mathbf{v}_m, \boldsymbol{\varepsilon}_k\rangle \boldsymbol{\varepsilon}_k}{\left\| \mathbf{v}_m - \sum_{k=1}^{m-1}\langle \mathbf{v}_m, \boldsymbol{\varepsilon}_k\rangle \boldsymbol{\varepsilon}_k \right\|} \qquad (\text{ where } 1 \leq m \leq n) \tag{19.6}$$

Proof: We shall use a process, called the **Gram–Schmidt orthogonalization procedure**, to construct the orthonormal basis B'.

Step 1. We set

$$\boldsymbol{\varepsilon}_1 = \frac{\mathbf{v}_1}{\|\mathbf{v}_1\|}$$

so $\|\boldsymbol{\varepsilon}_1\| = 1$. We are guaranteed that $\|\mathbf{v}_1\| \neq 0$, since the set B is linearly independent, and no such set can contain the zero vector.

Step 2. The vector $\boldsymbol{\varepsilon}_1$ spans a one-dimensional subspace S_1 of V. So, to obtain $\boldsymbol{\varepsilon}_2$, we find the vector component $\mathbf{v}_2$ that is orthogonal to S_1, and then normalize that component:

$$\boldsymbol{\varepsilon}_2 = \frac{\mathbf{v}_2 - proj_{S_1}\mathbf{v}_2}{\|\mathbf{v}_2 - proj_{S_1}\mathbf{v}_2\|} = \frac{\mathbf{v}_2 - \langle \mathbf{v}_2, \boldsymbol{\varepsilon}_1 \rangle \boldsymbol{\varepsilon}_1}{\|\mathbf{v}_2 - \langle \mathbf{v}_2, \boldsymbol{\varepsilon}_1 \rangle \boldsymbol{\varepsilon}_1\|}$$

Observe that this normalization is not possible if $\mathbf{v}_2 - \langle \mathbf{v}_2, \boldsymbol{\varepsilon}_1 \rangle \boldsymbol{\varepsilon}_1 = \mathbf{0}$; but this can never happen since

$$\mathbf{v}_2 - \langle \mathbf{v}_2, \boldsymbol{\varepsilon}_1 \rangle \boldsymbol{\varepsilon}_1 = 0 \implies \mathbf{v}_2 = \frac{\langle \mathbf{v}_2, \boldsymbol{\varepsilon}_1 \rangle}{\|\mathbf{v}_1\|}\mathbf{v}_1$$

and then, the set of vectors $\{\mathbf{v}_1, \mathbf{v}_2\}$ would be linearly dependent, contrary to hypothesis.

Step 3. Now, the vectors $\{\boldsymbol{\varepsilon}_1, \boldsymbol{\varepsilon}_2\}$ span a two-dimensional subspace S_2 of V. So, to obtain $\boldsymbol{\varepsilon}_3$, we find the vector component $\mathbf{v}_3$ that is orthogonal to S_2, and then normalize it:

$$\boldsymbol{\varepsilon}_3 = \frac{\mathbf{v}_3 - proj_{S_2}\mathbf{v}_3}{\|\mathbf{v}_3 - proj_{S_2}\mathbf{v}_3\|} = \frac{\mathbf{v}_3 - \langle \mathbf{v}_3, \boldsymbol{\varepsilon}_1 \rangle \boldsymbol{\varepsilon}_1 - \langle \mathbf{v}_3, \boldsymbol{\varepsilon}_2 \rangle \boldsymbol{\varepsilon}_2}{\|\mathbf{v}_3 - \langle \mathbf{v}_3, \boldsymbol{\varepsilon}_1 \rangle \boldsymbol{\varepsilon}_1 - \langle \mathbf{v}_3, \boldsymbol{\varepsilon}_2 \rangle \boldsymbol{\varepsilon}_2\|}$$

Once again, we are guaranteed that the denominator is non-zero due to the linear independence of the vectors in B.

Step n. We continue in this fashion, obtaining the set of orthonormal vectors $\{\boldsymbol{\varepsilon}_1, \boldsymbol{\varepsilon}_2, \ldots, \boldsymbol{\varepsilon}_{n-1}\}$ that span an $(n-1)$-dimensional subspace S_{n-1} of V. So, to obtain $\boldsymbol{\varepsilon}_n$, we find the vector component $\mathbf{v}_n$ that is orthogonal to S_{n-1}, and then normalize that vector to obtain

$$\boldsymbol{\varepsilon}_n = \frac{\mathbf{v}_n - proj_{S_{n-1}}\mathbf{v}_n}{\|\mathbf{v}_n - proj_{S_{n-1}}\mathbf{v}_n\|} = \frac{\mathbf{v}_n - \sum_{k=1}^{n-1} \langle \mathbf{v}_n, \boldsymbol{\varepsilon}_k \rangle \boldsymbol{\varepsilon}_k}{\left\| \mathbf{v}_n - \sum_{k=1}^{n-1} \langle \mathbf{v}_n, \boldsymbol{\varepsilon}_k \rangle \boldsymbol{\varepsilon}_k \right\|}$$

which is (19.6). Once again, we are assured that the denominator is not zero due to the linear independence of the set B. □.

Example-2: Consider the space $\mathbb{R}^3$, equipped with the Euclidean inner product. It's easy to show that the vectors

$$\mathbf{v}_1 = \begin{bmatrix} 1 \\ -1 \\ 1 \end{bmatrix}, \quad \mathbf{v}_2 = \begin{bmatrix} 1 \\ 0 \\ 1 \end{bmatrix}, \quad \mathbf{v}_3 = \begin{bmatrix} 0 \\ 1 \\ 1 \end{bmatrix}$$

are linearly independent, and so, form a basis B for $\mathbb{R}^3$. Use this basis to find an orthornormal basis $B' = \{\boldsymbol{\varepsilon}_1, \boldsymbol{\varepsilon}_2, \boldsymbol{\varepsilon}_3\}$ for $\mathbb{R}^3$.

Solution: We carry out the Gram-Schmidt orthogonalization procedure.

Step 1. We have

$$\boldsymbol{\varepsilon}_1 = \frac{\mathbf{v}_1}{\|\mathbf{v}_1\|} = \frac{1}{\sqrt{3}}\begin{bmatrix} 1 \\ -1 \\ 1 \end{bmatrix} = \begin{bmatrix} \frac{1}{3}\sqrt{3} \\ -\frac{1}{3}\sqrt{3} \\ \frac{1}{3}\sqrt{3} \end{bmatrix}$$

Step 2.

$$\mathbf{v}_2 - \langle \mathbf{v}_2, \boldsymbol{\varepsilon}_1 \rangle \boldsymbol{\varepsilon}_1 = \begin{bmatrix} 1 \\ 0 \\ 1 \end{bmatrix} - \left(\frac{2}{3}\sqrt{3}\right)\begin{bmatrix} \frac{\sqrt{3}}{3} \\ -\frac{\sqrt{3}}{3} \\ \frac{\sqrt{3}}{3} \end{bmatrix} = \begin{bmatrix} \frac{1}{3} \\ \frac{2}{3} \\ \frac{1}{3} \end{bmatrix}$$

$$\|\mathbf{v}_2 - \langle \mathbf{v}_2, \boldsymbol{\varepsilon}_1 \rangle \boldsymbol{\varepsilon}_1\| = \frac{\sqrt{6}}{3}$$

So,

$$\boldsymbol{\varepsilon}_2 = \frac{\mathbf{v}_2 - \langle \mathbf{v}_2, \boldsymbol{\varepsilon}_1 \rangle \boldsymbol{\varepsilon}_1}{\|\mathbf{v}_2 - \langle \mathbf{v}_2, \boldsymbol{\varepsilon}_1 \rangle \boldsymbol{\varepsilon}_1\|} = \frac{1}{\left(\frac{1}{3}\sqrt{6}\right)}\begin{bmatrix} \frac{1}{3} \\ \frac{2}{3} \\ \frac{1}{3} \end{bmatrix} = \begin{bmatrix} \frac{1}{6}\sqrt{6} \\ \frac{1}{3}\sqrt{6} \\ \frac{1}{6}\sqrt{6} \end{bmatrix}$$

Step 3.

$$\mathbf{v}_3 - \langle \mathbf{v}_3, \boldsymbol{\varepsilon}_1 \rangle \boldsymbol{\varepsilon}_1 - \langle \mathbf{v}_3, \boldsymbol{\varepsilon}_2 \rangle \boldsymbol{\varepsilon}_2 = \begin{bmatrix} 0 \\ 1 \\ 1 \end{bmatrix} - (\mathbf{0}) - \left(\frac{\sqrt{6}}{2}\right)\begin{bmatrix} \frac{\sqrt{6}}{6} \\ \frac{\sqrt{6}}{3} \\ \frac{\sqrt{6}}{6} \end{bmatrix} = \begin{bmatrix} -\frac{1}{2} \\ 0 \\ \frac{1}{2} \end{bmatrix}$$

$$\|\mathbf{v}_3 - \langle \mathbf{v}_3, \boldsymbol{\varepsilon}_1 \rangle \boldsymbol{\varepsilon}_1 - \langle \mathbf{v}_3, \boldsymbol{\varepsilon}_2 \rangle \boldsymbol{\varepsilon}_2 \| = \frac{1}{\sqrt{2}}$$

Consequently,

$$\boldsymbol{\varepsilon}_3 = \frac{\mathbf{v}_3 - \langle \mathbf{v}_3, \boldsymbol{\varepsilon}_1 \rangle \boldsymbol{\varepsilon}_1 - \langle \mathbf{v}_3, \boldsymbol{\varepsilon}_2 \rangle \boldsymbol{\varepsilon}_2}{\|\mathbf{v}_3 - \langle \mathbf{v}_3, \boldsymbol{\varepsilon}_1 \rangle \boldsymbol{\varepsilon}_1 - \langle \mathbf{v}_3, \boldsymbol{\varepsilon}_2 \rangle \boldsymbol{\varepsilon}_2 \|} = \begin{bmatrix} -\frac{1}{2}\sqrt{2} \\ 0 \\ \frac{1}{2}\sqrt{2} \end{bmatrix}$$

and an orthonormal basis for $\mathbb{R}^3$ is

$$\boldsymbol{\varepsilon}_1 = \begin{bmatrix} \frac{1}{3}\sqrt{3} \\ -\frac{1}{3}\sqrt{3} \\ \frac{1}{3}\sqrt{3} \end{bmatrix}, \quad \boldsymbol{\varepsilon}_2 = \begin{bmatrix} \frac{1}{6}\sqrt{6} \\ \frac{1}{3}\sqrt{6} \\ \frac{1}{6}\sqrt{6} \end{bmatrix}, \quad \boldsymbol{\varepsilon}_3 = \begin{bmatrix} -\frac{1}{2}\sqrt{2} \\ 0 \\ \frac{1}{2}\sqrt{2} \end{bmatrix}$$

The reader should check that these vectors are, in fact, orthonormal by direct calculation.

Example-3: Consider the space P_2 of all polynomials of degree ≤ 2, equipped with the inner product

$$\langle \mathbf{p}, \mathbf{q} \rangle = \int_{-1}^{+1} p(x)q(x)dx$$

Starting with the standard basis for P_2:

$$\mathbf{p}_1(x) = 1, \quad \mathbf{p}_2(x) = x, \quad \mathbf{p}_3(x) = x^2$$

construct an orthonormal basis.

Solution: As in the previous example, we will use the Gram-Schmidt orthogonalization procedure.

Step 1.

$$\boldsymbol{\varepsilon}_1 = \frac{\mathbf{p}_1}{\|\mathbf{p}_1\|} \quad \text{where} \quad \|\mathbf{p}_1\| = \left[\int_{-1}^{+1} (1)dx\right]^{1/2} = \sqrt{2}$$

Thus, $\boldsymbol{\varepsilon}_1 = 1/\sqrt{2}$

Step 2.

$$\mathbf{p}_2 - \langle \mathbf{p}_2, \boldsymbol{\varepsilon}_1 \rangle \boldsymbol{\varepsilon}_1 = x - \frac{1}{\sqrt{2}} \int_{-1}^{+1} \frac{1}{\sqrt{2}} x dx = x$$

$$\|\mathbf{p}_2 - \langle \mathbf{p}_2, \boldsymbol{\varepsilon}_1 \rangle \boldsymbol{\varepsilon}_1 \| = \left[\int_{-1}^{+1} x^2 dx\right]^{1/2} = \frac{\sqrt{6}}{3}$$

$$\varepsilon_2 = \frac{x}{\sqrt{6}/3} = \frac{\sqrt{6}}{2}x$$

Step 3. Similarly, we have

$$\mathbf{p}_3 - \langle \mathbf{p}_3, \varepsilon_1 \rangle \varepsilon_1 - \langle \mathbf{p}_3, \varepsilon_2 \rangle \varepsilon_2 = x^2 - \frac{1}{\sqrt{2}} \int_{-1}^{+1} \frac{1}{\sqrt{2}} x^2 dx - \frac{\sqrt{6}}{2} x \int_{-1}^{+1} \frac{\sqrt{6}}{2} x^3 dx$$
$$= x^2 - \frac{1}{3}$$

$$\| \mathbf{p}_3 - \langle \mathbf{p}_3, \varepsilon_1 \rangle \varepsilon_1 - \langle \mathbf{p}_3, \varepsilon_2 \rangle \varepsilon_2 \| = \left[\int_{-1}^{+1} \left(x^2 - \frac{1}{3} \right)^2 dx \right]^{1/2} = \frac{2}{15}\sqrt{10}$$

$$\varepsilon_3 = \frac{\mathbf{p}_3 - \langle \mathbf{p}_3, \varepsilon_1 \rangle \varepsilon_1 - \langle \mathbf{p}_3, \varepsilon_2 \rangle \varepsilon_2}{\| \mathbf{p}_3 - \langle \mathbf{p}_3, \varepsilon_1 \rangle \varepsilon_1 - \langle \mathbf{p}_3, \varepsilon_2 \rangle \varepsilon_2 \|} = \frac{3\sqrt{10}}{4}\left(x^2 - \frac{1}{3} \right)$$

Thus, the vectors

$$\varepsilon_1 = \frac{1}{\sqrt{2}}, \quad \varepsilon_2 = \frac{\sqrt{6}}{2}x, \quad \varepsilon_3 = \frac{3\sqrt{10}}{4}\left(x^2 - \frac{1}{3} \right)$$

form an orthonormal basis for P_2; they are called the first three *normalized Legendre polynomials.*

We now state and prove a theorem which is a generalization of Lemma 19.3, and which shows the connection between the concept of orthogonal decomposition and the notion of a direct sum that we studied previously.

Theorem 19.5: *(Projection Theorem)*

If S is a subspace of a finite-dimensional Euclidean space V, then

$$V = S \oplus S^{\perp} \tag{19.7}$$

and for every vector $\mathbf{v} \in V$, there exist vectors $\mathbf{s} \in S$ and $\mathbf{s}_{\perp} \in S^{\perp}$ such that

$$\mathbf{v} = \mathbf{s} + \mathbf{s}_{\perp}$$

where this decomposition is *unique*.

Proof: We first show that $V = S \oplus S^{\perp}$. Now, by Lemma 19.1, we already know that for any $\mathbf{v} \in V$, there exist vectors $\mathbf{s} \in S$ and $\mathbf{s}_{\perp} \in S^{\perp}$ such that

$$\mathbf{v} = \mathbf{s} + \mathbf{s}_{\perp}$$

where

$$\mathbf{s} = proj_S\,\mathbf{v} = \sum_{k=1}^{r}\langle \mathbf{v}, \boldsymbol{\varepsilon}_k\rangle \boldsymbol{\varepsilon}_k \quad \text{and} \quad \mathbf{s}_\perp = \mathbf{v} - \mathbf{s}$$

Consequently, we have $V = S + S^\perp$, and by Theorem 15.3, it suffices to show that $S \cap S^\perp = \{\mathbf{0}\}$. To this end, assume that $\mathbf{x} \in S \cap S^\perp$, so $\mathbf{x} \in S^\perp$ and $\mathbf{x} \in S$. Observe that

$$\mathbf{x} \in S^\perp \Rightarrow \langle \mathbf{x}, \mathbf{s}'\rangle = 0 \text{ for every vector } \mathbf{s}' \in S$$

But then $\mathbf{x} \in S$ implies $\langle \mathbf{x}, \mathbf{x}\rangle = 0$ as well, so $\mathbf{x} = \mathbf{0}$. Thus, we conclude that $V = S \oplus S^\perp$.

Secondly, the fact that for any $\mathbf{v} \in V$, the decomposition $\mathbf{v} = \mathbf{s} + \mathbf{s}_\perp$ is *unique* follows directly from the *definition* of the direct sum. □.

The previous theorem guarantees that a Euclidean space V can be written as the direct sum of any subspace S of V, and its orthogonal complement $S^\perp$. An immediate consequence of the theorem is the following useful corollary.

Corollary 19.6: *(Dimension of an Orthogonal Complement)*

If S is a subspace of a finite-dimensional Euclidean space V, then

$$\dim(S^\perp) = \dim(V) - \dim(S) \tag{19.8}$$

Proof: The proof follows immediately from the previous theorem, and Corollary 15.1. □.

The Projection Theorem has an interesting interpretation. Assume that S is a subspace of a finite-dimensional Euclidean space V, and we want to approximate a fixed vector $\mathbf{v} \in V$ with a some vector $\mathbf{s} \in S$. Although infinitely many choices are possible for $\mathbf{s}$, it turns out that if we choose

$$\mathbf{s} = proj_S\mathbf{v}$$

then the error E of our approximation, i.e.,

$$E = \|\mathbf{v} - \mathbf{s}\| = \|\mathbf{v} - proj_S\mathbf{v}\|$$

will be *minimized.*

Thus, as long as we confine ourselves to only choosing vectors from S to approximate $\mathbf{v}$, then the **best approximation** is given by the orthogonal projection $\mathbf{s}$ onto the subspace S. This is the content of the last theorem of the lesson.

Theorem 19.7: *(Best Approximation Theorem)*

Let S be a subspace of a finite-dimensional Euclidean space V. Then for any vector $\mathbf{v} \in V$, the orthogonal projection of $\mathbf{v}$ onto S

$$\mathbf{s} = proj_S \mathbf{v}$$

provides the **best approximation** of $\mathbf{v}$ among all of the possible vectors that can be chosen from S; that is, for any other vector $\mathbf{s}' \in S$, we have

$$\|\mathbf{v} - \mathbf{s}\| < \|\mathbf{v} - \mathbf{s}'\|$$

Proof: Let $\mathbf{v} \in V$ be chosen, and kept fixed. By Theorem 19.5, $V = S \oplus S^{\perp}$, so there exist vectors $\mathbf{s} \in S$ and $\mathbf{s}_{\perp} \in S^{\perp}$ such that

$$\mathbf{v} = \mathbf{s} + \mathbf{s}_{\perp}$$

where the decomposition is unique. Now, observe that for any $\mathbf{s}' \in S$, which is assumed to be *distinct* from $\mathbf{s}$, we have

$$\mathbf{v} - \mathbf{s}' = (\mathbf{s} + \mathbf{s}_{\perp}) - \mathbf{s}' = (\mathbf{s} - \mathbf{s}') + \mathbf{s}_{\perp}$$

where $(\mathbf{s} - \mathbf{s}') \in S$. Clearly, the vectors $(\mathbf{s} - \mathbf{s}')$ and $\mathbf{s}_{\perp}$ are orthogonal to each other, so by the Pythagorean Theorem, we can write

$$\begin{aligned} \|\mathbf{v} - \mathbf{s}'\|^2 &= \|\mathbf{s} - \mathbf{s}'\|^2 + \|\mathbf{s}_{\perp}\|^2 \\ &= \|\mathbf{s} - \mathbf{s}'\|^2 + \|\mathbf{v} - \mathbf{s}\|^2 \\ &< \|\mathbf{v} - \mathbf{s}\|^2 \end{aligned}$$

where the last step follows from the fact that $\|\mathbf{s} - \mathbf{s}'\|^2 > 0$. The result then follows upon taking square roots. □.

Exercise Set 19

---

1. Given $\mathbb{R}^2$ equipped with the Euclidean inner product, use the Gram-Schmidt process to convert each given basis $\{\mathbf{v}_1, \mathbf{v}_2\}$ for $\mathbb{R}^2$ into an orthonormal basis:

(a) $\mathbf{v}_1 = [1,1]^T$, $\mathbf{v}_2 = [1,-1]^T$

(b) $\mathbf{v}_1 = [0,-1]^T$, $\mathbf{v}_2 = [2,0]^T$

2. Consider the space $\mathbb{R}^3$, equipped with the Euclidean inner product. Use the Gram-Schmidt process to convert each basis $\{\mathbf{v}_1, \mathbf{v}_2, \mathbf{v}_3\}$ into an orthonormal basis:

(a) $\mathbf{v}_1 = [1,0,0]^T$, $\mathbf{v}_2 = [1,0,1]^T$, $\mathbf{v}_3 = [0,1,1]^T$

(b) $\mathbf{v}_1 = [0,0,1]^T$, $\mathbf{v}_2 = [1,1,1]^T$, $\mathbf{v}_3 = [1,0,1]^T$

3. Given the Euclidean space P_2 equipped with the inner product

$$\langle \mathbf{p}, \mathbf{q} \rangle = \int_0^1 p(x)q(x)dx$$

Starting with the basis for P_2:

$$\mathbf{p}_1(x) = 1, \ \mathbf{p}_2(x) = x, \ \mathbf{p}_3(x) = x^2$$

use the Gram-Schmidt process to construct an orthonormal basis.

4. Let V be a Euclidean space with an orthonormal basis $\{\varepsilon_1, \varepsilon_2, \ldots, \varepsilon_n\}$. Show that for any vectors $\mathbf{u}, \mathbf{v} \in V$:

(a)

$$\|\mathbf{v}\|^2 = \sum_{k=1}^{n} \langle \mathbf{v}, \boldsymbol{\varepsilon}_k \rangle^2 = \langle \mathbf{v}, \boldsymbol{\varepsilon}_1 \rangle^2 + \langle \mathbf{v}, \boldsymbol{\varepsilon}_2 \rangle^2 + \ldots + \langle \mathbf{v}, \boldsymbol{\varepsilon}_n \rangle^2$$

(b)

$$\langle \mathbf{u}, \mathbf{v} \rangle = \sum_{k=1}^{n} \langle \mathbf{u}, \boldsymbol{\varepsilon}_k \rangle \langle \boldsymbol{\varepsilon}_k, \mathbf{v} \rangle = \langle \mathbf{u}, \boldsymbol{\varepsilon}_1 \rangle \langle \boldsymbol{\varepsilon}_1, \mathbf{v} \rangle + \langle \mathbf{u}, \boldsymbol{\varepsilon}_2 \rangle \langle \boldsymbol{\varepsilon}_2, \mathbf{v} \rangle + \ldots + \langle \mathbf{u}, \boldsymbol{\varepsilon}_n \rangle \langle \boldsymbol{\varepsilon}_n, \mathbf{v} \rangle$$

5. *(Alternative Version of the Best Approximation Theorem)*
Let $B = \{\boldsymbol{\varepsilon}_1, \boldsymbol{\varepsilon}_2,,, \boldsymbol{\varepsilon}_r\}$ be an orthonormal set of vectors in a Euclidean space V, and let $S = span\{\boldsymbol{\varepsilon}_1, \boldsymbol{\varepsilon}_2,,, \boldsymbol{\varepsilon}_r\}$ so S is an r-dimensional subspace of V. Show that for any $\mathbf{v} \in V$, that if we approximate $\mathbf{v}$ by an expression of the form

$$\sum_{k=1}^{r} a_k \boldsymbol{\varepsilon}_k$$

where the coefficients a_k are real numbers, then the error

$$E = \left\| \mathbf{v} - \sum_{k=1}^{r} a_k \boldsymbol{\varepsilon}_k \right\|$$

is *minimized* when

$$a_k = \langle \mathbf{v}, \boldsymbol{\varepsilon}_k \rangle \quad \text{for} \quad k = 1, 2, \ldots, r.$$

[**Hint**: Assume that $a_k = \langle \mathbf{v}, \boldsymbol{\varepsilon}_k \rangle + \Delta_k$, and show that E is *smallest* when we take $\Delta_k = 0$.]

6. Let S be the subspace of $\mathbb{R}^3$ given by

$$S = \left\{ \begin{bmatrix} a \\ -a \\ a \end{bmatrix} : a \in R \right\}$$

so S is a line through the origin. Find the distance between S and the vector $\mathbf{v} = [1,0,-1]^T$ in $\mathbb{R}^3$. [**Hint:** Use Theorem 19.7.]

7. Let S be the subspace of $\mathbb{R}^3$ given by the plane

$$S = \left\{ \begin{bmatrix} x \\ y \\ z \end{bmatrix} \in \mathbb{R}^3 : x + 2y + z = 0 \right\}$$

(a) Find a basis for S.
(b) Use the Gram-Schmidt process to obtain an orthonormal basis for S.
(c) Find the distance between the plane and the vector $\mathbf{v} = [1,1,1]^T$.

Unit V: Linear Transformations

Oliver Heaviside (1850-1925)

One of the first mathematicians to systematically study vector-valued functions was Oliver Heaviside, an English engineer, mathematician, and physicist who was born in Camden Town, London to a family of modest means. Heaviside was largely self-taught, as he had no formal education beyond the age of 16.

In 1873, he took his first look at James Clerk Maxwell's famous *Treatise on Electricity and Magnetism*. Although this work was to have a profound effect on Heaviside, he couldn't understand it at first. Consequently, he spent the next several years studying the requisite mathematics that enabled him to master Maxwell's work.

Maxwell's work was largely formulated in the language of quaternions. Later on, Heaviside and Josiah Willard Gibbs (1839-1903) streamlined Maxwell's notation, and made three-dimensional vector analysis a separate subject in its own right.

Unfortunately, for much of his reclusive life, Heaviside's work was largely unrecognized by the members of the scientific elite; nevertheless, he went on to make major contributions to the fields of engineering, physics, and the use of operational methods in mathematics.

20. Introduction to Linear Transformations

In this lesson, we begin the study of functions that map one vector space into another vector space. If V and W are vector spaces, then a mapping

$$F : V \to W,$$

is a function which assigns a *unique* vector $\mathbf{w} \in W$ to each vector $\mathbf{v} \in V$. Thus, a mapping is nothing more than a *rule* which assigns a *unique* vector $\mathbf{w}$ in W to each vector $\mathbf{v}$ in V.

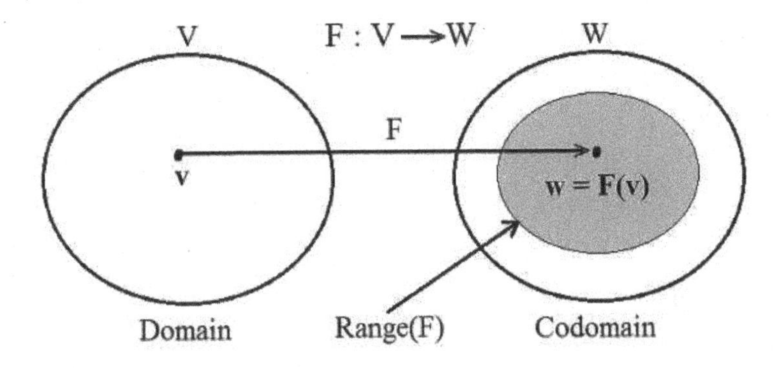

Figure 20.1: A typical mapping, $F : V \to W$.

We call V the **domain** of F, and W the **codomain** of F. If, in particular, we have $\mathbf{w} = F(\mathbf{v})$, then $\mathbf{w}$ is said to be the **image** of $\mathbf{v}$ under F. The set of all images of vectors in V is called the **range** of F; that is,

$$range(F) = \{\mathbf{w} \in W : \mathbf{w} = F(\mathbf{v}), \text{ where } \mathbf{v} \in V\} \tag{20.1}$$

A mapping $F : V \to W$ is said to be **one-to-one** (or **injective**), if for any $\mathbf{v}_1, \mathbf{v}_2 \in V$ with $\mathbf{v}_1 \neq \mathbf{v}_2$, we have $F(\mathbf{v}_1) \neq F(\mathbf{v}_2)$, while the mapping is said to be **onto** (or **surjective**), if $range(F) = W$. Thus, a mapping $F : V \to W$ is *one-to-one* if distinct vectors in V have distinct images in W, while it is *onto* if for any vector $\mathbf{w} \in W$ there exists some $\mathbf{v} \in V$ such that $\mathbf{w} = F(\mathbf{v})$.

Whenever we want to show that a mapping $F : V \to W$ is one-to-one, it suffices to show that $F(\mathbf{v}_1) = F(\mathbf{v}_2)$ implies $\mathbf{v}_1 = \mathbf{v}_2$ for all vectors $\mathbf{v}_1, \mathbf{v}_2 \in V$. Here, we are just using the contrapositive of the original statement in the definition.

We now turn our attention to the study of a special family of mappings, called *linear transformations*. As we shall see, linear transformations have many applications to the physical sciences, and to mathematics itself, so they are certainly worthy of our attention.

Definition: *(Linear Transformation & Linear Operator)*

Let $T : V \to W$ be a mapping from a vector space V into a vector space W. We say that T is a **linear transformation**, if

$$T(\mathbf{u}+\mathbf{v}) = T\mathbf{u} + T\mathbf{v} \quad \text{for all vectors } \mathbf{u},\mathbf{v} \in V. \tag{20.2}$$

and

$$T(\lambda\mathbf{u}) = \lambda(T\mathbf{u}) \quad \text{for all vectors } \mathbf{u} \in V \text{ and for all scalars } \lambda. \tag{20.3}$$

If, in addition, $V = W$, then T is called a **linear operator** on V.

Example-1: *(A Linear Operator on $\mathbb{R}^2$)*
Consider the mapping $T : \mathbb{R}^2 \to \mathbb{R}^2$ defined by

$$T\begin{bmatrix} x \\ y \end{bmatrix} = \begin{bmatrix} ax+by \\ cx+dy \end{bmatrix}$$

where a, b, c, d are scalars.

Clearly, T is a *linear operator* on $\mathbb{R}^2$ since if $\mathbf{u} = [x_1, y_1]^T$, $\mathbf{v} = [x_2, y_2]^T$, and λ is an arbitrary scalar, then

$$\begin{aligned} T(\mathbf{u}+\mathbf{v}) &= T\begin{bmatrix} x_1+x_2 \\ y_1+y_2 \end{bmatrix} = \begin{bmatrix} a(x_1+x_2)+b(y_1+y_2) \\ c(x_1+x_2)+d(y_1+y_2) \end{bmatrix} \\ &= \begin{bmatrix} ax_1+by_1 \\ cx_1+dy_1 \end{bmatrix} + \begin{bmatrix} ax_2+by_2 \\ cx_2+dy_2 \end{bmatrix} \\ &= T(\mathbf{u}) + T(\mathbf{v}) \end{aligned}$$

so (20.2) is satisfied, while

$$\begin{aligned} T(\lambda\mathbf{u}) &= T\begin{bmatrix} \lambda x_1 \\ \lambda y_1 \end{bmatrix} = \begin{bmatrix} a\lambda x_1+b\lambda y_1 \\ c\lambda x_1+d\lambda y_1 \end{bmatrix} \\ &= \lambda\begin{bmatrix} ax_1+by_1 \\ cx_1+dy_1 \end{bmatrix} \\ &= \lambda T(\mathbf{u}) \end{aligned}$$

and (20.3) is satisfied as well.

Example-2: *(Transformations Defined by Matrices)*
Let A be an $m \times n$ matrix. Assuming that A is fixed, we can define a mapping

$T : \mathbb{R}^n \to R^m$ by $T(\mathbf{x}) = A\mathbf{x}$ for all $\mathbf{x} \in \mathbb{R}^n$. Note that this definition makes sense, since if A is an $m \times n$ matrix, when we multiply it by the $n \times 1$ column vector $\mathbf{x}$, we get an $m \times 1$ column vector $A\mathbf{x} \in R^m$. Using the basic properties of matrices, it's easy to see that T is a linear, since for any $\mathbf{x}, \mathbf{y} \in \mathbb{R}^n$ and for any scalar λ, we have

$$T(\mathbf{x} + \mathbf{y}) = A(\mathbf{x} + \mathbf{y}) = A\mathbf{x} + A\mathbf{y}$$
$$T(\lambda\mathbf{x}) = A(\lambda\mathbf{x}) = \lambda A\mathbf{x}$$

Thus, T is a *linear transformation* from $\mathbb{R}^n$ into R^m.

Example-3: *(Rotation in the plane, $\mathbb{R}^2$)*
Let T be the linear operator on $\mathbb{R}^2$ defined by

$$T(\mathbf{x}) = R\mathbf{x} \text{ where } R = \begin{bmatrix} \cos\alpha & -\sin\alpha \\ \sin\alpha & \cos\alpha \end{bmatrix} \tag{20.4}$$

The operator T rotates any vector $x \in \mathbb{R}^2$ counterclockwise by the angle α, while keeping its length unchanged.

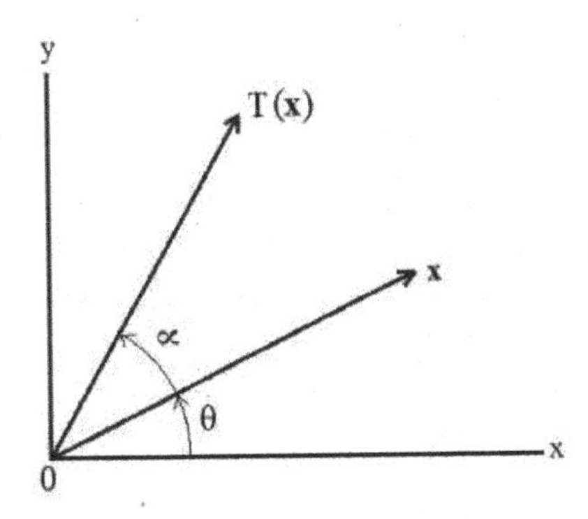

Figure 20.2: The vector $\mathbf{x}$ is rotated counterclockwise through an angle α.

To see this, let's use polar coordinates to write

$$\mathbf{x} = [r\cos\theta, r\sin\theta]^T$$

so that $\|\mathbf{x}\| = r$, and

$$T(\mathbf{x}) = \begin{bmatrix} \cos\alpha & -\sin\alpha \\ \sin\alpha & \cos\alpha \end{bmatrix} \begin{bmatrix} r\cos\theta \\ r\sin\theta \end{bmatrix} = \begin{bmatrix} r\cos\theta\cos\alpha - r\sin\theta\sin\alpha \\ r\cos\theta\sin\alpha + r\cos\alpha\sin\theta \end{bmatrix}$$
$$= \begin{bmatrix} r\cos(\theta + \alpha) \\ r\sin(\theta + \alpha) \end{bmatrix}$$

Thus, the vector $T(\mathbf{x})$ makes an angle $(\theta + \alpha)$ with the positive x-axis. Note also that $\|T(\mathbf{x})\| = r$ so the transformation T is **length-preserving**.

Example-4: *(The Derivative as a Linear Transformation)*
Let $C^1[a,b]$ be the set of all functions that have continuous derivatives on some interval $[a,b]$. Define the **differential operator** $D_x = d/dx$, so that for any function $\mathbf{f} \in C^1[a,b]$ we have

$$D_x(\mathbf{f}) = \frac{d}{dx}(\mathbf{f}) \qquad (20.5$$

It's easy to see that D_x is a linear transformation from $C^1[a,b]$ into $C[a,b]$, since for any functions $\mathbf{f},\mathbf{g} \in C^1[a,b]$ and for any scalar λ, we can write

$$D_x(\mathbf{f}+\mathbf{g}) = \frac{d}{dx}(\mathbf{f}+\mathbf{g}) = \frac{d}{dx}(\mathbf{f}) + \frac{d}{dx}(\mathbf{g}) = D_x(\mathbf{f}) + D_x(\mathbf{g})$$

and

$$D_x(\lambda\mathbf{f}) = \frac{d}{dx}(\lambda\mathbf{f}) = \lambda\frac{d}{dx}(\mathbf{f}) = \lambda D_x(\mathbf{f}),$$

so that (20.2) and (20.3) are satisfied

Exercise Set 20

In Exercises 1-5, determine whether the given mapping T is a linear transformation. If the given mapping is linear, show that properties (20.2) and (20.3) are satisfied.

1. $T : \mathbb{R}^2 \to \mathbb{R}^2$ where $T\left(\begin{bmatrix} x \\ y \end{bmatrix}\right) = \begin{bmatrix} 2x+3y \\ x-2y \end{bmatrix}$

2. $T : \mathbb{R}^2 \to \mathbb{R}^3$ where $T\left(\begin{bmatrix} x \\ y \end{bmatrix}\right) = \begin{bmatrix} x+y \\ x-y \\ 1 \end{bmatrix}$

3. $T : \mathbb{R}^3 \to \mathbb{R}^3$ where $T\left(\begin{bmatrix} x \\ y \\ z \end{bmatrix}\right) = \begin{bmatrix} 2x+3y+z \\ 2x-y+z \\ z \end{bmatrix}$

4. $T : P_2 \to P_1$ where $T(a_0 + a_1x + a_2x^2) = a_1 + 2a_2x$

5. $T : M(2,2) \to M(2,2)$ where $T(A) = A^T$ for all $A \in M(2,2)$.

6. Let V be a vector space, and consider the **identity transformation**

$T : V \to V$ defined by $T(\mathbf{x}) = \mathbf{x}$ for all vectors $\mathbf{x} \in V$. Show that the identity transformation is a linear operator on V.

7. If $T : V \to W$ is a linear transformation, then show that:

(a) $T(-\mathbf{v}) = -T(\mathbf{v})$ for all $\mathbf{v} \in V$
(b) $T(\mathbf{u} - \mathbf{v}) = T(\mathbf{u}) - T(\mathbf{v})$ for all $\mathbf{u}, \mathbf{v} \in V$
(c) $T(\mathbf{0}) = \mathbf{0}$

8. Let $T : \mathbb{R}^n \to \mathbb{R}^n$ be given by $T(\mathbf{x}) = A\mathbf{x}$ where A is a fixed square matrix of order n. Show that:

(a) T is a linear operator on $\mathbb{R}^n$
(b) T is a one-to-one transformation if $\det(A) \neq 0$.

9. Let V be a vector space, and consider the **zero transformation** $T : V \to V$ defined by $T(\mathbf{x}) = \mathbf{0}$ for all vectors $\mathbf{x} \in V$. Show that the zero transformation is a linear operator on V.

10. Consider the tranformation $T : P_n(x) \to P_{n+1}(x)$ defined by

$$T(\mathbf{p}) = \int_0^x p(t)dt$$

for all polynomials $\mathbf{p} \in P_n(x)$.

(a) Show that T is a linear transformation
(b) Find $T(x^2 + 1)$
(c) Find $T(2x + 3)$.

11. Assume that T is a linear operator on $P_2(x)$ such that

$$T(1) = 2x + 1, \; T(x) = x^2, \text{ and } T(x^2) = 1$$

(a) Find $T(2x + 1)$
(b) Find $T(2x^2 + 3x + 4)$

12. Let $T : V \to V$ be a linear operator on V. A vector $\mathbf{v} \in V$ is said to be a **fixed point** of T if $T(\mathbf{v}) = \mathbf{v}$. Find all fixed points of the linear operator $T : \mathbb{R}^2 \to \mathbb{R}^2$ which is given by

$$T\left(\begin{bmatrix} x \\ y \end{bmatrix}\right) = \begin{bmatrix} 2x + y \\ x + y \end{bmatrix}$$

21. Isomorphisms of Vector Spaces

We now turn our attention to the notion of **isomorphism** of vector spaces. The basic idea is that two vector spaces are *isomorphic* if one space "looks like" the other one. Admittedly, we need to make this idea more precise, and we shall do so in the following definition.

Definition: *(Isomorphism of Vector Spaces)*

If V and V' are vector spaces over the *same* field of scalars, then an **isomorphism** is a linear transformation $\varphi : V \to V'$ that is both 1-1 and onto. Furthermore, if such a mapping φ exists from V to V', then we say that the vector space V is **isomorphic** to V', and write $V \cong V'$.

It follows immediately that if $\varphi : V \to V'$ is a isomorphism from V to V' then for any vectors $\mathbf{u}, \mathbf{v} \in V$, and for any scalar λ, we have both

$$\varphi(\mathbf{u} + \mathbf{v}) = \varphi(\mathbf{u}) + \varphi(\mathbf{v}) \tag{21.1a}$$

$$\varphi(\lambda \mathbf{u}) = \lambda \varphi(\mathbf{u}) \tag{21.1b}$$

By requiring the mapping φ to be linear, we are basically requiring that φ preserves or respects the operations of vector addition and scalar multiplication in V.

Also, since the mapping $\varphi : V \to V'$ is both 1-1 and onto, we say that φ sets up a **one-to-one correspondence** between these sets. Thus, whenever $V \cong V'$, to each vector $\mathbf{v} \in V$ there corresponds a *unique* vector $\mathbf{v}' \in V'$, and conversely to each vector $\mathbf{v}' \in V$ there corresponds a *unique* vector $\mathbf{v} \in V$.

Theorem 21.1: *(Isomorphism of Any Two Spaces of the Same Finite Dimension)*

If V and V' are both finite dimensional vector spaces of the *same* dimension, then these spaces are isomorphic; i.e., $V \cong V'$.

Proof: Let V and V' have bases $\{\mathbf{v}_1, \mathbf{v}_2, \ldots, \mathbf{v}_n\}$ and $\{\mathbf{v}'_1, \mathbf{v}'_2, \ldots, \mathbf{v}'_n\}$ respectively. Given any vector $\mathbf{v} \in V$, there exist scalars $a_1, a_2, .., a_n$ such that

$$\mathbf{v} = a_1\mathbf{v}_1 + a_2\mathbf{v}_2 + \ldots + a_n\mathbf{v}_n$$

Consider the mapping $\varphi : V \to V'$ defined by

$$\varphi(a_1\mathbf{v}_1 + a_2\mathbf{v}_2 + \ldots + a_n\mathbf{v}_n) = a_1\mathbf{v}'_1 + a_2\mathbf{v}'_2 + \ldots + a_n\mathbf{v}'_n$$

so that, in particular, we have

$$\varphi(\mathbf{v}_1) = \mathbf{v}_1', \ \varphi(\mathbf{v}_2) = \mathbf{v}_2', \ldots, \varphi(\mathbf{v}_n) = \mathbf{v}_n'$$

We now show that the mapping φ is linear. If $\mathbf{w} = b_1\mathbf{v}_1 + b_2\mathbf{v}_2 + \ldots + b_n\mathbf{v}_n$ is also an arbitrary vector in V, then

$$\begin{aligned}
\varphi(\mathbf{v} + \mathbf{w}) &= \varphi((a_1 + b_1)\mathbf{v}_1 + (a_2 + b_2)\mathbf{v}_2 + \ldots + (a_n + b_n)\mathbf{v}_n) \\
&= (a_1 + b_1)\mathbf{v}_1' + (a_2 + b_2)\mathbf{v}_2' + \ldots + (a_n + b_n)\mathbf{v}_n' \\
&= (a_1\mathbf{v}_1' + a_2\mathbf{v}_2' + \ldots + a_n\mathbf{v}_n') + (b_1\mathbf{v}_1' + b_2\mathbf{v}_2' + \ldots + b_n\mathbf{v}_n') \\
&= \varphi(\mathbf{v}) + \varphi(\mathbf{w})
\end{aligned}$$

Also, for any scalar λ, we have

$$\begin{aligned}
\varphi(\lambda\mathbf{v}) &= \varphi\left(\lambda a_1\mathbf{v}_1 + \lambda a_2\mathbf{v}_2 + \ldots + \lambda a_n\mathbf{v}_n\right) \\
&= \lambda a_1\mathbf{v}_1' + \lambda a_2\mathbf{v}_2' + \ldots + \lambda a_n\mathbf{v}_n' \\
&= \lambda\left(a_1\mathbf{v}_1' + a_2\mathbf{v}_2' + \ldots + a_n\mathbf{v}_n'\right) \\
&= \lambda\varphi(\mathbf{v})
\end{aligned}$$

So, $\varphi : V \to V'$ is indeed a linear mapping. The mapping φ is *onto*, since given any vector $\mathbf{v}' \in V'$, there exist scalars $a_1, a_2, \ldots, a_n$ such that

$$\begin{aligned}
\mathbf{v}' &= a_1\mathbf{v}_1' + a_2\mathbf{v}_2' + \ldots + a_n\mathbf{v}_n' \\
&= \varphi\left(a_1\mathbf{v}_1 + a_2\mathbf{v}_2 + \ldots + a_n\mathbf{v}_n\right) \\
&= \varphi(\mathbf{v})
\end{aligned}$$

Finally, we show that the mapping φ is *one-to-one*. Suppose that $\varphi(\mathbf{v}) = \varphi(\mathbf{w})$ for some vectors $\mathbf{v}, \mathbf{w} \in V$ where

$$\mathbf{v} = a_1\mathbf{v}_1 + a_2\mathbf{v}_2 + \ldots + a_n\mathbf{v}_n \text{ and } \mathbf{w} = b_1\mathbf{v}_1 + b_2\mathbf{v}_2 + \ldots + b_n\mathbf{v}_n$$

Clearly, $\varphi(\mathbf{v}) = \varphi(\mathbf{w})$ implies

$$a_1\mathbf{v}_1' + a_2\mathbf{v}_2' + \ldots + a_n\mathbf{v}_n' = b_1\mathbf{v}_1' + b_2\mathbf{v}_2' + \ldots + b_n\mathbf{v}_n'$$

so that

$$(a_1 - b_1)\mathbf{v}_1' + (a_2 - b_2)\mathbf{v}_2' + \ldots + (a_n - b_n)\mathbf{v}_n' = \mathbf{0}$$

from which we find $a_1 = b_1, a_2 = b_2, \ldots, a_n = b_n$, owing to the linear independence of the set of basis vectors in V'. We conclude that $\mathbf{v} = \mathbf{w}$ so the mapping is one-to-one. This completes the proof. □.

From the numerous examples given in the previous lessons, we have seen that any n-dimensional vector space behaves like $\mathbb{R}^n$. In other words, any n-dimensional vector space should be isomorphic to $\mathbb{R}^n$. This is an immediate corollary of the previous theorem.

Corollary 21.2: *(Every n-Dimensional Space is Isomorphic to $\mathbb{R}^n$)*

If V is a real n-dimensional vector space, then $V \cong \mathbb{R}^n$.

Theorem 21.3: *(The Converse of Theorem 21.1)*

If any two *finite* dimensional vector spaces V and V' are isomorphic, then

$$\dim(V) = \dim(V')$$

Proof: Assume that $V \cong V'$ so there exists an isomorphism $\varphi : V \to V'$. Let the set of vectors $B = \{\mathbf{v}_1, \mathbf{v}_2, \ldots, \mathbf{v}_n\}$ comprise a basis for V so that $\dim(V) = n$. We will show that the set of n-vectors $B' = \{\varphi(\mathbf{v}_1), \varphi(\mathbf{v}_2), \ldots, \varphi(\mathbf{v}_n)\}$ is a basis for V'. Since φ is onto, then given any vector $\mathbf{v}' \in V'$ we know that there exists a vector $\mathbf{v} \in V$ such that $\varphi(\mathbf{v}) = \mathbf{v}'$. Since $\{\mathbf{v}_1, \mathbf{v}_2, \ldots, \mathbf{v}_n\}$ is a basis for V there exist scalars $a_1, a_2, \ldots, a_n$ such that

$$\mathbf{v} = a_1\mathbf{v}_1 + a_2\mathbf{v}_2 + \ldots + a_n\mathbf{v}_n$$

But then,

$$\varphi(\mathbf{v}) = a_1\varphi(\mathbf{v}_1) + a_2\varphi(\mathbf{v}_2) + .. + a_n\varphi(\mathbf{v}_n) = \mathbf{v}'$$

so the set of vectors $\{\varphi(\mathbf{v}_1), \varphi(\mathbf{v}_2), \ldots, \varphi(\mathbf{v}_n)\}$ spans V'. Now suppose that there exist scalars $c_1, c_2, \ldots, c_n$ such that

$$c_1\varphi(\mathbf{v}_1) + c_2\varphi(\mathbf{v}_2) + .. + c_n\varphi(\mathbf{v}_n) = \mathbf{0}$$

Consequently,

$$\varphi\left(c_1\mathbf{v}_1 + c_2\mathbf{v}_2 + .. + c_n\mathbf{v}_n\right) = \mathbf{0}$$

Now, we must have $c_1\mathbf{v}_1 + c_2\mathbf{v}_2 + .. + c_n\mathbf{v}_n = \mathbf{0}$ since we already know that $\varphi(\mathbf{0}) = \mathbf{0}$ and φ is a one-to-one mapping. Consequently, the set B' forms a basis for V' and

$$\dim(V') = n = \dim(V)$$

as claimed. $\square$.

Exercise Set 21

---

1. Show that P_1 is isomorphic to $\mathbb{R}^2$.

2. Show that $\mathbb{R}^4$ is isomorphic to $M(2,2)$.

3. Let φ be the mapping $\varphi : M(2,2) \rightarrow M(2,2)$ defined by

$$\varphi(A) = A^T$$

Show that φ is an isomorphism.

4. Let $\varphi : \mathbb{R}^3 \rightarrow P_2$ such that

$$\varphi\left(\begin{bmatrix} a_1 \\ a_2 \\ a_3 \end{bmatrix}\right) = a_1 + a_2t + a_3t^2 \text{ for all } \mathbf{a} = [a_1, a_2, a_3]^T \in \mathbb{R}^3$$

Show that φ is an isomorphism.

5. Let $A_0 \in M(n,n)$ be a fixed invertible matrix. Show that the mapping $\varphi : M(n,n) \rightarrow M(n,n)$ given by

$$\varphi(X) = A_0X \text{ for all } X \in M(n,n)$$

is an isomorphism.

6. Assume that φ is an isomorphism from V to V'. If S is a subspace of V, show that the set of vectors given by

$$\varphi(S) = \{\varphi(\mathbf{v}) : \mathbf{v} \in S\}$$

is a subspace of V', so φ preserves subspaces.

7. Show that every vector space V is isomorphic to itself.

8. Show that if $\varphi : V \rightarrow V'$ is an isomorphism from V to V', then φ^{-1} exists, and is an isomorphism from V' to V.

9. Show that if $U \cong V$ and $V \cong W$, then $U \cong W$.

10. Let S_1 and S_2 be subspaces of a vector space V and assume that $\varphi : V \rightarrow V'$ is an isomorphism. Show that

(a)

$$\varphi(S_1 \cap S_2) = \varphi(S_1) \cap \varphi(S_2)$$

(b)

$$\varphi(S_1 + S_2) = \varphi(S_1) + \varphi(S_2)$$

22. The Kernel and Range of a Linear Transformation

If $T : V \to W$ is a linear transformation from V into W, then T must map the zero vector $\mathbf{0} \in V$ into the zero vector of W. To see this, observe that for any vector $\mathbf{v} \in V$ we can write

$$T(\mathbf{0}) = T(\mathbf{v} - \mathbf{v}) = T(\mathbf{v}) - T(\mathbf{v}) = \mathbf{0}$$

so $T(\mathbf{0}) = \mathbf{0}$. Sometimes it may happen that additional vectors are mapped to the zero vector of W; the set of all such vectors is called the **kernel** of T.

Definition: *(Kernel)*

If $T : V \to W$ is a linear transformation, then the **kernel** of T, denoted by $\ker(T)$, is the set of all vectors in V that are mapped into the zero vector of W; i.e.,

$$\ker(T) = \{\mathbf{v} \in V : T(\mathbf{v}) = \mathbf{0}\} \tag{22.1}$$

Example-1: Find the kernel of the linear transformation $T : \mathbb{R}^2 \to \mathbb{R}^2$ given by

$$T\left(\begin{bmatrix} x \\ y \end{bmatrix}\right) = \begin{bmatrix} 4 & 2 \\ -2 & -1 \end{bmatrix}\begin{bmatrix} x \\ y \end{bmatrix}$$

Solution: By definition, the kernel of T is set of all vectors $\mathbf{x} = [x,y]^T \in \mathbb{R}^2$ such that

$$T(\mathbf{x}) = \mathbf{0}$$

Thus, we are led to find the solutions of the homogeneous system

$$\begin{aligned} 4x + 2y &= 0 \\ -2x - y &= 0 \end{aligned}$$

whose general solution is

$$\begin{bmatrix} x \\ y \end{bmatrix} = t\begin{bmatrix} 1 \\ -2 \end{bmatrix} \quad \text{where } t \text{ is a real parameter}$$

Consequently, the kernel of T is given by

$$\ker(T) = \left\{ t\begin{bmatrix} 1 \\ -2 \end{bmatrix} \mid t \in R \right\} = span\left\{\begin{bmatrix} 1 \\ -2 \end{bmatrix}\right\}$$

In the previous example, we see that the kernel of T contains infinitely many vectors. In terms of geometry, the kernel consists of a line through the origin of $\mathbb{R}^2$, and hence it is a vector subspace of $\mathbb{R}^2$. This state of affairs holds in general, as shown in the following theorem.

Theorem 22.1: *(The Kernel of a Linear Transformation is a Subspace)*

If $T : V \to W$ is a linear transformation, then the ker(T) is a subspace of V.

Proof: Since $T(\mathbf{0}) = \mathbf{0}$, we know that the ker($T$) contains at least one element, and therefore, it is not empty. By Theorem 10.1, it suffices to show that for any $\mathbf{u}, \mathbf{v} \in$ ker(T), we must have both $\mathbf{u} + \mathbf{v} \in$ ker(T) and $a\mathbf{u} \in$ ker(T) for all scalars a.

So assume that $\mathbf{u}, \mathbf{v} \in$ ker(T), and let a be an arbitrary scalar. Now, owing to the linearity of T, we can write

$$T(\mathbf{u} + \mathbf{v}) = T(\mathbf{u}) + T(\mathbf{v}) = \mathbf{0} + \mathbf{0} = \mathbf{0}$$
$$T(a\mathbf{u}) = aT(\mathbf{u}) = a(\mathbf{0}) = \mathbf{0}$$

Consequently, we are guaranteed that both $\mathbf{u} + \mathbf{v} \in$ ker(T) and $a\mathbf{u} \in$ ker(T). Thus, ker(T) is a subspace of the domain V. □.

Since the kernel of any given linear transformation is a subspace of the domain of that transformation, it makes sense to look for a basis for the kernel. How we can accomplish this task is shown in the next example.

Example-2: Find a basis for the kernel of the linear transformation $T : \mathbb{R}^4 \to \mathbb{R}^4$ given by

$$T\left(\begin{bmatrix} x \\ y \\ z \\ w \end{bmatrix}\right) = \begin{bmatrix} 1 & 0 & -1 & 2 \\ 2 & 1 & 1 & 3 \\ 1 & 0 & 1 & 2 \\ 0 & -1 & -1 & 1 \end{bmatrix} \begin{bmatrix} x \\ y \\ z \\ w \end{bmatrix}$$

Solution: Using the same reasoning as the previous example, we see that ker(T) will consist of all solutions of the homogeneous system:

$$\begin{bmatrix} 1 & 0 & -1 & 2 \\ 2 & 1 & 1 & 3 \\ 3 & 1 & 0 & 5 \\ -1 & -1 & -2 & -1 \end{bmatrix} \begin{bmatrix} x \\ y \\ z \\ w \end{bmatrix} = \begin{bmatrix} 0 \\ 0 \\ 0 \\ 0 \end{bmatrix}$$

Using the Gauss-Jordan method, we find that the solution of this system is:

$$\begin{bmatrix} x \\ y \\ z \\ w \end{bmatrix} = s\begin{bmatrix} 1 \\ -3 \\ 1 \\ 0 \end{bmatrix} + t\begin{bmatrix} -2 \\ 1 \\ 0 \\ 1 \end{bmatrix} \quad \text{where } s,t \text{ are real parameters.}$$

We conclude that one basis for ker(T) is given by

$$B = \{[1,-3,1,0]^T,[-2,1,0,1]^T\}$$

We now turn our attention to the range of a linear transformation $T : V \to W$. From a previous lesson, recall that the **range** of T, denoted by *range*(T), is the set of vectors in W given by

$$range(T) = \{\mathbf{w} \in W : \mathbf{w} = T(\mathbf{v}), \text{ where } \mathbf{v} \in V\} \tag{22.2}$$

In other words, the range of T consists of those vectors in W that are the images of vectors in the domain of T.

Theorem 22.2: *(The Range of a Linear Transformation is a Subspace)*

If $T : V \to W$ is a linear transformation, then the *range*(T) is a subspace of W.

Proof: Since $T(\mathbf{0}) = \mathbf{0}$, then clearly $\mathbf{0} \in range(T)$, so the range of T is a non-empty subset of W. Once again, using Theorem 10.1, we must show that for any $\mathbf{w}_1, \mathbf{w}_2 \in range(T)$, both $\mathbf{w}_1 + \mathbf{w}_2 \in range(T)$ and $a\mathbf{w}_1 \in range(T)$ for all scalars a.

Let $\mathbf{w}_1, \mathbf{w}_2 \in range(T)$, and let a be an arbitrary scalar. Now, there must exist vectors $\mathbf{v}_1, \mathbf{v}_2 \in V$ such that $\mathbf{w}_1 = T(\mathbf{v}_1)$ and $\mathbf{w}_2 = T(\mathbf{v}_2)$, so we can write

$$\mathbf{w}_1 + \mathbf{w}_2 = T(\mathbf{v}_1) + T(\mathbf{v}_2) = T(\mathbf{v}_1 + \mathbf{v}_2)$$
$$a\mathbf{w}_1 = aT(\mathbf{v}_1) = T(a\mathbf{v}_1)$$

Since V is a vector space, $\mathbf{v}_1 + \mathbf{v}_2 \in V$ and $a\mathbf{v}_1 \in V$ for any scalar a. Consequently, $\mathbf{w}_1 + \mathbf{w}_2 \in range(T)$ and $a\mathbf{w}_1 \in range(T)$. □.

We now examine how we can characterize the kernel and range of a linear transformation $T : \mathbb{R}^n \to R^m$ which is defined by $T(\mathbf{x}) = A\mathbf{x}$, where A is an $m \times n$ matrix. From the above examples, it is obvious that the kernel of T is *identical* to the **solution space** of the homogeneous system $A\mathbf{x} = \mathbf{0}$.

On the other hand, the range of T must consist of all vectors $\mathbf{b} \in R^m$ for which the linear system $A\mathbf{x} = \mathbf{b}$ is *consistent*. As in the past, let's write this system in matrix form:

$$\begin{bmatrix} a_{11} & a_{12} & \cdots & a_{1n} \\ a_{21} & a_{22} & \cdots & a_{2n} \\ \vdots & \vdots & & \vdots \\ a_{m1} & a_{m2} & \cdots & a_{mn} \end{bmatrix} \begin{bmatrix} x_1 \\ x_2 \\ \vdots \\ x_n \end{bmatrix} = \begin{bmatrix} b_1 \\ b_2 \\ \vdots \\ b_n \end{bmatrix}$$

We can rewrite the last expression in the form

$$A\mathbf{x} = x_1\mathbf{c}_1 + x_2\mathbf{c}_2 + \ldots + x_n\mathbf{c}_n = \mathbf{b}$$

where $\mathbf{c}_1, \mathbf{c}_2, \ldots, \mathbf{c}_n$ are the column vectors of A. The last equation implies that $\mathbf{b}$ is in the range of T if and only if $\mathbf{b} \in span\{\mathbf{c}_1, \mathbf{c}_2, \ldots, \mathbf{c}_n\}$. We conclude that the range of T is equal to the **column space** of A. We summarize the results of our analysis in the next theorem.

Theorem 22.3: *(Range and Kernel of Matrix Transformations)*

If $T : \mathbb{R}^n \to R^m$ is a linear transformation defined by $T(\mathbf{x}) = A\mathbf{x}$ where A is an $m \times n$ matrix, then
(**a**) the kernel of T is equal to the solution space of the system $A\mathbf{x} = \mathbf{0}$, and
(**b**) the range of T is equal to the column space of A.

Example-3: Assume that $T : \mathbb{R}^4 \to \mathbb{R}^4$ is given by $T(\mathbf{x}) = A\mathbf{x}$ where

$$A = \begin{bmatrix} 1 & 0 & -1 & 2 \\ 2 & 1 & 1 & 3 \\ 3 & 1 & 0 & 5 \\ -1 & -1 & -2 & -1 \end{bmatrix}$$

Find a basis for the range of T.

Solution: As shown in Lesson 13, we must first find the transpose A^T of A, and then use the Gauss-Jordan method to transform A^T to reduced-row echelon form. In doing so, we obtain

$$A^T = \begin{bmatrix} 1 & 2 & 3 & -1 \\ 0 & 1 & 1 & -1 \\ -1 & 1 & 0 & -2 \\ 2 & 3 & 5 & -1 \end{bmatrix} \quad \overset{\textit{Gauss-Jordan}}{\underset{\textit{Reduction}}{\Longrightarrow}} \quad \begin{bmatrix} 1 & 0 & 1 & 1 \\ 0 & 1 & 1 & -1 \\ 0 & 0 & 0 & 0 \\ 0 & 0 & 0 & 0 \end{bmatrix}$$

We conclude that a basis for the range of T is given by

$$B = \left\{ \begin{bmatrix} 1 \\ 0 \\ 1 \\ 1 \end{bmatrix}, \begin{bmatrix} 0 \\ 1 \\ 1 \\ -1 \end{bmatrix} \right\}$$

Definition: *(Rank and Nullity)*

If $T : V \to W$ is a linear transformation, then the dimension of the range of T is called the **rank** of T, while the dimension of the kernel of T is called the **nullity** of T.

Theorem 22.4: *(A Useful Isomorphism)*

If $T : V \to W$ is a linear transformation from V into W, then the quotient space $V/\ker(T)$ is isomorphic to the range of T, i.e.,

$$V/\ker(T) \cong range(T) \tag{22.3}$$

Proof: For the sake of brevity, let $K = \ker(T)$. We must construct a linear map $\varphi : V/K \to range(T)$ which is both one-to-one and onto. Let φ be given by

$$\varphi(\mathbf{u} + K) = T(\mathbf{u})$$

where $\mathbf{u} + K$ is a typical coset in the quotient space V/K. Clearly, the map φ is linear, since for all $\mathbf{u} + K, \mathbf{v} + K \in V/K$, we have

$$\begin{aligned} \varphi((\mathbf{u} + K) + (\mathbf{v} + K)) &= \varphi((\mathbf{u} + \mathbf{v}) + K) = T(\mathbf{u} + \mathbf{v}) \\ &= T(\mathbf{u}) + T(\mathbf{v}) \\ &= \varphi(\mathbf{u} + K) + \varphi(\mathbf{v} + K) \end{aligned}$$

and for any scalar a,

$$\begin{aligned} \varphi(a(\mathbf{u} + K)) &= \varphi(a\mathbf{u} + K) = T(a\mathbf{u}) = aT(\mathbf{u}) \\ &= a\varphi(\mathbf{u} + K) \end{aligned}$$

Next, we show that φ is one-to-one. Now, if $\varphi(\mathbf{u} + K) = \varphi(\mathbf{v} + K)$ then $T(\mathbf{u}) = T(\mathbf{v})$, so $T(\mathbf{u} - \mathbf{v}) = \mathbf{0}$. But $T(\mathbf{u} - \mathbf{v}) = \mathbf{0}$ implies that, $\mathbf{u} - \mathbf{v} \in K$ so $\mathbf{u} + K = \mathbf{v} + K$.

Finally, we show that φ is onto the *range*(T). Let $\mathbf{w}$ be an arbitrary vector in the range of T. Since $\mathbf{w} \in range(T)$ there must exist a vector $\mathbf{u} \in V$ such that $T(\mathbf{u}) = \mathbf{w}$. But then we have $\varphi(\mathbf{u} + K) = T(\mathbf{u}) = \mathbf{w}$, where the coset $\mathbf{u} + K \in V/K$. Thus, φ maps V/K onto the *range*(T). □.

Theorem 22.5: *(Dimension Theorem)*

Let V be an n-dimensional vector space. If $T : V \rightarrow W$ is a linear transformation from V into a finite-dimensional vector space W, then

$$n = (\text{rank of } T) + (\text{nullity of } T) \tag{22.4}$$

Proof: By the previous theorem, we have $V/\ker(T) \cong range(T)$. Since these spaces are isomorphic, then by Theorem 21.3, they must have the same dimension, so we can write

$$\dim(V/\ker(T)) = \dim(range(T)) = (\text{rank of } T)$$

However, by equation (16.7), we have

$$\dim(V/\ker(T)) = \dim V - \dim(\ker(T)) = n - (\text{nullity of } T)$$

Combining the last two equations gives

$$n - (\text{nullity of } T) = (\text{rank of } T)$$

from which (22.4) follows immediately. □.

Example-4: Verify the Dimension Theorem for the linear transformation $T : \mathbb{R}^4 \rightarrow \mathbb{R}^4$ is given by $T(\mathbf{x}) = A\mathbf{x}$ where

$$A = \begin{bmatrix} 1 & 0 & -1 & 2 \\ 2 & 1 & 1 & 3 \\ 3 & 1 & 0 & 5 \\ -1 & -1 & -2 & -1 \end{bmatrix},$$

which was discussed previously in Example-2 and Example-3.

Solution: From Example-2, we know that

$$\text{nullity of } T = \dim(\text{null space of } T) = 2$$

while from the results of Example-3, we have

$$\text{rank of } T = \dim(range(T)) = 2$$

Observe that

$$\dim(\text{domain of } T) = 4 = 2 + 2 = (\text{rank of } T) + (\text{nullity of } T)$$

as it should.

Exercise Set 22

1. Consider the linear transformation $T : \mathbb{R}^3 \to \mathbb{R}^3$ is given by $T(\mathbf{x}) = A\mathbf{x}$ where

$$A = \begin{bmatrix} 1 & 0 & -1 \\ 2 & 1 & 0 \\ -1 & -1 & -1 \end{bmatrix}$$

Which of the following vectors are in the range of T?

(a) $\begin{bmatrix} -3 \\ -2 \\ -1 \end{bmatrix}$ **(b)** $\begin{bmatrix} 2 \\ 0 \\ 1 \end{bmatrix}$ **(c)** $\begin{bmatrix} -1 \\ 3 \\ -4 \end{bmatrix}$ **(d)** $\begin{bmatrix} 0 \\ 0 \\ 2 \end{bmatrix}$

2. Given the linear transformation in Exercise-1, determine
(a) A basis for the null space of A.
(b) A basis for the range space of A.
(c) Verify the Dimension Theorem.

3. Given each linear transformation, find the nullity of T.
(a) $T : \mathbb{R}^3 \to \mathbb{R}^4$ has rank 3.
(b) $T : \mathbb{R}^2 \to \mathbb{R}^3$ has rank 1.
(c) $T : P_3 \to P_3$ has rank 4.

4. The linear transformation $T : \mathbb{R}^3 \to \mathbb{R}^2$ is given by $T(\mathbf{x}) = A\mathbf{x}$ where

$$A = \begin{bmatrix} 1 & 0 & -1 \\ 2 & 1 & 0 \end{bmatrix}$$

(a) Find a basis for the range of T.
(b) Find a basis for the kernel of T.
(c) What are the rank and nullity of T?

5. Let $T : \mathbb{R}^n \to \mathbb{R}^n$ be given by $T(\mathbf{x}) = A\mathbf{x}$ where A is an $n \times n$ matrix. If $\det(A) \neq 0$, show that T is both one-to-one and onto.

6. Given a linear transformation $T : V \to W$, show that T is one-to-one if and only if $\ker(T) = \{\mathbf{0}\}$.

7. Consider the linear operator $T : M(n,n) \to M(n,n)$ which is given by $T(A) = A - A^T$. Determine the kernel of T.

23. Matrices of Linear Transformations

Recall that if A is any $m \times n$ matrix, then we can define a mapping $T : \mathbb{R}^n \to R^m$ by $T(\mathbf{x}) = A\mathbf{x}$ for all $\mathbf{x} \in \mathbb{R}^n$, and this matrix transformation is *linear*. On the other hand, it turns out that every linear transformation $T : \mathbb{R}^n \to R^m$ can be represented by an $m \times n$ matrix.

To see this, let's consider a linear transformation $T : \mathbb{R}^n \to R^m$, and assume that $E = \{\mathbf{e}_1, \mathbf{e}_2, \ldots, \mathbf{e}_n\}$ is the **standard basis** for $\mathbb{R}^n$; that is,

$$\mathbf{e}_1 = \begin{bmatrix} 1 \\ 0 \\ 0 \\ \vdots \\ 0 \end{bmatrix}, \ \mathbf{e}_2 = \begin{bmatrix} 0 \\ 1 \\ 0 \\ \vdots \\ 0 \end{bmatrix}, \ \mathbf{e}_3 = \begin{bmatrix} 0 \\ 0 \\ 1 \\ \vdots \\ 0 \end{bmatrix}, \ldots, \mathbf{e}_n = \begin{bmatrix} 0 \\ 0 \\ 0 \\ \vdots \\ 1 \end{bmatrix} \tag{23.1}$$

Now, since E is a basis for $\mathbb{R}^n$, given any vector $\mathbf{v} \in \mathbb{R}^n$ we can write

$$\mathbf{v} = v_1\mathbf{e}_1 + v_2\mathbf{e}_2 + \ldots + v_n\mathbf{e}_n$$

where the numbers $v_1, v_2, \ldots, v_n$ are just the components of $\mathbf{v}$ relative to the standard basis. Consequently,

$$T(\mathbf{v}) = T(v_1\mathbf{e}_1 + v_2\mathbf{e}_2 + \ldots + v_n\mathbf{e}_n) = v_1T(\mathbf{e}_1) + v_2T(\mathbf{e}_2) + \ldots + v_nT(\mathbf{e}_n)$$

and we see that the action of T on any vector $\mathbf{v} \in \mathbb{R}^n$ is completely determined by its action on the standard basis vectors in E. So let

$$T(\mathbf{e}_1) = \begin{bmatrix} a_{11} \\ a_{21} \\ \vdots \\ a_{m1} \end{bmatrix}, \ T(\mathbf{e}_2) = \begin{bmatrix} a_{12} \\ a_{22} \\ \vdots \\ a_{m2} \end{bmatrix}, \ldots, T(\mathbf{e}_n) = \begin{bmatrix} a_{1n} \\ a_{2n} \\ \vdots \\ a_{mn} \end{bmatrix}$$

where each column matrix $T(\mathbf{e}_k)$ is a vector in R^m by hypothesis. Now observe that

$$\begin{aligned} T(\mathbf{v}) &= v_1T(\mathbf{e}_1) + v_2T(\mathbf{e}_2) + \ldots + v_nT(\mathbf{e}_n) \\ &= \begin{bmatrix} a_{11}v_1 \\ a_{21}v_1 \\ \vdots \\ a_{m1}v_1 \end{bmatrix} + \begin{bmatrix} a_{12}v_2 \\ a_{22}v_2 \\ \vdots \\ a_{m2}v_2 \end{bmatrix} + \ldots + \begin{bmatrix} a_{1n}v_n \\ a_{2n}v_n \\ \vdots \\ a_{mn}v_n \end{bmatrix} \end{aligned}$$

The last equation can be rewritten neatly as

$$T(\mathbf{v}) = \begin{bmatrix} a_{11} & a_{12} & \cdots & a_{1n} \\ a_{21} & a_{22} & \cdots & a_{2n} \\ \vdots & \vdots & & \vdots \\ a_{m1} & a_{m2} & \cdots & a_{mn} \end{bmatrix} \begin{bmatrix} v_1 \\ v_2 \\ \vdots \\ v_n \end{bmatrix} \equiv A\mathbf{v} \tag{23.2}$$

where A is an $m \times n$ matrix. In other words, we have shown that any given linear transformation $T : \mathbb{R}^n \rightarrow R^m$ is equivalent to multiplication by a matrix. Let's summarize this result as our first theorem.

Theorem 23.1: *(Standard Matrix for a Linear Transformation)*

Let $T : \mathbb{R}^n \rightarrow R^m$ be any linear transformation where the action of T on the standard basis vectors of $\mathbb{R}^n$ is given by the column vectors

$$T(\mathbf{e}_1) = \begin{bmatrix} a_{11} \\ a_{21} \\ \vdots \\ a_{m1} \end{bmatrix}, \quad T(\mathbf{e}_2) = \begin{bmatrix} a_{12} \\ a_{22} \\ \vdots \\ a_{m2} \end{bmatrix}, \ldots, T(\mathbf{e}_n) = \begin{bmatrix} a_{1n} \\ a_{2n} \\ \vdots \\ a_{mn} \end{bmatrix} \tag{23.3}$$

and let A be the $m \times n$ matrix whose columns are formed from the column vectors in (23.3); that is,

$$A = \begin{bmatrix} T(\mathbf{e}_1) \mid T(\mathbf{e}_2) \mid \cdots \mid T(\mathbf{e}_n) \end{bmatrix} = \begin{bmatrix} a_{11} & a_{12} & \cdots & a_{1n} \\ a_{21} & a_{22} & \cdots & a_{2n} \\ \vdots & \vdots & & \vdots \\ a_{m1} & a_{m2} & \cdots & a_{mn} \end{bmatrix} \tag{23.4}$$

Then the action of T on any vector $\mathbf{v} \in \mathbb{R}^n$ is given by

$$T(\mathbf{v}) = A\mathbf{v} \tag{23.4}$$

We shall call A the **standard matrix** for T.

Example-1: Consider the linear transformation $T : \mathbb{R}^2 \rightarrow \mathbb{R}^3$ which is given by

$$T\left(\begin{bmatrix} x \\ y \end{bmatrix}\right) = \begin{bmatrix} x + y \\ x - y \\ 2x + y \end{bmatrix}$$

Find the standard matrix for T.

Solution: According to the previous theorem, we must find the action of T on each of the standard basis vectors for $\mathbb{R}^2$. We find that

$$T(\mathbf{e}_1) = T\left(\begin{bmatrix} 1 \\ 0 \end{bmatrix}\right) = \begin{bmatrix} 1 \\ 1 \\ 2 \end{bmatrix} \text{ and } T(\mathbf{e}_2) = T\left(\begin{bmatrix} 0 \\ 1 \end{bmatrix}\right) = \begin{bmatrix} 1 \\ -1 \\ 1 \end{bmatrix}$$

So the standard matrix for T is the matrix whose columns are $T(\mathbf{e}_1)$ and $T(\mathbf{e}_2)$:

$$A = \begin{bmatrix} 1 & 1 \\ 1 & -1 \\ 2 & 1 \end{bmatrix}$$

We can check our work by noting that

$$T\left(\begin{bmatrix} x \\ y \end{bmatrix}\right) = \begin{bmatrix} 1 & 1 \\ 1 & -1 \\ 2 & 1 \end{bmatrix}\begin{bmatrix} x \\ y \end{bmatrix} = \begin{bmatrix} x+y \\ x-y \\ 2x+y \end{bmatrix}$$

so $T(\mathbf{x}) = A\mathbf{x}$ for all vectors $\mathbf{x} = [x,y]^T$ as it should.

Example-2: (*Reflection About the x-Axis*)
As shown in Figure 23.1, let $T : \mathbb{R}^2 \to \mathbb{R}^2$ be the linear transformation that maps each plane vector $\mathbf{x}$ into its reflection about the x-axis.

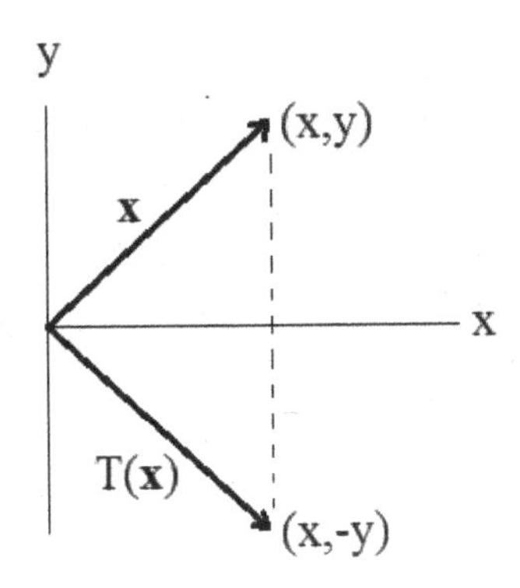

Figure 23.1: Reflection about the x-axis.

Find the standard matrix for T.

Solution: According to Theorem 23.1, we must find the action of T on each of the standard basis vectors for $\mathbb{R}^2$. We find

$$T(\mathbf{e}_1) = T\left(\begin{bmatrix} 1 \\ 0 \end{bmatrix}\right) = \begin{bmatrix} 1 \\ 0 \end{bmatrix} \text{ and } T(\mathbf{e}_2) = T\left(\begin{bmatrix} 0 \\ 1 \end{bmatrix}\right) = \begin{bmatrix} 0 \\ -1 \end{bmatrix}$$

so the standard matrix for T is given by

$$A = \begin{bmatrix} 1 & 0 \\ 0 & -1 \end{bmatrix}$$

Once again, we can check our work by noting that

$$T\left(\begin{bmatrix} x \\ y \end{bmatrix}\right) = \begin{bmatrix} 1 & 0 \\ 0 & -1 \end{bmatrix}\begin{bmatrix} x \\ y \end{bmatrix} = \begin{bmatrix} x \\ -y \end{bmatrix}$$

which agrees with the image vector $T(\mathbf{x})$ in Figure 23.1.

We now extend the result of Theorem 23.1 to handle any linear transformation $T : V \to V'$ where V and V' are vector spaces of dimension n and m, respectively. We will show that we can still represent such a transformation by an $m \times n$ matrix.

To this end, let $B = \{\mathbf{v}_1, \mathbf{v}_2, \ldots, \mathbf{v}_n\}$ and $B' = \{\mathbf{v}'_1, \mathbf{v}'_2, \ldots, \mathbf{v}'_m\}$ be bases for V and V', respectively. Given any vector $\mathbf{v} \in V$ we have

$$\mathbf{v} = v_1\mathbf{v}_1 + v_2\mathbf{v}_2 + \ldots + v_n\mathbf{v}_n = \begin{bmatrix} v_1 \\ v_2 \\ \vdots \\ v_n \end{bmatrix}_B \equiv [\mathbf{v}]_B$$

where the column matrix $[\mathbf{v}]_B$ is called the **component matrix** of $\mathbf{v}$ relative to the basis B. Now, we find that

$$T(\mathbf{v}) = T(v_1\mathbf{v}_1 + v_2\mathbf{v}_2 + \ldots + v_n\mathbf{v}_n) = v_1T(\mathbf{v}_1) + v_2T(\mathbf{v}_2) + \ldots + v_nT(\mathbf{v}_n)$$

so, once again, the action of T on any vector $\mathbf{v} \in V$ is completely determined by its action on the basis vectors in B. Now let

$$T(\mathbf{v}_1) = \begin{bmatrix} a_{11} \\ a_{21} \\ \vdots \\ a_{m1} \end{bmatrix}_{B'}, \; T(\mathbf{v}_2) = \begin{bmatrix} a_{12} \\ a_{22} \\ \vdots \\ a_{m2} \end{bmatrix}_{B'}, \ldots, T(\mathbf{v}_n) = \begin{bmatrix} a_{1n} \\ a_{2n} \\ \vdots \\ a_{mn} \end{bmatrix}_{B'}$$

where the column vectors represent the **component matrices** of the image vectors $T(\mathbf{v}_1), T(\mathbf{v}_2), \ldots, T(\mathbf{v}_n)$ relative to the basis B'. Now, we can write

$$T(\mathbf{v}) = v_1T(\mathbf{v}_1) + v_2T(\mathbf{v}_2) + \ldots + v_nT(\mathbf{v}_n)$$

or equivalently,

$$T(\mathbf{v}) = \begin{bmatrix} a_{11}v_1 \\ a_{21}v_1 \\ \vdots \\ a_{m1}v_1 \end{bmatrix}_{B'} + \begin{bmatrix} a_{12}v_2 \\ a_{22}v_2 \\ \vdots \\ a_{m2}v_2 \end{bmatrix}_{B'} + \ldots + \begin{bmatrix} a_{1n}v_n \\ a_{2n}v_n \\ \vdots \\ a_{mn}v_n \end{bmatrix}_{B'}$$

so that

$$[T(\mathbf{v})]_{B'} = \begin{bmatrix} a_{11} & a_{12} & \cdots & a_{1n} \\ a_{21} & a_{22} & \cdots & a_{2n} \\ \vdots & \vdots & & \vdots \\ a_{m1} & a_{m2} & \cdots & a_{mn} \end{bmatrix} \begin{bmatrix} v_1 \\ v_2 \\ \vdots \\ v_n \end{bmatrix} \equiv A[\mathbf{v}]_B$$

where A is an $m \times n$ matrix. This result is completely analogous to (23.2). We have shown that any linear transformation $T : V \to V'$ where V and V' are vector spaces of dimension n and m respectively, is equivalent to.simple matrix multiplication by an appropriately constructed $m \times n$ matrix.

Theorem 23.2: *(Transformation Matrix for Non-Standard Bases)*

Let $T : V \to V'$ be a linear transformation where V and V' are finite-dimensional vector spaces.with bases $B = \{\mathbf{v}_1, \mathbf{v}_2, \ldots, \mathbf{v}_n\}$ and $B' = \{\mathbf{v}'_1, \mathbf{v}'_2, \ldots, \mathbf{v}'_m\}$ respectively. Let the action of T on the basis vectors of V be given by

$$T(\mathbf{v}_1) = \begin{bmatrix} a_{11} \\ a_{21} \\ \vdots \\ a_{m1} \end{bmatrix}_{B'}, \quad T(\mathbf{v}_2) = \begin{bmatrix} a_{12} \\ a_{22} \\ \vdots \\ a_{m2} \end{bmatrix}_{B'}, \ldots, T(\mathbf{v}_n) = \begin{bmatrix} a_{1n} \\ a_{2n} \\ \vdots \\ a_{mn} \end{bmatrix}_{B'} \tag{23.5}$$

where the column vectors represent the components of the corresponding image vectors $T(\mathbf{v}_1), T(\mathbf{v}_2), \ldots, T(\mathbf{v}_n)$ relative to the basis B'. Furthermore, let A be the $m \times n$ matrix given by

$$A = \left[\ T(\mathbf{v}_1) \mid T(\mathbf{v}_2) \mid \cdots \mid T(\mathbf{v}_n)\ \right] = \begin{bmatrix} a_{11} & a_{12} & \cdots & a_{1n} \\ a_{21} & a_{22} & \cdots & a_{2n} \\ \vdots & \vdots & & \vdots \\ a_{m1} & a_{m2} & \cdots & a_{mn} \end{bmatrix} \tag{23.6}$$

Then the action of T on any vector $\mathbf{v} \in V$ is given by

$$[T(\mathbf{v})]_{B'} = A[\mathbf{v}]_B \tag{23.7}$$

We shall call A the **matrix of T relative to the bases B and B'.**

Example-3: Let $T : \mathbb{R}^2 \to \mathbb{R}^2$ be the linear operator

$$T\left(\begin{bmatrix} x \\ y \end{bmatrix}\right) = \begin{bmatrix} x+y \\ x-y \end{bmatrix}$$

Find the matrix of T relative to the bases

$$B = \{\mathbf{v}_1, \mathbf{v}_2\} = \left\{\begin{bmatrix} 1 \\ -1 \end{bmatrix}, \begin{bmatrix} 1 \\ 1 \end{bmatrix}\right\}$$

$$B' = \{\mathbf{v}'_1, \mathbf{v}'_2\} = \left\{\begin{bmatrix} 1 \\ 0 \end{bmatrix}, \begin{bmatrix} 0 \\ 2 \end{bmatrix}\right\}$$

Solution: We have

$$T(\mathbf{v}_1) = \begin{bmatrix} 0 \\ 2 \end{bmatrix} = (0)\mathbf{v}'_1 + (1)\mathbf{v}'_2 = \begin{bmatrix} 0 \\ 1 \end{bmatrix}_{B'}$$

$$T(\mathbf{v}_2) = \begin{bmatrix} 2 \\ 0 \end{bmatrix} = (2)\mathbf{v}'_1 + (0)\mathbf{v}'_2 = \begin{bmatrix} 2 \\ 0 \end{bmatrix}_{B'}$$

So, the matrix of T relative to the bases B and B' is given by

$$A = \begin{bmatrix} 0 & 2 \\ 1 & 0 \end{bmatrix}$$

As a check on our work, we first observe that for any $\mathbf{x} \in \mathbb{R}^2$we have

$$\mathbf{x} = \begin{bmatrix} x \\ y \end{bmatrix} = \begin{bmatrix} (x-y)/2 \\ (x+y)/2 \end{bmatrix}_B$$

and

$$A[\mathbf{x}]_B = \begin{bmatrix} 0 & 2 \\ 1 & 0 \end{bmatrix}\begin{bmatrix} (x-y)/2 \\ (x+y)/2 \end{bmatrix}_B = \begin{bmatrix} x+y \\ \frac{1}{2}(x-y) \end{bmatrix}_{B'}$$

$$= \begin{bmatrix} x+y \\ x-y \end{bmatrix}$$

as it should.

Example-4: Consider the linear operator $T : \mathbb{R}^2 \to \mathbb{R}^2$ defined by

$$T(\mathbf{x}) = \lambda \mathbf{x}$$

where λ is a real constant. Find the matrix of T with respect to the bases E and B where

$$E = \{\mathbf{e}_1, \mathbf{e}_2\} = \left\{ \begin{bmatrix} 1 \\ 0 \end{bmatrix}, \begin{bmatrix} 0 \\ 1 \end{bmatrix} \right\} \quad \text{(Standard Basis for } \mathbb{R}^2\text{)}$$

$$B = \{\mathbf{v}_1, \mathbf{v}_2\} = \left\{ \begin{bmatrix} 1 \\ -1 \end{bmatrix}, \begin{bmatrix} 1 \\ 2 \end{bmatrix} \right\}$$

Solution: We find that

$$T(\mathbf{e}_1) = \lambda \begin{bmatrix} 1 \\ 0 \end{bmatrix} = \begin{bmatrix} \lambda \\ 0 \end{bmatrix} = \begin{bmatrix} \frac{2}{3}\lambda \\ \frac{1}{3}\lambda \end{bmatrix}_B$$

$$T(\mathbf{e}_2) = \lambda \begin{bmatrix} 0 \\ 1 \end{bmatrix} = \begin{bmatrix} 0 \\ \lambda \end{bmatrix} = \begin{bmatrix} -\frac{1}{3}\lambda \\ \frac{1}{3}\lambda \end{bmatrix}_B$$

Consequently, the matrix of T relative to the bases E and B is

$$A = \begin{bmatrix} \frac{2}{3}\lambda & -\frac{1}{3}\lambda \\ \frac{1}{3}\lambda & \frac{1}{3}\lambda \end{bmatrix}$$

Finally, checking our work, we find that

$$\begin{aligned} A[\mathbf{x}]_E &= \begin{bmatrix} \frac{2}{3}\lambda & -\frac{1}{3}\lambda \\ \frac{1}{3}\lambda & \frac{1}{3}\lambda \end{bmatrix} \begin{bmatrix} x \\ y \end{bmatrix} = \begin{bmatrix} \frac{2}{3}x\lambda - \frac{1}{3}y\lambda \\ \frac{1}{3}x\lambda + \frac{1}{3}y\lambda \end{bmatrix}_B \\ &= \lambda \begin{bmatrix} \frac{2}{3}x - \frac{1}{3}y \\ \frac{1}{3}x + \frac{1}{3}y \end{bmatrix}_B \\ &= \begin{bmatrix} \lambda x \\ \lambda y \end{bmatrix} \end{aligned}$$

which agrees with the definition of T.

Exercise Set 23

1. Determine the standard matrix for each of the following linear transformations:

(**a**) $T : \mathbb{R}^2 \to \mathbb{R}^2$ where

$$T\left(\begin{bmatrix} x \\ y \end{bmatrix}\right) = \begin{bmatrix} ax + by \\ cx + dy \end{bmatrix}$$ where a, b, c, d are real constants.

(**b**) $T : \mathbb{R}^2 \to \mathbb{R}^3$ given by

$$T\left(\begin{bmatrix} x \\ y \end{bmatrix}\right) = \begin{bmatrix} 2x + y \\ x - y \\ x + 3y \end{bmatrix}$$

(**c**) $T : \mathbb{R}^2 \to \mathbb{R}^2$ where

$$T\left(\begin{bmatrix} x \\ y \end{bmatrix}\right) = \begin{bmatrix} -x \\ y \end{bmatrix}$$

Describe what T does geometrically.

2. As shown in Figure 23.2, the linear operator $T : \mathbb{R}^2 \to \mathbb{R}^2$ rotates any vector $\mathbf{x} \in \mathbb{R}^2$ counterclockwise through an angle θ.

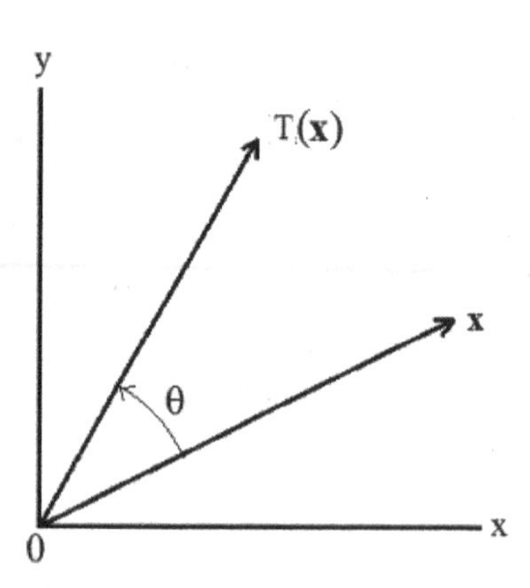

Figure 23.2: Rotation through an angle θ.

Find the standard matrix for T.

3. As shown in Figure 23.3, a linear operator $T : \mathbb{R}^2 \to \mathbb{R}^2$ reflects any vector $\mathbf{x} \in \mathbb{R}^2$ through the line $y = x$.

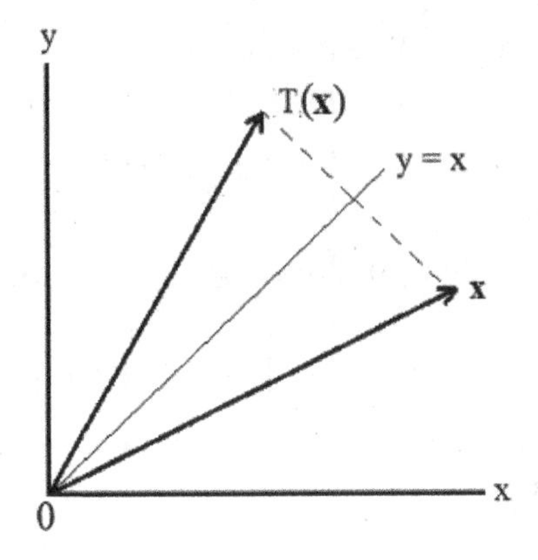

Figure 23.3: Reflection through the line $y = x$.

Find the standard matrix for T.

4. Let $T : P_2 \rightarrow P_1$ be the linear transformation given by

$$T(a_0 + a_1x + a_2x^2) = a_1 + 2a_2x$$

Find the matrix of T relative to the bases $B = \{1,x,x^2\}$ and $B' = \{1,x\}$.

5. As shown in Figure 23.4, a linear operator $T : \mathbb{R}^2 \rightarrow \mathbb{R}^2$ maps any vector $\mathbf{x} \in \mathbb{R}^2$ to its orthogonal projection on the x-axis.

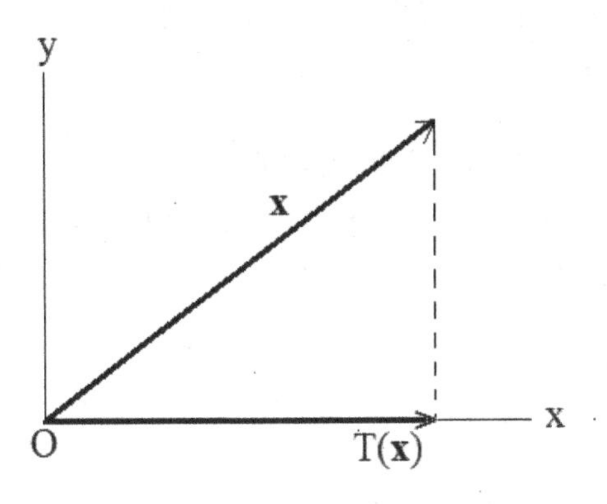

Figure 23.4: Orthogonal projection of $\mathbf{x}$ onto the x-axis.

Determine the standard matrix of T.

6. Given the linear transformation $T : P_1 \rightarrow P_2$ where

$$T(a_0 + a_1x) = a_0x + a_1x^2$$

Find the matrix of T relative to the standard bases $B = \{1,x\}$ and $B' = \{1,x,x^2\}$.

7. Consider the linear operator $T : \mathbb{R}^3 \to \mathbb{R}^3$ defined by

$$T\left(\begin{bmatrix} x \\ y \\ z \end{bmatrix}\right) = \begin{bmatrix} 2x+y \\ y+z \\ x+2z \end{bmatrix}$$

(**a**) Find the matrix of T with respect to the basis $B = \{\mathbf{v}_1, \mathbf{v}_2, \mathbf{v}_3\}$ where

$$\mathbf{v}_1 = \begin{bmatrix} 1 \\ 1 \\ 0 \end{bmatrix}, \ \mathbf{v}_2 = \begin{bmatrix} 1 \\ 0 \\ 1 \end{bmatrix}, \ \mathbf{v}_3 = \begin{bmatrix} 0 \\ 1 \\ 1 \end{bmatrix}$$

(**b**) Use the matrix found in (a) to calculate

$$T\left(\begin{bmatrix} 1 \\ 1 \\ 1 \end{bmatrix}\right)$$

8. Let the linear operator $T : \mathbb{R}^3 \to \mathbb{R}^3$ be the **zero operator**, i.e., for any vector $\mathbf{x} \in \mathbb{R}^3$, we have $T(\mathbf{x}) = \mathbf{0}$. Show that the standard matrix for T is the 3×3 zero matrix.

9. Let $D : P_2 \to P_2$ be the differentiation operator $D(\mathbf{p}) = \mathbf{p}'$.
(**a**) Find the matrix of D with respect to the basis $B = \{1, x, x^2\}$.
(**b**) Use the matrix obtained in (a) to find $D(1 + 2x + x^2)$.

10. Let the linear operator $T : \mathbb{R}^2 \to \mathbb{R}^2$ be the **identity operator**, i.e., for any vector $\mathbf{x} \in \mathbb{R}^2$, we have $T(\mathbf{x}) = \mathbf{x}$. Show that the standard matrix for T is the 2×2 identity matrix.

11. Consider the linear operator $T : \mathbb{R}^n \to \mathbb{R}^n$ where $T(\mathbf{x}) = \lambda\mathbf{x}$ for any vector $\mathbf{x} \in \mathbb{R}^n$. Show that the standard matrix for T is a diagonal matrix.

12. Let $T : P_1 \to P_2$ be the linear transformation defined by

$$T(x^n) = \int_0^x t^n dt$$

where n is a non-negative integer.
(**a**) Find the matrix of T relative to the standard bases $B = \{1, x\}$ and $B' = \{1, x, x^2\}$.
(**b**) Use the matrix found in (a) to find $T(1 + 2x)$.

24. Similar Matrices

Let $T : V \rightarrow V$ be a linear operator on a finite dimensional vector space V. In the previous lesson, we saw that the matrix A of T clearly depends upon the underlying basis chosen for V. An important problem in linear algebra is to find a basis for V which somehow makes the matrix A as simple as possible.

Before we can fully discuss this problem, we need to know how the matrix of a linear operator behaves under a change of basis. This is the content of our first theorem.

Theorem 24.1: *(Matrix of an Operator Under a Change of Basis)*

Let $T : V \rightarrow V$ be a linear operator on a finite dimensional vector space V. Let A and A' be the matrices of T relative to the bases B and B', respectively. Furthermore, let P be the transition matrix from the basis B' to the basis B; i.e.,

$$[\mathbf{x}]_B = P[\mathbf{x}]_{B'} \qquad (\text{for all } \mathbf{x} \in V) \tag{24.1}$$

Then

$$A' = P^{-1}AP \tag{24.2}$$

Proof: Let $\mathbf{x}$ be an *arbitrary* vector in V, and assume that

$$A[\mathbf{x}]_B = [T(\mathbf{x})]_B \tag{a}$$

$$A'[\mathbf{x}]_{B'} = [T(\mathbf{x})]_{B'} \tag{b}$$

By (24.1), we can write

$$[\mathbf{x}]_{B'} = P^{-1}[\mathbf{x}]_B \tag{c}$$

$$[T(\mathbf{x})]_{B'} = P^{-1}[T(\mathbf{x})]_B \tag{d}$$

Substituting expressions (c) and (d) into (b), we obtain

$$A'P^{-1}[\mathbf{x}]_B = P^{-1}[T(\mathbf{x})]_B$$

so that

$$(PA'P^{-1})[\mathbf{x}]_B = [T(\mathbf{x})]_B \tag{e}$$

Comparing (a) and (e), we see that the matrix $PA'P^{-1}$ has the same effect on *any* vector $\mathbf{x} \in V$ as the matrix A, so we must have $A = PA'P^{-1}$, and consequently, $A' = P^{-1}AP$ as claimed. □.

The content of the previous theorem is summarized in Figure 24.1.

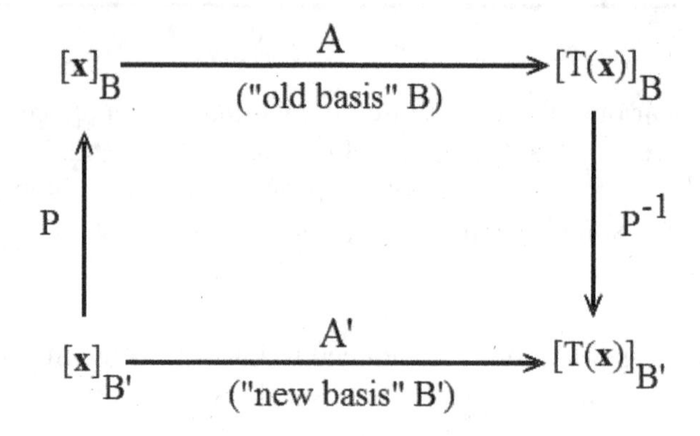

Figure 24.1: Pictorial representation of Theorem 24.1

If we start in the lower left-hand corner of the diagram with the coordinate matrix $[\mathbf{x}]_{B'}$, we see that there are two paths we can follow to obtain the matrix $[T(\mathbf{x})]_{B'}$. That is, we can go straight across the bottom path to get

$$A'[\mathbf{x}]_{B'} = [T(\mathbf{x})]_{B'}$$

or alternatively, we can move up the left-hand side, then across the top, and finally, down the right-hand side to get

$$(P^{-1}AP)[\mathbf{x}]_{B'} = [T(\mathbf{x})]_{B'}$$

Since the last two expressions hold for any vector $\mathbf{x} \in V$, we conclude that

$$A' = P^{-1}AP$$

which is just (24.2) again.

Example-1: Consider the linear operator $T : \mathbb{R}^2 \to \mathbb{R}^2$ given by

$$T\left(\begin{bmatrix} x \\ y \end{bmatrix}\right) = \begin{bmatrix} 2x+y \\ x+3y \end{bmatrix}$$

(a) Find the matrix A for T relative to the standard basis $E = \{\mathbf{e}_1, \mathbf{e}_2\}$ where

$$\mathbf{e}_1 = \begin{bmatrix} 1 \\ 0 \end{bmatrix} \text{ and } \mathbf{e}_2 = \begin{bmatrix} 0 \\ 1 \end{bmatrix}$$

(b) Use Theorem 24.1 to find the matrix A' for T relative to the basis

$$\mathbf{b}_1' = \begin{bmatrix} 1 \\ -1 \end{bmatrix} \text{ and } \mathbf{b}_2' = \begin{bmatrix} 1 \\ 1 \end{bmatrix}$$

(c) Check your answer in (b) by directly computing A'.

Solution:

(a) To find the matrix of T relative to the standard basis E, we must first find the action of T on each of the standard basis vectors. We find that

$$T(\mathbf{e}_1) = T\left(\begin{bmatrix} 1 \\ 0 \end{bmatrix}\right) = \begin{bmatrix} 2 \\ 1 \end{bmatrix}$$

$$T(\mathbf{e}_2) = T\left(\begin{bmatrix} 0 \\ 1 \end{bmatrix}\right) = \begin{bmatrix} 1 \\ 3 \end{bmatrix}$$

So, the matrix A for T relative to the standard basis E is

$$A = \begin{bmatrix} 2 & 1 \\ 1 & 3 \end{bmatrix}$$

(b) Before we can use Theorem 24.1, we need to find the transition matrix P from the "new basis" B' to the "old basis" E. To do this, we must express the new basis vectors $\mathbf{b}_1'$ and $\mathbf{b}_2'$ in terms of the standard basis vectors. By inspection,

$$\mathbf{b}_1' = (1)\mathbf{e}_1 + (-1)\mathbf{e}_2$$
$$\mathbf{b}_2' = (1)\mathbf{e}_1 + (1)\mathbf{e}_2$$

so the required transition matrix is

$$P = \begin{bmatrix} 1 & 1 \\ -1 & 1 \end{bmatrix}$$

Also, it's easy to see that

$$P^{-1} = \begin{bmatrix} 1/2 & -1/2 \\ 1/2 & 1/2 \end{bmatrix}$$

By Theorem 24.1, we are guaranteed that the matrix A' for T relative to the new basis B' is given by

$$A' = P^{-1}AP = \begin{bmatrix} 1/2 & -1/2 \\ 1/2 & 1/2 \end{bmatrix}\begin{bmatrix} 2 & 1 \\ 1 & 3 \end{bmatrix}\begin{bmatrix} 1 & 1 \\ -1 & 1 \end{bmatrix} = \begin{bmatrix} \frac{3}{2} & -\frac{1}{2} \\ -\frac{1}{2} & \frac{7}{2} \end{bmatrix}$$

(c) We can directly compute A' by simply expressing each of the image vectors $T(\mathbf{b}_1')$ and $T(\mathbf{b}_2')$ in terms of the new basis (this amounts to moving left to right, along the bottom of Figure 24.1). We find

$$T(\mathbf{b}_1') = T\left(\begin{bmatrix} 1 \\ -1 \end{bmatrix}\right) = \begin{bmatrix} 1 \\ -2 \end{bmatrix} = \left(\tfrac{3}{2}\right)\mathbf{b}_1' + \left(-\tfrac{1}{2}\right)\mathbf{b}_2'$$

$$T(\mathbf{b}_2') = T\left(\begin{bmatrix} 1 \\ 1 \end{bmatrix}\right) = \begin{bmatrix} 3 \\ 4 \end{bmatrix} = \left(-\tfrac{1}{2}\right)\mathbf{b}_1' + \left(\tfrac{7}{2}\right)\mathbf{b}_2'$$

so that

$$A' = \begin{bmatrix} \frac{3}{2} & -\frac{1}{2} \\ -\frac{1}{2} & \frac{7}{2} \end{bmatrix},$$

which agrees with the result that we found in (b).

Example-2: Given that the matrix A of the linear operator $T : \mathbb{R}^2 \to \mathbb{R}^2$ relative to the standard basis $E = \{\mathbf{e}_1, \mathbf{e}_2\}$ is given by

$$A = \begin{bmatrix} -3 & -4 \\ 1 & 2 \end{bmatrix}$$

Find the matrix A' for T relative to the basis

$$\mathbf{b}_1' = \begin{bmatrix} -1 \\ 1 \end{bmatrix} \quad \text{and} \quad \mathbf{b}_2' = \begin{bmatrix} -4 \\ 1 \end{bmatrix}$$

Solution: Once again, we will use Theorem 24.1. We first find the transition matrix P from the "new basis" B' to the "old basis" E by expressing each of the new basis vectors $\mathbf{b}_1'$ and $\mathbf{b}_2'$ in terms of the standard basis vectors. We have

$$\begin{aligned} \mathbf{b}_1' &= (-1)\mathbf{e}_1 + (1)\mathbf{e}_2 \\ \mathbf{b}_2' &= (-4)\mathbf{e}_1 + (1)\mathbf{e}_2 \end{aligned}$$

so the required transition matrix is

$$P = \begin{bmatrix} -1 & -4 \\ 1 & 1 \end{bmatrix}$$

and its inverse P^{-1} is simply

$$P^{-1} = \begin{bmatrix} \frac{1}{3} & \frac{4}{3} \\ -\frac{1}{3} & -\frac{1}{3} \end{bmatrix}$$

By Theorem 24.1, the matrix A' for T relative to the new basis B' is

$$A' = P^{-1}AP = \begin{bmatrix} \frac{1}{3} & \frac{4}{3} \\ -\frac{1}{3} & -\frac{1}{3} \end{bmatrix} \begin{bmatrix} -3 & -4 \\ 1 & 2 \end{bmatrix} \begin{bmatrix} -1 & -4 \\ 1 & 1 \end{bmatrix} = \begin{bmatrix} 1 & 0 \\ 0 & -2 \end{bmatrix}$$

The previous example illustrates that sometimes, through a proper choice of basis, we can express the matrix of a linear operator as a *diagonal* matrix relative to the new basis. Whenever this can be done, it is highly beneficial since diagonal matrices are extremely easy to work with. In the next lesson, we will return to this topic, and discuss a general procedure that will enable us to do this in most practical cases.

In general, any two square matrices A and A' of order n are said to be **similar** if there exists an invertible $n \times n$ matrix P such that $A' = P^{-1}AP$. We formalize this concept in the next definition.

Definition: *(Similar Matrices)*

If A and A' are square matrices of order n, then we say that A' is **similar** to A, and write

$$A' \approx A,$$

if and only if there is an *invertible* $n \times n$ matrix P such that

$$A' = P^{-1}AP$$

We see that if two matrices A and A' represent the *same* linear operator $T : V \to V$ with respect to the bases B and B' for a vector space V, then the matrix A' must be similar to A.

Theorem 24.2: *(Properties of Similar Matrices)*

Let A, B, C be square matrices of order n. Then the following properties hold true:

(a) $A \approx A$ (Reflexive Property)
(b) If $A \approx B$ then $B \approx A$. (Symmetric Property)
(c) If $A \approx B$ and $B \approx C$ then $A \approx C$. (Transitive Property)
(d) If $A \approx B$ then $\det(A) = \det(B)$.
(e) If $A \approx B$ then $A^k \approx B^k$ for any positive integer k.
(f) If $A \approx B$ then $Tr(A) = Tr(B)$

Proof: We prove only properties (a), (b), and (d); the remaining properties are left as exercises. To establish property (a), we simply observe that if

I is the $n \times n$ identity matrix, then

$$A = I^{-1}AI$$

so every square matrix is similar to itself. To prove property (b), assume that $A \approx B$ so there exists an invertible square matrix P such that

$$A = P^{-1}BP$$

but then

$$B = PAP^{-1} = (P^{-1})^{-1}A(P^{-1})$$

so $B \approx A$ as well. Finally, to establish property (d), assume that $A \approx B$ so $A = P^{-1}BP$ where P is an invertible square matrix of order n. Observe that

$$\begin{aligned} \det(A) &= \det(P^{-1}BP) \\ &= \det(P^{-1})\det(B)\det(P) \\ &= \frac{1}{\det(P)}\det(B)\det(P) \\ &= \det(B) \end{aligned}$$

as claimed. □.

We can summarize properties (a),(b), and (c) of the previous theorem by simply saying that **similarity** is an **equivalence relation** on the set of square matrices of order n. Finally, observe that property (b) permits us to simply say that two matrices A and B are **similar** without specifying whether $A \approx B$ or $B \approx A$.

Exercise Set 24

---

1. Given the linear operator $T : \mathbb{R}^2 \to \mathbb{R}^2$ where

$$T\left(\begin{bmatrix} x \\ y \end{bmatrix}\right) = \begin{bmatrix} x + 2y \\ 2x + y \end{bmatrix}$$

Let $B = \{\mathbf{e}_1, \mathbf{e}_2\}$ be the *standard basis* for $\mathbb{R}^2$, and let $B' = \{\mathbf{b}_1', \mathbf{b}_2'\}$ where

$$\mathbf{b}_1' = \begin{bmatrix} 1 \\ 1 \end{bmatrix} \quad \text{and} \quad \mathbf{b}_2' = \begin{bmatrix} -1 \\ 1 \end{bmatrix}$$

Find the matrix A' of T with respect to the basis B'.

2. If $T : \mathbb{R}^2 \to \mathbb{R}^2$ is the orthogonal projection onto the x-axis, and the bases B and B' are the same as Exercise-1, then find the matrix A' of T with respect to the basis B'.

3. The linear operator $T : \mathbb{R}^3 \to \mathbb{R}^3$ is defined by

$$T\left(\begin{bmatrix} x \\ y \\ z \end{bmatrix}\right) = \begin{bmatrix} 3 & -1 & 1 \\ -2 & 4 & 2 \\ -1 & 1 & 5 \end{bmatrix}\begin{bmatrix} x \\ y \\ z \end{bmatrix}$$

relative to the **standard basis** $B = \{\mathbf{e}_1, \mathbf{e}_2, \mathbf{e}_3\}$ for $\mathbb{R}^3$. Find the matrix A' of T relative to the basis $B' = \{\mathbf{b}_1', \mathbf{b}_2', \mathbf{b}_3'\}$ where

$$\mathbf{b}_1' = \begin{bmatrix} 1 \\ 1 \\ 0 \end{bmatrix}, \quad \mathbf{b}_2' = \begin{bmatrix} 1 \\ 0 \\ 1 \end{bmatrix}, \quad \mathbf{b}_3' = \begin{bmatrix} 0 \\ 1 \\ 1 \end{bmatrix}$$

4. Prove that if $A \approx B$ then $A^2 \approx B^2$.

5. Generalize the result of Exercise-4, by showing that if $A \approx B$ then $A^k \approx B^k$ for any positive integer k.

6. Show that if $A \approx B$ then $Tr(A) = Tr(B)$.

7. Consider the linear operator $T : \mathbb{R}^2 \to \mathbb{R}^2$ defined by $T(\mathbf{x}) = 2\mathbf{x}$. Using the same bases as in Exercise-1, find the matrix of T relative to B'.

8. The matrix A of a linear operator $T : \mathbb{R}^2 \to \mathbb{R}^2$ is given by

$$A = \begin{bmatrix} 1/\sqrt{2} & 1/\sqrt{2} \\ 1/\sqrt{2} & -1/\sqrt{2} \end{bmatrix}$$

relative to the standard basis $B = \{\mathbf{e}_1, \mathbf{e}_2\}$ for $\mathbb{R}^2$. Let the new basis $B' = \{\mathbf{b}_1', \mathbf{b}_2'\}$ where

$$\mathbf{b}_1' = \begin{bmatrix} 1 - \sqrt{2} \\ 1 \end{bmatrix} \text{ and } \mathbf{b}_2' = \begin{bmatrix} 1 + \sqrt{2} \\ 1 \end{bmatrix}$$

Find the matrix A' of T with respect to the basis B'.

Unit VI: Matrix Diagonalization

James Joseph Sylvester (1814-1897)

James Joseph Sylvester was a talented English mathematician who made many contributions to mathematics. He was born into a modest family in London, and at the age of 14, he entered the University of London, where he studied under Augustus DeMorgan.

Sylvester then attended St. John's College, Cambridge, where he began his study of mathematics. Although he had a high degree of academic success, and even finished second in Cambridge's Mathematical Tripos, he was not awarded a degree. This was due to the extant policy that all graduates were required to formally state their acceptance of the Thirty Nine Articles of the Church of England, and as a Jew, Sylvester could not do so in good conscience. This unfortunate injustice was later corrected in 1872, when Sylvester finally received both his B.A. and M.A. degrees from Cambridge.

During his productive life, Sylvester held numerous positions in both England and the United States. These positions included professorships at the University of Virginia, Johns Hopkins University, and finally, at Oxford University, an appointment he held until his death.

Upon his return from the University of Virginia, Sylvester studied law. While doing so, he met the English mathematician Arthur Cayley, and the two became close friends and collaborators. During the remainder of his life, Sylvester went on to make many contributions to the theory of matrices.

25. Eigenvalues and Eigenvectors

Given a linear operator T on an n-dimensional space V, we know that we can always find a square matrix A of order n such that $T(\mathbf{x}) = A\mathbf{x}$ for every vector $\mathbf{x}$ in V. In many applications, however, it is often useful to find those scalars λ such that $A\mathbf{x} = \lambda\mathbf{x}$ has *non-zero* solution vectors $\mathbf{x}$. Whenever this is possible, we achieve a great simplification, since the action of the operator T on any vector $\mathbf{x}$ is reduced to scalar multiplication.

Definition: *(Eigenvalue and Eigenvector)*

Let A be a square matrix of order n. A scalar λ is said to be an **eigenvalue** of A if there exists a *non-zero* vector $\mathbf{x}$ such that

$$A\mathbf{x} = \lambda\mathbf{x} \tag{25.1}$$

The vector $\mathbf{x}$ is called an **eigenvector** of A corresponding to the eigenvalue λ.

Two important points should be made at this juncture. First, we must remember that *an eigenvector can never be the zero vector*. In fact, if we were to allow $\mathbf{x} = \mathbf{0}$ in (25.1), we would get $A(\mathbf{0}) = \lambda(\mathbf{0})$, which is an equation that would hold for *all* scalars λ, and the above definition would be useless.

Second, if $\mathbf{x}$ is an eigenvector of A corresponding to the eigenvalue λ, then any *non-zero* scalar multiple $k\mathbf{x}$ is also an eigenvector of A corresponding to the eigenvalue λ, since

$$A(k\mathbf{x}) = k(A\mathbf{x}) = k(\lambda\mathbf{x}) = \lambda(k\mathbf{x})$$

So, *an eigenvector corresponding to a given eigenvalue is not unique.*

Now, in order to find the eigenvalues of a square matrix A, we first rewrite (25.1) in the form

$$A\mathbf{x} = \lambda I\mathbf{x}$$

so that

$$(A - \lambda I)\mathbf{x} = \mathbf{0} \tag{25.2}$$

Here (25.2) represents a *homogeneous* system of n-equations in n-unknowns, i.e., the n-components of $\mathbf{x}$. From the theory of such systems, we know that it will have *non-trivial* solutions if and only if

$$\det(A - \lambda I) = 0 \tag{25.3}$$

This equation is called the **characteristic equation** of the matrix A, and its solutions are the required eigenvalues of A. If we expand the determinant on the left-hand side of (25.3), we obtain a polynomial $\Delta(\lambda)$ of degree n having the form

$$\Delta(\lambda) = \det(A - \lambda I) = (-1)^n(\lambda^n + a_{n-1}\lambda^{n-1} + a_{n-2}\lambda^{n-2} + \ldots + a_1\lambda + a_0) \quad (25.4$$

This is called the **characteristic polynomial** of the matrix A. Thus, the eigenvalues of A are simply the roots of the characteristic polynomial. Now, we know that every polynomial of degree n has exactly n roots - some of which may be repeated roots. Thus, a square matrix A of order n can have at most n distinct eigenvalues.

From the above discussion, we can formulate a general procedure for finding the eigenvalues and eigenvectors of an $n \times n$ matrix A.

Procedure for Finding Eigenvalues and Eigenvectors

Given a square matrix A of order n, the following procedure gives us the eigenvalues and eigenvectors of A:

1. First find the characteristic equation by setting

$$\det(A - \lambda I) = 0$$

This will be a polynomial equation of degree n in the unknown λ. It will have n-solutions, say

$$\lambda = \lambda_1, \lambda_2, \ldots, \lambda_n$$

not necessarily distinct, and these solutions will be the eigenvalues of the matrix A.

2. Substitute each distinct solution, say $\lambda = \lambda_k$, of the characteristic equation into (25.2) to obtain the homogeneous system

$$(A - \lambda_k I)\mathbf{x} = \mathbf{0}$$

Use the Gauss-Jordan method (or any convenient method) to solve the homogeneous system. Any *non-trivial* solution vector $\mathbf{x}_k$ of this system will be an eigenvector corresponding to the eigenvalue $\lambda = \lambda_k$.

Let's see exactly how this general procedure works in a few examples.

Example-1: Find the eigenvalues and eigenvectors of the matrix A where

$$A = \begin{bmatrix} 1 & 2 \\ 3 & 0 \end{bmatrix}$$

Solution: We first observe that

$$\det(A - \lambda I) = \det\begin{bmatrix} 1-\lambda & 2 \\ 3 & -\lambda \end{bmatrix} = \lambda^2 - \lambda - 6$$

So, the characteristic equation of A is

$$\lambda^2 - \lambda - 6 = 0$$

The solutions of the characteristic equation are $\lambda_1 = -2$, and $\lambda_2 = 3$; these are the required eigenvalues of A.

In order to find an eigenvector corresponding to $\lambda_1 = -2$, we substitute $\lambda = -2$ into (25.2) to obtain the homogeneous system

$$\begin{bmatrix} 3 & 2 \\ 3 & 2 \end{bmatrix}\begin{bmatrix} x_1 \\ x_2 \end{bmatrix} = \begin{bmatrix} 0 \\ 0 \end{bmatrix}$$

from which we conclude that $3x_1 + 2x_2 = 0$, or $x_2 = (-3/2)x_1$. For convenience, let $x_1 = 2$, so then $x_2 = -3$, and an eigenvector corresponding to $\lambda = -2$ is

$$\mathbf{x}_1 = \begin{bmatrix} 2 \\ -3 \end{bmatrix}$$

Substituting $\lambda = \lambda_2 = 3$ into (25.2), we obtain the homogeneous system

$$\begin{bmatrix} -2 & 2 \\ 3 & -3 \end{bmatrix}\begin{bmatrix} x_1 \\ x_2 \end{bmatrix} = \begin{bmatrix} 0 \\ 0 \end{bmatrix}$$

from which we find that $x_1 - x_2 = 0$, or $x_2 = x_1$. As a matter of convenience, set $x_1 = 1$, so $x_2 = 1$. Thus, an eigenvector corresponding to $\lambda_2 = 3$ is

$$\mathbf{x}_2 = \begin{bmatrix} 1 \\ 1 \end{bmatrix}$$

Example-2: Determine the eigenvalues and eigenvectors of the matrix

$$A = \begin{bmatrix} 1 & 0 & 0 \\ 0 & 2 & 0 \\ 0 & 1 & 4 \end{bmatrix}$$

Solution: Proceeding as in the previous example, we first observe that

$$\det(A - \lambda I) = \det\begin{bmatrix} 1-\lambda & 0 & 0 \\ 0 & 2-\lambda & 0 \\ 0 & 1 & 4-\lambda \end{bmatrix} = -\lambda^3 + 7\lambda^2 - 14\lambda + 8$$

So, the characteristic equation is

$$\lambda^3 - 7\lambda^2 + 14\lambda - 8 = 0$$

Factoring the left-hand side of this equation, we obtain

$$(\lambda - 1)(\lambda - 2)(\lambda - 4) = 0$$

so the eigenvalues of A are $\lambda_1 = 1$, $\lambda_2 = 2$, and $\lambda_3 = 4$

In order to find an eigenvector corresponding to $\lambda_1 = 1$, we substitute $\lambda = 1$ into (25.2) to obtain the homogeneous system

$$\begin{bmatrix} 0 & 0 & 0 \\ 0 & 1 & 0 \\ 0 & 1 & 3 \end{bmatrix}\begin{bmatrix} x_1 \\ x_2 \\ x_3 \end{bmatrix} = \begin{bmatrix} 0 \\ 0 \\ 0 \end{bmatrix}$$

whose solution is

$$\mathbf{x} = t\begin{bmatrix} 1 \\ 0 \\ 0 \end{bmatrix}$$

where t is a real parameter. Now, we can choose any non-zero value of t we like, so for convenience, set $t = 1$, then an eigenvector $\mathbf{x}_1$ corresponding to $\lambda_1 = 1$ is

$$\mathbf{x}_1 = \begin{bmatrix} 1 \\ 0 \\ 0 \end{bmatrix}$$

To find an eigenvector corresponding to $\lambda_2 = 2$, we substitute $\lambda = 2$ into (25.2) to obtain the homogeneous system

$$\begin{bmatrix} -1 & 0 & 0 \\ 0 & 0 & 0 \\ 0 & 1 & 2 \end{bmatrix}\begin{bmatrix} x_1 \\ x_2 \\ x_3 \end{bmatrix} = \begin{bmatrix} 0 \\ 0 \\ 0 \end{bmatrix}$$

whose solution is

$$\mathbf{x} = t\begin{bmatrix} 0 \\ -2 \\ 1 \end{bmatrix}$$

where t is again a real parameter. For convenience, let $t = 1$, so an eigenvector $\mathbf{x}_2$corresponding to $\lambda_2 = 2$ is

$$\mathbf{x}_2 = \begin{bmatrix} 0 \\ -2 \\ 1 \end{bmatrix}$$

Finally, we substitute $\lambda = \lambda_3 = 4$ into (25.2) to obtain the system

$$\begin{bmatrix} -3 & 0 & 0 \\ 0 & -2 & 0 \\ 0 & 1 & 0 \end{bmatrix}\begin{bmatrix} x_1 \\ x_2 \\ x_3 \end{bmatrix} = \begin{bmatrix} 0 \\ 0 \\ 0 \end{bmatrix}$$

whose solution is:

$$\mathbf{x} = t\begin{bmatrix} 0 \\ 0 \\ 1 \end{bmatrix}$$

where t is a real parameter. Set $t = 1$, so an eigenvector of A corresponding to the eigenvalue $\lambda_3 = 4$ is

$$\mathbf{x}_3 = \begin{bmatrix} 0 \\ 0 \\ 1 \end{bmatrix}$$

Theorem 25.1: *(Two Important Properties of Eigenvalues)*

Let A be a square matrix of order n. If $\lambda_1, \lambda_2, \lambda_3, \ldots, \lambda_n$ are the n eigenvalues of A, then

$$\det A = \lambda_1\lambda_2\lambda_3 \cdots \lambda_n \tag{25.5}$$

and

$$Tr(A) = \lambda_1 + \lambda_2 + \lambda_3 + \ldots + \lambda_n \tag{25.6}$$

Proof: In order to establish (25.5), observe that by factoring the right-hand side of (25.4), we can write

$$\det(A - \lambda I) = (-1)^n(\lambda - \lambda_1)(\lambda - \lambda_2)(\lambda - \lambda_3)\cdots(\lambda - \lambda_n) \tag{25.7}$$

Since this expression is an *identity* in λ, let $\lambda = 0$, so (25.7) reduces to

$$\det A = \lambda_1\lambda_2\lambda_3\cdots\lambda_n$$

as claimed. To establish (25.6), we first rewrite (25.7) in the form

$$\det(A - \lambda I) = \begin{vmatrix} (a_{11} - \lambda) & a_{12} & a_{13} & \cdots & a_{1n} \\ a_{21} & (a_{22} - \lambda) & a_{23} & \cdots & a_{2n} \\ a_{31} & a_{32} & (a_{33} - \lambda) & \cdots & a_{3n} \\ \vdots & \vdots & \vdots & & \vdots \\ a_{n1} & a_{n2} & a_{n3} & \cdots & (a_{nn} - \lambda) \end{vmatrix} \tag{25.8}$$
$$= (-1)^n(\lambda - \lambda_1)(\lambda - \lambda_2)(\lambda - \lambda_3)\cdots(\lambda - \lambda_n)$$

If we expand the left-hand side of (25.8) by minors, it's easy to show that the

$$\text{coefficient of } \lambda^{n-1} = (-1)^{n-1}(a_{11} + a_{22} + \cdots + a_{nn})$$

On the other hand, if we expand the right-hand side of (25.7), we find that the

$$\text{coefficient of } \lambda^{n-1} = (-1)^{n-1}(\lambda_1 + \lambda_2 + \lambda_3 + \ldots + \lambda_n)$$

Consequently, if we equate coefficients, then we obtain

$$Tr(A) = a_{11} + a_{22} + \cdots + a_{nn} = \lambda_1 + \lambda_2 + \lambda_3 + \ldots + \lambda_n$$

which is precisely (25.6). □.

Note that the previous theorem provides us with an easy and useful computational check whenever we are finding the eigenvalues of a square matrix.

Example-3: Find the eigenvalues of the matrix

$$A = \begin{bmatrix} 1 & 0 & 1 \\ 1 & 2 & 3 \\ 0 & 0 & 1 \end{bmatrix}$$

and check your answer by using Theorem 25.1.

Solution: We have

$$\det(A - \lambda I) = \begin{vmatrix} 1-\lambda & 0 & 1 \\ 1 & 2-\lambda & 3 \\ 0 & 0 & 1-\lambda \end{vmatrix} = 0$$

so that

$$\lambda^3 - 4\lambda^2 + 5\lambda - 2 = (\lambda - 1)^2(\lambda - 2) = 0$$

Thus, the eigenvalues of A are $\lambda_1 = 1, \lambda_2 = 1$, and $\lambda_3 = 2$. It's easy to see that our work is correct since

$$\begin{aligned} Tr(A) &= 4 = 1 + 1 + 2 \\ \det(A) &= 2 = 1 \cdot 1 \cdot 2 \end{aligned}$$

as required by the previous theorem.

Now, if A is a square matrix of order n, and $\mathbf{x}_1$.and $\mathbf{x}_2$ are any eigenvectors of A corresponding to the *same* eigenvalue λ, then any *non-zero* linear combination $c_1\mathbf{x}_1 + c_2\mathbf{x}_2 \neq \mathbf{0}$ is also an eigenvector of A since

$$\begin{aligned} A(c_1\mathbf{x}_1 + c_2\mathbf{x}_2) &= c_1A\mathbf{x}_1 + c_2A\mathbf{x}_2 = c_1\lambda\mathbf{x}_1 + c_2\lambda\mathbf{x}_2 \\ &= \lambda(c_1\mathbf{x}_1 + c_2\mathbf{x}_2) \end{aligned}$$

Thus, the set of all eigenvectors corresponding to a given eigenvalue λ, *together with the zero vector*, forms a *subspace* of $\mathbb{R}^n$. The subspace of $\mathbb{R}^n$ which corresponds to an eigenvalue λ is called the **eigenspace** of λ. We have proved the following theorem:

Theorem 25.2: *(Eigenspace of λ is a Subspace of $\mathbb{R}^n$)*

Let A be a square matrix of order n, and let λ be an eigenvalue of A. Then the set of all eigenvectors corresponding to the *same* eigenvalue λ, combined with the zero vector, i.e.,

$$\{\mathbf{0}\} \cup \{\mathbf{x} \in \mathbb{R}^n : A\mathbf{x} = \lambda\mathbf{x}\}$$

is a subspace of $\mathbb{R}^n$, called the **eigenspace** of λ.

We have already seen that in some cases, a given eigenvalue may be a multiple root of the characteristic equation. In general, if A is a square matrix of order n, then its characteristic equation can be written in the form

$$\det(A - \lambda I) = (-1)^n(\lambda - \lambda_1)^{m_1}(\lambda - \lambda_2)^{m_2}(\lambda - \lambda_3)^{m_3} \cdots (\lambda - \lambda_r)^{m_r} = 0 \tag{25.9}$$

Here $\lambda_1, \lambda_2, \lambda_3, \ldots, \lambda_r$ are the *distinct* eigenvalues of A; the exponents $m_1, m_2, m_3, \ldots, m_r$ are called the **algebraic multiplicities** of each root, and they satisfy the condition:

$$m_1 + m_2 + m_3 + \ldots + m_r = n.$$

If the algebraic multiplicity of an eigenvalue is one, then we say that the eigenvalue is **simple**. On the other hand, if the algebraic multiplicity of an eigenvalue is greater than one, we say that the eigenvalue is **degenerate**.

Now, if $\lambda = \lambda_r$ is a degenerate eigenvalue of A with algebraic multiplicity $m_r > 1$, then we are not guaranteed that there will exist m_r linearly independent eigenvectors corresponding to λ_r. The *maximum* number of linearly independent eigenvectors corresponding to the same eigenvalue is called the **geometric multiplicity** of that eigenvalue. Stated differently, the geometric multiplicity of an eigenvalue λ_r is simply the *dimension* of the eigenspace of λ_r. In general, the algebraic and geometric multiplicities of a degenerate eigenvalue may differ.

Example-4: Given the matrix

$$A = \begin{bmatrix} 1 & 1 & 0 \\ 0 & 2 & 0 \\ 0 & 0 & 1 \end{bmatrix}$$

Find suitable bases for the eigenspaces of A, and for each eigenvalue of A, determine its algebraic and geometric multiplicity.

Solution: We find immediately that the characteristic equation of A is

$$(\lambda - 1)^2(\lambda - 2) = 0$$

Thus, eigenvalue $\lambda = 1$ has an algebraic multiplicity of 2 while the eigenvalue $\lambda = 2$ is a simple eigenvalue with an algebraic multiplicity of unity.

Let's find the eigenvectors corresponding to $\lambda = 1$. We obtain the homogeneous system

$$\begin{bmatrix} 0 & 1 & 0 \\ 0 & 1 & 0 \\ 0 & 0 & 0 \end{bmatrix} \begin{bmatrix} x_1 \\ x_2 \\ x_3 \end{bmatrix} = \begin{bmatrix} 0 \\ 0 \\ 0 \end{bmatrix}$$

whose general solution is

$$\mathbf{x} = s \begin{bmatrix} 1 \\ 0 \\ 0 \end{bmatrix} + t \begin{bmatrix} 0 \\ 0 \\ 1 \end{bmatrix}$$

where s and t are real parameters. Since the above vectors on the right-hand side of the last equation are linearly independent, then a suitable basis for the eigenspace of $\lambda = 1$ is

$$B_1 = \left\{ \begin{bmatrix} 1 \\ 0 \\ 0 \end{bmatrix}, \begin{bmatrix} 0 \\ 0 \\ 1 \end{bmatrix} \right\}$$

We see that $\lambda = 1$ has a geometric multiplicity of 2.

Finally, let's find the eigenvector corresponding to $\lambda = 2$. We obtain the homogeneous system

$$\begin{bmatrix} -1 & 1 & 0 \\ 0 & 0 & 0 \\ 0 & 0 & -1 \end{bmatrix} \begin{bmatrix} x_1 \\ x_2 \\ x_3 \end{bmatrix} = \begin{bmatrix} 0 \\ 0 \\ 0 \end{bmatrix}$$

whose general solution is

$$\mathbf{x} = s \begin{bmatrix} 1 \\ 1 \\ 0 \end{bmatrix}$$

where s is a real parameter. We see that a suitable basis for the eigenspace of $\lambda = 2$ is

$$B_2 = \left\{ \begin{bmatrix} 1 \\ 1 \\ 0 \end{bmatrix} \right\}$$

so the eigenvalue $\lambda = 2$ has a geometric multiplicity of unity.

Exercise Set 25

1. Find the eigenvalues and eigenvectors of the following matrices:

(a) $\begin{bmatrix} 1 & 2 \\ 2 & 1 \end{bmatrix}$ **(b)** $\begin{bmatrix} -1 & 0 \\ 1 & 1 \end{bmatrix}$ **(c)** $\begin{bmatrix} 3 & -1 \\ -1 & 2 \end{bmatrix}$

2. Determine the eigenvalues and eigenvectors of the following matrices:

(a) $\begin{bmatrix} 4 & -2 & -2 \\ 0 & 1 & 0 \\ 1 & 0 & 1 \end{bmatrix}$ (b) $\begin{bmatrix} 1 & 0 & 0 \\ -2 & 0 & 1 \\ 1 & 0 & 1 \end{bmatrix}$ (c) $\begin{bmatrix} -1 & 0 & 0 \\ 3 & 1 & 0 \\ 2 & 0 & 1 \end{bmatrix}$

3. Find bases for the eigenspaces of the matrices given in Exercise 2.

4. Show that the eigenvalues of a triangular matrix are given by the elements on its main diagonal.

5 Show that a square matrix A is *invertible* if and only if each of its eigenvalues is not zero.

6. Let A be a square matrix. Show that if λ is an eigenvalue of A, then λ^n is an eigenvalue of A^n for any positive integer n.

7. Given an *invertible* square matrix A, show that if λ is an eigenvalue of A, then $1/\lambda$ is an eigenvalue of A^{-1}.

8. Prove that for any square matrix A, both A and its transpose A^T have the same eigenvalues.

9. A square matrix A is said to be **idempotent** if $A^2 = A$. Show that the eigenvalues of an idempotent matrix must be $\lambda = 0$ or $\lambda = 1$.

10. If A is a square matrix of order 2, show that its characteristic equation can be written in the form

$$\lambda^2 - Tr(A)\lambda + \det(A) = 0$$

11. Use the result in Exercise 4 to find the eigenvalues and eigenvectors of the matrix

$$A = \begin{bmatrix} 1 & 0 & 2 \\ 0 & 2 & -1 \\ 0 & 0 & 4 \end{bmatrix}$$

26. Diagonalization of Square Matrices

From Theorem 24.1, recall that if we are given a linear operator $T : V \to V$ on a finite dimensional vector space V, and if A and A' are the matrices of T relative to the bases B and B' respectively, then the matrix of T relative to a new basis B' is given by

$$A' = P^{-1}AP \tag{24.2}$$

where P is the transition matrix from the new basis B' to the old basis B; i.e.,

$$[\mathbf{x}]_B = P[\mathbf{x}]_{B'} \quad (\text{for all } \mathbf{x} \in V) \tag{24.1}$$

Now, a question which naturally arises is can we choose a new basis B' such that the new matrix A' of the operator T is a *diagonal* matrix? Clearly, if this is possible, then we will achieve a great simplification, since diagonal matrices are particularly easy to work with. We shall answer this question shortly, but let's first define our terms carefully.

Definition: *(Diagonalization)*

A square matrix A of order n is said to be **diagonalizable** if there exists a *non-singular* matrix P of order n such that the matrix

$$A' = P^{-1}AP \tag{26.1}$$

is a *diagonal* matrix. Furthermore, the linear transformation S given by

$$S(A) = P^{-1}AP \tag{26.2}$$

which maps the matrix A to the diagonal matrix $P^{-1}AP$ is called a **similarity transformation**.

Now, it turns out that not all matrices can be diagonalized. The next theorem, however, gives us a useful criterion for determining when this is possible.

Theorem 26.1: *(Criterion for Diagonalizability)*

Let A be a square matrix of order n. Then A is diagonalizable if and only if A has n linearly independent eigenvectors.

Proof: Assume that A has n linearly independent eigenvectors $\mathbf{p}_1, \mathbf{p}_2, \ldots, \mathbf{p}_n$ with corresponding eigenvalues $\lambda_1, \lambda_2, \ldots, \lambda_n$. We construct the matrix P

$$P = [\mathbf{p}_1|\mathbf{p}_2|\cdots|\mathbf{p}_n] = \begin{bmatrix} p_{11} & p_{12} & \cdots & p_{1n} \\ p_{21} & p_{22} & \cdots & p_{2n} \\ \vdots & \vdots & & \vdots \\ p_{n1} & p_{n2} & \cdots & p_{nn} \end{bmatrix}$$

whose columns are formed from the column vectors $\mathbf{p}_1, \mathbf{p}_2, \ldots, \mathbf{p}_n$. Now since $A\mathbf{p}_k = \lambda_k \mathbf{p}_k$ for $k = 1$ to n, we can write

$$\begin{aligned} AP &= [A\mathbf{p}_1|A\mathbf{p}_2|\cdots|A\mathbf{p}_n] = [\lambda_1\mathbf{p}_1|\lambda_2\mathbf{p}_2|\cdots|\lambda_n\mathbf{p}_n] \\ &= [\mathbf{p}_1|\mathbf{p}_2|\cdots|\mathbf{p}_n]\begin{bmatrix} \lambda_1 & 0 & \cdots & 0 \\ 0 & \lambda_2 & \cdots & 0 \\ \vdots & \vdots & & \vdots \\ 0 & 0 & \ldots & \lambda_n \end{bmatrix} \\ &= PD \end{aligned} \qquad (26.3)$$

where D is the diagonal matrix

$$D = \begin{bmatrix} \lambda_1 & 0 & \cdots & 0 \\ 0 & \lambda_2 & \cdots & 0 \\ \vdots & \vdots & & \vdots \\ 0 & 0 & \ldots & \lambda_n \end{bmatrix}$$

Since the column vectors of P are linearly independent, then $\det(P) \neq 0$ and therefore, P^{-1} exists. Multiplying both sides of (26.3) on the left by P^{-1} gives

$$P^{-1}AP = D$$

so A has been diagonalized.

Now assume that A is diagonalizable, so there exists a non-singular matrix

$$P = \begin{bmatrix} p_{11} & p_{12} & \cdots & p_{1n} \\ p_{21} & p_{22} & \cdots & p_{2n} \\ \vdots & \vdots & & \vdots \\ p_{n1} & p_{n2} & \cdots & p_{nn} \end{bmatrix} = [\mathbf{p}_1|\mathbf{p}_2|\cdots|\mathbf{p}_n]$$

whose columns consist of the column vectors $\mathbf{p}_1, \mathbf{p}_2, \ldots, \mathbf{p}_n$, such that $P^{-1}AP = D$ is a diagonal matrix. But then $AP = PD$, so that

$$AP = [A\mathbf{p}_1|A\mathbf{p}_2|\cdots|A\mathbf{p}_n] = PD = [\lambda_1\mathbf{p}_1|\lambda_2\mathbf{p}_2|\cdots|\lambda_n\mathbf{p}_n]$$

and consequently,

$$A\mathbf{p}_1 = \lambda_1\mathbf{p}_1,\ A\mathbf{p}_2 = \lambda_2\mathbf{p}_2, \ldots, A\mathbf{p}_n = \lambda_n\mathbf{p}_n \tag{26.4}$$

Since P is non singular, its column vectors must be *linearly independent*, and in particular, they are all *non-zero*. But then from (26.4), we see that A has the n linearly independent eigenvectors $\mathbf{p}_1, \mathbf{p}_2, \ldots, \mathbf{p}_n$ corresponding to the eigenvalues $\lambda_1, \lambda_2, \ldots, \lambda_n$. □.

An alternative but only *sufficient condition* for diagonalizability is given in the next theorem which states that if a square matrix of order n has exactly n *distinct* eigenvalues, i.e., its characteristic equation doesn't have any multiple roots, then it can be diagonalized.

Theorem 26.2: *(A Sufficient Condition for Diagonalizability)*

Let A be a square matrix of order n. If A has n *distinct* eigenvalues, then its corresponding eigenvectors are linearly independent so that A is diagonalizable.

Proof: Assume that A has n distinct eigenvalues $\lambda_1, \lambda_2, \ldots, \lambda_n$ with corresponding eigenvectors $\mathbf{x}_1, \mathbf{x}_2, \ldots, \mathbf{x}_n$. We must show that the equation

$$c_1\mathbf{x}_1 + c_2\mathbf{x}_2 + \ldots + c_n\mathbf{x}_n = \mathbf{0}$$

implies that $c_1 = c_2 = \ldots = c_n = 0$. We shall use mathematical induction. For $n = 1$, we have $c_1\mathbf{x}_1 = \mathbf{0}$. Since $\mathbf{x}_1$ is an eigenvector, however, we are guaranteed that $\mathbf{x}_1 \neq \mathbf{0}$, so $c_1 = 0$, and the result holds for $n = 1$. Next, we assume that the theorem holds for $n = k$ so that the eigenvectors $\mathbf{x}_1, \mathbf{x}_2, \ldots, \mathbf{x}_k$ are linearly independent, and

$$c_1\mathbf{x}_1 + c_2\mathbf{x}_2 + \ldots + c_k\mathbf{x}_k = \mathbf{0} \tag{A}$$

implies $c_1 = c_2 = \ldots = c_k = 0$. Now, let's assume that

$$c_1\mathbf{x}_1 + c_2\mathbf{x}_2 + \ldots + c_{k+1}\mathbf{x}_{k+1} = \mathbf{0} \tag{B}$$

Multiplying (B) by A we obtain

$$c_1(A\mathbf{x}_1) + c_2(A\mathbf{x}_2) + \ldots + c_{k+1}(A\mathbf{x}_{k+1}) = \mathbf{0}$$

which, in turn, implies

$$c_1(\lambda_1\mathbf{x}_1) + c_2(\lambda_2\mathbf{x}_2) + \ldots + c_{k+1}(\lambda_{k+1}\mathbf{x}_{k+1}) = \mathbf{0} \tag{C}$$

Multiplying (B) by λ_{k+1} and subtracting the result from (C) gives

$$c_1(\lambda_1 - \lambda_{k+1})\mathbf{x}_1 + c_2(\lambda_2 - \lambda_{k+1})\mathbf{x}_2 + \ldots + c_k(\lambda_k - \lambda_{k+1})\mathbf{x}_k = \mathbf{0} \tag{D}$$

Since by hypothesis, the eigenvectors $\mathbf{x}_1, \mathbf{x}_2, \ldots \mathbf{x}_k$ are linearly independent and the eigenvalues are all distinct, we must have $c_1 = c_2 = \ldots = c_k = 0$. Substituting $c_1 = c_2 = \ldots = c_k = 0$ into (B) implies $c_{k+1} = 0$ as well, and the result follows. □.

The proof of Theorem 26.1 provides us with a general method for diagonalizing a square matrix whenever this is possible.

Procedure for Diagonalizing a Square Matrix

Given a square matrix A of order n,

1. We first attempt to find n linearly independent eigenvectors $\mathbf{p}_1, \mathbf{p}_2, \ldots, \mathbf{p}_n$ of A. If this is not possible, then by Theorem 26.1, A is not diagonalizable.
2. If A has n linearly independent eigenvectors $\mathbf{p}_1, \mathbf{p}_2, \ldots, \mathbf{p}_n$ we construct the matrix

$$P = [\mathbf{p}_1 | \mathbf{p}_2 | \cdots | \mathbf{p}_n]$$

whose columns are formed from the eigenvectors $\mathbf{p}_1, \mathbf{p}_2, \ldots, \mathbf{p}_n$ of A.
3. The matrix $P^{-1}AP = D$ will then be a *diagonal* matrix, where

$$D = \begin{bmatrix} \lambda_1 & 0 & \cdots & 0 \\ 0 & \lambda_2 & \cdots & 0 \\ \vdots & \vdots & & \vdots \\ 0 & 0 & \ldots & \lambda_n \end{bmatrix}$$

and the diagonal elements of D are the eigenvalues of A. Here, the order of the eigenvalues along the main diagonal of D will correspond to the order of the eigenvectors in the columns of P.

Let's take a look at some examples.

Example-1: Given the matrix

$$A = \begin{bmatrix} 1 & 2 \\ 2 & 1 \end{bmatrix}$$

show that A is diagonalizable, and find a matrix P such that $P^{-1}AP$ is a diagonal matrix.

Solution: The characteristic equation for A is

$$\det(A - \lambda I) = \begin{vmatrix} 1-\lambda & 2 \\ 2 & 1-\lambda \end{vmatrix} = \lambda^2 - 2\lambda - 3 = 0$$

so its eigenvalues are $\lambda_1 = -1$, and $\lambda_2 = 3$. In order to find the corresponding eigenvectors, we must find a non-trivial solution of the equation:

$$(A - \lambda I)\mathbf{x} = \mathbf{0}. \tag{a}$$

Substituting $\lambda = \lambda_1 = -1$ into (a), we obtain

$$\begin{bmatrix} 2 & 2 \\ 2 & 2 \end{bmatrix}\begin{bmatrix} x_1 \\ x_2 \end{bmatrix} = \begin{bmatrix} 0 \\ 0 \end{bmatrix}$$

which implies that we can take

$$\mathbf{x}_1 = \begin{bmatrix} -1 \\ 1 \end{bmatrix}$$

as the corresponding eigenvector. Substituting $\lambda = \lambda_2 = 3$ into (a), we get

$$\begin{bmatrix} -2 & 2 \\ 2 & -2 \end{bmatrix}\begin{bmatrix} x_1 \\ x_2 \end{bmatrix} = \begin{bmatrix} 0 \\ 0 \end{bmatrix}$$

so that

$$\mathbf{x}_2 = \begin{bmatrix} 1 \\ 1 \end{bmatrix}$$

Since the eigenvalues of A are distinct, then by Theorem 26.2, we are guaranteed that its eigenvectors are linearly independent. So, we construct the matrix

$$P = \begin{bmatrix} -1 & 1 \\ 1 & 1 \end{bmatrix}$$

and we are assured that $P^{-1}AP$ is a diagonal matrix. In fact, a simple calculation gives

$$P^{-1}AP = \begin{bmatrix} -\frac{1}{2} & \frac{1}{2} \\ \frac{1}{2} & \frac{1}{2} \end{bmatrix}\begin{bmatrix} 1 & 2 \\ 2 & 1 \end{bmatrix}\begin{bmatrix} -1 & 1 \\ 1 & 1 \end{bmatrix} = \begin{bmatrix} -1 & 0 \\ 0 & 3 \end{bmatrix}$$

Example-2: Show that the matrix

$$A = \begin{bmatrix} 1 & 0 & 0 \\ 0 & 2 & 0 \\ 0 & 1 & -4 \end{bmatrix}$$

can be diagonalized, and find a matrix P such that $P^{-1}AP$ is a diagonal matrix.

Solution: Proceeding as in the previous example, we find that the eigenvalues of A are $\lambda_1 = 1, \lambda_2 = 2$, and $\lambda_3 = -4$, with corresponding eigenvectors

$$\mathbf{x}_1 = \begin{bmatrix} 1 \\ 0 \\ 0 \end{bmatrix}, \ \mathbf{x}_2 = \begin{bmatrix} 0 \\ 6 \\ 1 \end{bmatrix}, \text{ and } \mathbf{x}_3 = \begin{bmatrix} 0 \\ 0 \\ 1 \end{bmatrix}$$

Since the eigenvalues of A are distinct, its eigenvectors must be linearly independent. So we construct the matrix

$$P = \begin{bmatrix} 1 & 0 & 0 \\ 0 & 6 & 0 \\ 0 & 1 & 1 \end{bmatrix}$$

and we are guaranteed that $P^{-1}AP = D$ is the diagonal matrix given by

$$P^{-1}AP = \begin{bmatrix} 1 & 0 & 0 \\ 0 & 2 & 0 \\ 0 & 0 & -4 \end{bmatrix}$$

Exercise Set 26

---

1. For each matrix A, find an invertible matrix P such that $P^{-1}AP$ is a diagonal matrix, with the eigenvalues of A along its main diagonal.

(a) $A = \begin{bmatrix} -1 & 0 \\ 3 & 2 \end{bmatrix}$ **(b)** $A = \begin{bmatrix} 1 & 1 \\ 0 & -2 \end{bmatrix}$

(c) $A = \begin{bmatrix} 1 & 0 & 0 \\ 0 & 2 & 1 \\ 0 & 1 & 3 \end{bmatrix}$ **(d)** $A = \begin{bmatrix} 3 & -1 & 1 \\ -2 & 4 & 2 \\ -1 & 1 & 5 \end{bmatrix}$

2. Show that the matrix

$$\begin{bmatrix} 1 & 1 & 1 \\ 0 & 2 & 1 \\ 0 & 0 & 2 \end{bmatrix}$$

is not diagonalizable because it only has two linearly independent eigenvectors.

3. Given the matrix

$$A = \begin{bmatrix} 1 & 0 & 0 \\ 1 & 2 & 0 \\ 0 & 0 & 1 \end{bmatrix}$$

(a) Find the eigenvalues of A and show that they are not all distinct.
(b) Show that we still can find three linearly independent eigenvectors of A.
(c) Find an invertible matrix P such that $P^{-1}AP$ is a diagonal matrix.
(d) Does this result contradict Theorem 26.2? Explain.

4. Show that the matrix

$$A = \begin{bmatrix} a & b \\ c & d \end{bmatrix}$$

is diagonalizable if $(a - d)^2 + 4bc \neq 0$.

5. *(Powers of a Square Matrix)* Assume that A is a square matrix of order n which is diagonalizable, so there exists a non-singular matrix P such that $P^{-1}AP = D$ is a diagonal matrix, where the elements of D along its main diagonal are the eigenvalues $\lambda_1, \lambda_2, \ldots, \lambda_n$ of A. Show that

$$A^k = P \begin{bmatrix} \lambda_1^k & 0 & \cdots & 0 \\ 0 & \lambda_2^k & \cdots & 0 \\ \vdots & \vdots & & \vdots \\ 0 & 0 & \ldots & \lambda_n^k \end{bmatrix} P^{-1}$$

for any positive integer k.

6. Use the result of Exercise 5 to compute

$$\begin{bmatrix} -3 & -4 \\ 1 & 2 \end{bmatrix}^4$$

7. Show that if A and B are square matrices of order n, and $A \approx B$, then A and B have the same characteristic equation. In other words, show that similar matrices have the same eigenvalues.

27. Diagonalizing Symmetric Matrices

We have already seen that in some cases, a square matrix A of order n may have fewer than n linearly independent eigenvectors, and consequently, it is not diagonalizable. It turns out, however, that any real symmetric matrix of order n always has n linearly independent eigenvectors, so it is always diagonalizable.

Recall that a square matrix A is said to be **symmetric** if $A = A^T$. We shall first examine some of the nice properties possessed by such matrices.

Theorem 27.1: *(Symmetric Matrices Have Real Eigenvalues)*

If A is a real symmetric matrix, then all of its eigenvalues are real.

Proof: Assume that λ is an eigenvalue of A with corresponding eigenvector $\mathbf{x}$, so we have

$$A\mathbf{x} = \lambda\mathbf{x} \tag{27.1}$$

Taking the transpose of both sides this equation, while keeping in mind that A is symmetric, so $A^T = A$, we get

$$\mathbf{x}^T A = \lambda\mathbf{x}^T \tag{27.2}$$

Since the elements of A are real numbers, if we take the complex conjugate of both sides of the last equation, we obtain

$$\bar{\mathbf{x}}^T A = \bar{\lambda}\bar{\mathbf{x}}^T \tag{27.3}$$

where we have used an overbar to denote the operation of complex conjugation.

Next, we multiply both sides of (27.1) on the left by $\bar{\mathbf{x}}^T$, and both sides of (27.3) on the right by $\mathbf{x}$, to obtain respectively,

$$\bar{\mathbf{x}}^T A\mathbf{x} = \lambda\bar{\mathbf{x}}^T\mathbf{x}$$

$$\bar{\mathbf{x}}^T A\mathbf{x} = \bar{\lambda}\bar{\mathbf{x}}^T\mathbf{x}$$

Subtracting the last two equations, gives

$$(\lambda - \bar{\lambda})\bar{\mathbf{x}}^T\mathbf{x} = 0 \tag{27.4}$$

Finally, let's explore the implications of equation (27.4). We first observe that if we set

$$\mathbf{x} = \begin{bmatrix} x_1 \\ x_2 \\ \vdots \\ x_n \end{bmatrix}$$

then

$$\bar{\mathbf{x}}^T\mathbf{x} = \begin{bmatrix} \bar{x}_1 & \bar{x}_2 & \cdots & \bar{x}_n \end{bmatrix} \begin{bmatrix} x_1 \\ x_2 \\ \vdots \\ x_n \end{bmatrix} = |x_1|^2 + |x_2|^2 + \ldots + |x_n|^2$$

But since $\mathbf{x}$ is an eigenvector, we must have $\mathbf{x} \neq \mathbf{0}$, so $\bar{\mathbf{x}}^T\mathbf{x} \neq 0$ as well, and we see that (27.4) implies $\lambda = \bar{\lambda}$, so λ must be a real number. □.

Theorem 27.2: *(Eigenvectors of a Symmetric Matrix)*

If A is a real symmetric matrix, then the eigenvectors that correspond to distinct eigenvalues of A are *orthogonal*.

Proof: Let λ_1 and λ_2 be distinct eigenvalues of A with eigenvectors $\mathbf{x}_1$ and $\mathbf{x}_2$ respectively, so that

$$A\mathbf{x}_1 = \lambda_1\mathbf{x}_1 \quad (27.5)$$

$$A\mathbf{x}_2 = \lambda_2\mathbf{x}_2 \quad (27.6)$$

To establish the result of the theorem, we must show that

$$\mathbf{x}_1^T\mathbf{x}_2 = \langle\mathbf{x}_1, \mathbf{x}_2\rangle = 0$$

To this end, we take the transpose of of (27.5) to obtain

$$\mathbf{x}_1^T A = \lambda_1\mathbf{x}_1^T \quad (27.7)$$

Next, we multiply (27.6) on the left by $\mathbf{x}_1^T$, and we multiply (27.7) on the right by $\mathbf{x}_2$ to get

$$\mathbf{x}_1^T A\mathbf{x}_2 = \lambda_2\mathbf{x}_1^T\mathbf{x}_2$$

$$\mathbf{x}_1^T A\mathbf{x}_2 = \lambda_1\mathbf{x}_1^T\mathbf{x}_2$$

Finally, subtracting the last two equations gives

$$(\lambda_2 - \lambda_1)\mathbf{x}_1^T\mathbf{x}_2 = (\lambda_2 - \lambda_1)\langle\mathbf{x}_1, \mathbf{x}_2\rangle = 0$$

By hypothesis $(\lambda_2 - \lambda_1) \neq 0$, so we must have $\langle\mathbf{x}_1, \mathbf{x}_2\rangle = 0$; that is, the eigenvectors $\mathbf{x}_1$ and $\mathbf{x}_2$ are orthogonal. □.

Example-1: Show that the eigenvalues of the symmetric matrix

$$A = \begin{bmatrix} 1 & 4 \\ 4 & 1 \end{bmatrix}$$

are real and distinct, and consequently, its eigenvectors are orthogonal.

Solution: We first find the eigenvalues of A by setting $|A - \lambda I| = 0$ to obtain

$$\begin{vmatrix} 1-\lambda & 4 \\ 4 & 1-\lambda \end{vmatrix} = \lambda^2 - 2\lambda - 15 = 0$$

whose solution is $\lambda_1 = -3$ and $\lambda_2 = 5$ so the eigenvalues are real, as guaranteed by Theorem 27.1.

As in the previous lesson, we can find the eigenvectors of A by solving the linear system $(A - \lambda I)\mathbf{x} = \mathbf{0}$ for each value of λ. In doing so, we find that the required eigenvectors are

$$\mathbf{x}_1 = \begin{bmatrix} -1 \\ 1 \end{bmatrix} \quad \text{and} \quad \mathbf{x}_2 = \begin{bmatrix} 1 \\ 1 \end{bmatrix}$$

Observe that the eigenvectors are orthogonal since

$$\langle \mathbf{x}_1, \mathbf{x}_2 \rangle = \mathbf{x}_1^T \mathbf{x}_2 = \begin{bmatrix} -1 & 1 \end{bmatrix} \begin{bmatrix} 1 \\ 1 \end{bmatrix} = 0$$

If A is a real symmetric matrix of order n such that its eigenvalues $\lambda_1, \lambda_2, \ldots, \lambda_n$ are *distinct*, then the previous theorem guarantees that A will have n pairwise *orthogonal* eigenvectors, say $\mathbf{x}_1, \mathbf{x}_2, \ldots, \mathbf{x}_n$. Now, we can normalize these eigenvectors by taking

$$\mathbf{p}_1 = \frac{1}{\|\mathbf{x}_1\|}\mathbf{x}_1,\ \mathbf{p}_2 = \frac{1}{\|\mathbf{x}_2\|}\mathbf{x}_2, \ldots, \mathbf{p}_n = \frac{1}{\|\mathbf{x}_n\|}\mathbf{x}_n,$$

so we will have an *orthonormal* set $\{\mathbf{p}_1, \mathbf{p}_2, \ldots, \mathbf{p}_n\}$ of eigenvectors of A that satisfy the relation

$$\langle \mathbf{p}_i, \mathbf{p}_j \rangle = \delta_{ij} \tag{27.8}$$

We are then guaranteed that if we form the $n \times n$ matrix P whose columns are formed from the eigenvectors $\mathbf{p}_1, \mathbf{p}_2, \ldots, \mathbf{p}_n$, then the matrix $P^{-1}AP$ will be diagonal. This matrix P has many useful properties, and is called an **orthogonal matrix**.

Definition: *(Orthogonal Matrix)*

An $n \times n$ matrix P is said to be **orthogonal** if it is non-singular, and

$$P^{-1} = P^T \tag{27.9}$$

Theorem 27.3: *(A Useful Property of Orthogonal Matrices)*

Let P be a square matrix of order n. Then P is orthogonal if and only if the column vectors of P form an orthonormal set.

Proof: First, assume that P is orthogonal so that $P^T P = I$, and let

$$P = \begin{bmatrix} \mathbf{p}_1 | & \mathbf{p}_2 | & \cdots & | \mathbf{p}_n \end{bmatrix} = \begin{bmatrix} p_{11} & p_{12} & \cdots & p_{1n} \\ p_{21} & p_{22} & \cdots & p_{2n} \\ \vdots & \vdots & & \vdots \\ p_{n1} & p_{n2} & \cdots & p_{nn} \end{bmatrix}$$

so we must have

$$P^T P = \begin{bmatrix} p_{11} & p_{21} & \cdots & p_{n1} \\ p_{12} & p_{22} & \cdots & p_{n2} \\ \vdots & \vdots & & \vdots \\ p_{1n} & p_{2n} & \cdots & p_{nn} \end{bmatrix} \begin{bmatrix} p_{11} & p_{12} & \cdots & p_{1n} \\ p_{21} & p_{22} & \cdots & p_{2n} \\ \vdots & \vdots & & \vdots \\ p_{n1} & p_{n2} & \cdots & p_{nn} \end{bmatrix}$$

$$= \begin{bmatrix} \langle \mathbf{p}_1, \mathbf{p}_1 \rangle & \langle \mathbf{p}_1, \mathbf{p}_2 \rangle & \cdots & \langle \mathbf{p}_1, \mathbf{p}_n \rangle \\ \langle \mathbf{p}_2, \mathbf{p}_1 \rangle & \langle \mathbf{p}_2, \mathbf{p}_2 \rangle & \cdots & \langle \mathbf{p}_2, \mathbf{p}_n \rangle \\ \vdots & \vdots & & \vdots \\ \langle \mathbf{p}_n, \mathbf{p}_1 \rangle & \langle \mathbf{p}_n, \mathbf{p}_2 \rangle & \cdots & \langle \mathbf{p}_n, \mathbf{p}_n \rangle \end{bmatrix} = I \tag{27.10}$$

Since the corresponding elements of equal matrices must themselves be equal, then

$$\langle \mathbf{p}_i, \mathbf{p}_j \rangle = \delta_{ij} \tag{27.11}$$

so column vectors of P form an orthonormal set. On the other hand, if the column vectors of P are orthogonal, then (27.11) is satisfied, and (27.10) is satisfied as well, so $P^T P = I$ and consequently, $P^{-1} = P^T$. □.

From the discussion immediately preceeding the last theorem, it's easy to see that if an $n \times n$ real symmetric matrix has n distinct eigenvalues, then it can always be diagonalized by an orthogonal matrix P whose columns consist of the *normalized* eigenvectors of A.

Now, let's suppose that an $n \times n$ real symmetric matrix does not have n distinct eigenvalues, so at least one of its eigenvalues, say λ, has an algebraic multiplicity of $k > 1$. In this case, it may be shown that there still exist k *linearly independent* eigenvectors corresponding to the eigenvalue λ.

In other words, if an eigenvalue of A has multiplicity k, then the eigenspace of λ will have dimension k. In this case, we can still use the Gram-Schmidt process to construct an *orthonormal* set of eigenvectors corresponding to λ. *This means that even if a symmetric matrix A doesn't have n distinct eigenvalues, we can still diagonalize A with an orthogonal matrix P.*

Theorem 27.4: *(All Symmetric Matrices are Orthogonally Diagonalizable)*

Every real $n \times n$ symmetric matrix A is **orthogonally diagonalizable**; that is, there exists an orthogonal matrix P such that

$$P^{-1}AP = D \tag{27.12}$$

is a diagonal matrix, where the eigenvalues of A lie along the main diagonal of D.

Example-2: Given the matrix

$$A = \begin{bmatrix} 1 & 2 & 2 \\ 2 & 1 & 2 \\ 2 & 2 & 1 \end{bmatrix}$$

Determine:
(**a**) the eigenvalues of A.
(**b**) the corresponding eigenvectors.
(**c**) an orthonormal set of eigenvectors of A.
(**d**) an orthogonal matrix P that diagonalizes A.

Solution:

(**a**) In order to find the eigenvalues of A, we set $\det(A - \lambda I) = 0$. We find that

$$\begin{vmatrix} 1-\lambda & 2 & 2 \\ 2 & 1-\lambda & 2 \\ 2 & 2 & 1-\lambda \end{vmatrix} = -\lambda^3 + 3\lambda^2 + 9\lambda + 5 = 0$$

Consequently,

$$\lambda^3 - 3\lambda^2 - 9\lambda - 5 = (\lambda - 5)(\lambda + 1)^2 = 0$$

So $\lambda_1 = 5$ is a simple eigenvalue, and $\lambda_2 = -1$ is an eigenvalue of multiplicity two.

(b) To find the eigenvectors of A, we simply solve the linear system $(A - \lambda I)\mathbf{x} = \mathbf{0}$ for each value of λ. In doing so, we find that the required eigenvectors are

$$\mathbf{x}_1 = \begin{bmatrix} 1 \\ 1 \\ 1 \end{bmatrix}, \quad \mathbf{x}_2^{(1)} = \begin{bmatrix} -1 \\ 1 \\ 0 \end{bmatrix}, \quad \text{and} \quad \mathbf{x}_2^{(2)} = \begin{bmatrix} -1 \\ 0 \\ 1 \end{bmatrix}$$

where $\mathbf{x}_1$ is the eigenvector corresponding to $\lambda_1 = 5$, while both of the eigenvectors $\mathbf{x}_2^{(1)}$ and $\mathbf{x}_2^{(2)}$ correspond to the eigenvector $\lambda_2 = -1$.

(c) We first normalize the vector $\mathbf{x}_1$ by setting

$$\mathbf{p}_1 = \frac{\mathbf{x}_1}{\|\mathbf{x}_1\|} = \begin{bmatrix} 1/\sqrt{3} \\ 1/\sqrt{3} \\ 1/\sqrt{3} \end{bmatrix}$$

By Theorem 27.2, we are guaranteed that the vector $\mathbf{p}_1$ will be orthogonal to each of the eigenvectors $\mathbf{x}_2^{(1)}$ and $\mathbf{x}_2^{(2)}$; but the eigenvectors $\mathbf{x}_2^{(1)}$ and $\mathbf{x}_2^{(2)}$ are not necessarily orthogonal, since they correspond to the *same eigenvalue* λ_2; therefore, we now use the Gram-Schmidt process on vectors $\mathbf{x}_2^{(1)}$ and $\mathbf{x}_2^{(2)}$ to obtain

$$\mathbf{p}_2 = \frac{\mathbf{x}_2^{(1)}}{\|\mathbf{x}_2^{(1)}\|} = \begin{bmatrix} -1/\sqrt{2} \\ 1/\sqrt{2} \\ 0 \end{bmatrix}$$

and

$$\mathbf{p}_3 = \frac{\mathbf{x}_2^{(2)} - \langle \mathbf{x}_2^{(2)}, \mathbf{p}_2 \rangle \mathbf{p}_2}{\|\mathbf{x}_2^{(2)} - \langle \mathbf{x}_2^{(2)}, \mathbf{p}_2 \rangle \mathbf{p}_2\|} = \begin{bmatrix} -\frac{1}{6}\sqrt{6} \\ -\frac{1}{6}\sqrt{6} \\ \frac{1}{3}\sqrt{6} \end{bmatrix}$$

Thus, an orthornormal set of eigenvectors is

$$\mathbf{p}_1 = \begin{bmatrix} 1/\sqrt{3} \\ 1/\sqrt{3} \\ 1/\sqrt{3} \end{bmatrix}, \quad \mathbf{p}_2 = \begin{bmatrix} -1/\sqrt{2} \\ 1/\sqrt{2} \\ 0 \end{bmatrix}, \quad \mathbf{p}_3 = \begin{bmatrix} -\frac{1}{6}\sqrt{6} \\ -\frac{1}{6}\sqrt{6} \\ \frac{1}{3}\sqrt{6} \end{bmatrix}$$

(d) The orthogonal matrix P is the matrix whose columns are formed from our orthonormal set of eigenvectors; that is,

$$P = \left[\ \mathbf{p}_1 | \ \mathbf{p}_2 | \ \mathbf{p}_3 \ \right] = \begin{bmatrix} 1/\sqrt{3} & -1/\sqrt{2} & -\frac{1}{6}\sqrt{6} \\ 1/\sqrt{3} & 1/\sqrt{2} & -\frac{1}{6}\sqrt{6} \\ 1/\sqrt{3} & 0 & \frac{1}{3}\sqrt{6} \end{bmatrix}$$

In fact, a direct calculation shows that

$$P^{-1}AP = \begin{bmatrix} 5 & 0 & 0 \\ 0 & -1 & 0 \\ 0 & 0 & -1 \end{bmatrix}$$

as it should.

Using the previous example as a guide, we can now write down a general procedure for diagonalizing any symmetric matrix.

Procedure for Diagonalizing a Symmetric Matrix

Given an $n \times n$ symmetric matrix A with real elements, to find an orthogonal matrix P that diagonalizes A, we proceed as follows:

1. Find the eigenvalues of A and note their respective multiplicities.

2. If λ is a **simple eigenvalue** of A (with multiplicity $k = 1$), and has a corresponding eigenvector $\mathbf{x}$, then simply normalize this eigenvector by taking

$$\mathbf{p} = \frac{\mathbf{x}}{\|\mathbf{x}\|}$$

Repeat this step for each of the remaining simple eigenvalues of A.

3. If λ is a **repeated eigenvalue** of A (with multiplicity $k > 1$) with eigenvectors $\mathbf{x}_1, \mathbf{x}_2, \ldots, \mathbf{x}_k$, then apply the Gram-Schmidt procedure to this set of eigenvectors to obtain an *orthonormal* set of eigenvectors $\mathbf{p}_1, \mathbf{p}_2, \ldots, \mathbf{p}_k$. Repeat this step for each of the remaining repeated eigenvalues of A.

4. Construct a matrix P whose columns are formed from the eigenvectors obtained in Step 2 and Step 3. The matrix P will be an *orthogonal matrix* that diagonalizes A; that is $P^{-1}AP$ will be a diagonal matrix whose elements along its main diagonal will be the eigenvalues of A.

Exercise Set 27

1. Find the eigenvalues of each symmetric matrix A:

(a) $A = \begin{bmatrix} 1 & 2 \\ 2 & 1 \end{bmatrix}$ **(b)** $A = \begin{bmatrix} 3 & 4 \\ 4 & 3 \end{bmatrix}$

(c) $A = \begin{bmatrix} 2 & 2 & 2 \\ 2 & 2 & 2 \\ 2 & 2 & 2 \end{bmatrix}$ **(d)** $A = \begin{bmatrix} 3 & 2 & 2 \\ 2 & 2 & 0 \\ 2 & 0 & 4 \end{bmatrix}$

2. For each matrix A in Exercise-1, find an orthogonal matrix P that diagonalizes A, i.e., such that $P^{-1}AP$ is a diagonal matrix.

3. Show that if A and B are symmetric matrices of order n that *commute*; that is, $AB = BA$, then AB and BA are also symmetric matrices.

4. Show that if P is an orthogonal matrix, then $\det(P) = \pm 1$.

5. Show that if P and Q are $n \times n$ orthogonal matrices, then both PQ and QP are orthogonal matrices.

6. Let P be an $n \times n$ orthogonal matrix so that $T(\mathbf{x}) = P\mathbf{x}$ is a linear transformation from $\mathbb{R}^n$ into $\mathbb{R}^n$, called an **orthogonal transformation**. Show that for any $\mathbf{x}, \mathbf{y} \in \mathbb{R}^n$ we have

$$\langle \mathbf{x}, \mathbf{y} \rangle = \langle P\mathbf{x}, P\mathbf{y} \rangle$$

In other words, show the inner product of any two vectors is **invariant** under an orthogonal transformation.

7. Show that the matrix

$$A = \begin{bmatrix} \cos\theta & \sin\theta & 0 \\ -\sin\theta & \cos\theta & 0 \\ 0 & 0 & 1 \end{bmatrix}$$

is orthogonal for any real number θ.

Unit VII: Complex Vector Spaces

Lesson 28 Complex Vector Spaces

Lesson 29 Unitary and Hermitian Matrices

William Rowan Hamilton (1805-1865)

William Rowan Hamilton was an important 19th century Irish mathematician and physicist who made important contributions to mathematical physics and pure mathematics. He entered Trinity College in Dublin at the age of 18, where he studied mathematics and the classics. In 1827, after much academic success, he was appointed Professor of Astronomy, a position which he held until his death in 1865.

In the field of mathematical physics, Hamilton made important contributions to mechanics, where he demonstrated that Newtonian mechanics could be derived using the Calculus of Variations, and by varying the "action" of a mechanical system. His approach naturally led to the equations of motion of any mechanical system and shed a bright light on the previously hidden symmetries of such problems. Today, his theory is known as Hamiltonian mechanics, and has proved invaluable to the study electromagnetism, optics, and the foundations of modern quantum mechanics.

In pure mathematics, Hamilton is best known as the creator of quaternions. Here, he generalized the notion of a complex number and defined a quaternion as a object of the form $\mathbf{q} = a + b\mathbf{i} + c\mathbf{j} + d\mathbf{k}$ where a is a scalar, and the unit vectors $\mathbf{i}, \mathbf{j}$, and $\mathbf{k}$ basically acted like the imaginary unit ($i = \sqrt{-1}$) does in the field of complex numbers. Although quaternions were never really embraced by physicists, their invention led indirectly to modern vector analysis.

28. Complex Vector Spaces

We now turn our attention to the study of complex vector spaces. Stated quite simply, a **complex vector space** is one in which we allow the scalars to be complex numbers. This means that the components of vectors can be complex numbers, and complex multiples of vectors are permitted.

The simplest example of a complex vector space is the space $\mathbb{C}^n$ which consists of all n-tuples of complex numbers. A typical vector $\mathbf{v} \in \mathbb{C}^n$ may be written in the form

$$\mathbf{v} = \begin{bmatrix} v_1 \\ v_2 \\ \vdots \\ v_n \end{bmatrix} \tag{28.1}$$

where each component of $\mathbf{v}$ is a complex number; that is, for $k = 1, 2, \ldots, n$, we have

$$v_k = a_k + ib_k \tag{28.2}$$

where $a_k, b_k \in \mathbb{R}$ and $i = \sqrt{-1}$. In a manner similar to the real vector space $\mathbb{R}^n$, we define the operations of vector addition and scalar multiplication in terms of components as

$$\mathbf{v} + \mathbf{w} = \begin{bmatrix} v_1 \\ v_2 \\ \vdots \\ v_n \end{bmatrix} + \begin{bmatrix} w_1 \\ w_2 \\ \vdots \\ w_n \end{bmatrix} = \begin{bmatrix} v_1 + w_1 \\ v_2 + w_2 \\ \vdots \\ v_n + w_n \end{bmatrix} \quad \text{for all } \mathbf{v}, \mathbf{w} \in C^n \tag{28.3}$$

and

$$c\mathbf{v} = \begin{bmatrix} cv_1 \\ cv_2 \\ \vdots \\ cv_n \end{bmatrix} \quad \text{for all } c \in \mathbb{C} \text{ and } \mathbf{v} \in \mathbb{C}^n. \tag{28.4}$$

Starting with (28.3) and (28.4), it's easy to show that all of the vector space axioms are satisfied, so $\mathbb{C}^n$ is indeed a vector space. Clearly, $\mathbb{C}^n$ is simply the complex analogue of the real vector space $\mathbb{R}^n$.

Definition: *(Complex Inner Product)*

Let V be a complex vector space. A **complex inner product** on V is a mapping

$$\langle\,,\rangle : V \times V \to \mathbb{C}$$

that assigns a complex number to each *ordered pair* of vectors in V and which satisfies the following properties:

(a) $\langle \mathbf{u}, \mathbf{u} \rangle \geq 0$

(b) $\langle \mathbf{u}, \mathbf{u} \rangle = 0$ if and only if $\mathbf{u} = \mathbf{0}$.

(c) $\langle \mathbf{u}, \mathbf{v} \rangle = \overline{\langle \mathbf{v}, \mathbf{u} \rangle}$

(d) $\langle c_1\mathbf{u} + c_2\mathbf{v}, \mathbf{w} \rangle = \overline{c_1}\,\langle \mathbf{u}, \mathbf{w} \rangle + \overline{c_2}\langle \mathbf{v}, \mathbf{w} \rangle$

A complex vector space which is equipped with an inner product is called a **unitary space**, or **complex inner product space.**

Example-1: Given any vectors $\mathbf{u}, \mathbf{v} \in \mathbb{C}^n$, we define the **Euclidean inner product** of $\mathbf{u}$ and $\mathbf{v}$ by

$$\langle \mathbf{u}, \mathbf{v} \rangle = \overline{u_1}v_1 + \overline{u_2}v_2 + \ldots + \overline{u_n}v_n \tag{28..}$$

We now show that $\langle \mathbf{u}, \mathbf{v} \rangle$ satisfies each of the properties (a)-(d), and thus, it is a bonefide complex inner product. First, observe that

$$\langle \mathbf{u}, \mathbf{u} \rangle = \overline{u_1}u_1 + \overline{u_2}u_2 + \ldots + \overline{u_n}u_n = |u_1|^2 + |u_2|^2 + \ldots + |u_n|^2$$

so $\langle \mathbf{u}, \mathbf{u} \rangle \geq 0$ and $\langle \mathbf{u}, \mathbf{u} \rangle = 0$ if and only if $\mathbf{u} = \mathbf{0}$. Also, property (c) is satisfied since

$$\begin{aligned}\overline{\langle \mathbf{v}, \mathbf{u} \rangle} &= \overline{\left(\overline{v_1}u_1 + \overline{v_2}u_2 + \ldots + \overline{v_n}u_n\right)} = \overline{u_1}v_1 + \overline{u_2}v_2 + \ldots + \overline{u_n}v_n \\ &= \langle \mathbf{u}, \mathbf{v} \rangle\end{aligned}$$

Finally, property (d) follows immediately, since

$$\begin{aligned}\langle c_1\mathbf{u} + c_2\mathbf{v}, \mathbf{w} \rangle &= \overline{(c_1u_1 + c_2v_1)}w_1 + \ldots + \overline{(c_1u_n + c_2v_n)}w_n \\ &= \overline{c_1}\left(\overline{u_1}w_1 + \ldots + \overline{u_n}w_n\right) + \overline{c_2}\left(\overline{u_1}w_1 + \ldots + \overline{u_n}w_n\right) \\ &= \overline{c_1}\langle \mathbf{u}, \mathbf{w} \rangle + \overline{c_2}\langle \mathbf{v}, \mathbf{w} \rangle\end{aligned}$$

Example-2: Let $\mathbb{C}[a,b]$ denote the set of all complex-valued functions of the form

$$\mathbf{f}(x) = \mathbf{u}(x) + i\mathbf{v}(x)$$

where $\mathbf{u}(x)$ and $\mathbf{v}(x)$ are real valued functions of a real variable that are continuous on the interval $[a,b]$. It's easy to show that $\mathbb{C}[a,b]$ is a complex vector space. We can define a complex inner product on $\mathbb{C}[a,b]$ by the expression

$$\langle \mathbf{f}(x), \mathbf{g}(x) \rangle = \int_a^b \overline{\mathbf{f}(x)}\mathbf{g}(x)dx \tag{28.6}$$

Properties (a) and (b) follow immediately upon observing that

$$\langle \mathbf{f}(x), \mathbf{f}(x) \rangle = \int_a^b |\mathbf{f}(x)|^2 dx = \int_a^b |\mathbf{u}(x)|^2 dx + \int_a^b |\mathbf{v}(x)|^2 dx$$

so $\langle \mathbf{f}(x), \mathbf{f}(x) \rangle \geq 0$ since each of the integrands is non-negative. At the same time, the last expression implies that $\langle \mathbf{f}(x), \mathbf{f}(x) \rangle = 0$ if and only if $\mathbf{u}(x) = 0$ and $\mathbf{v}(x) = 0$, or equivalently, $\mathbf{f}(x) = 0$.

In order to establish property (c), observe that

$$\overline{\langle \mathbf{g}(x), \mathbf{f}(x) \rangle} = \overline{\left(\int_a^b \overline{\mathbf{g}(x)}\mathbf{f}(x)dx \right)} = \int_a^b \overline{\mathbf{f}(x)}\mathbf{g}(x)dx = \langle \mathbf{f}(x), \mathbf{g}(x) \rangle$$

The proof of property (d) is also easy and is left as an exercise for the reader.

We now introduce a new matrix operation, which involves forming the **conjugate transpose** (or **Hermitian conjugate**) of a complex matrix.

Definition: *(Conjugate Transpose)*

The **conjugate transpose** (or **Hermitian conjugate**) of a complex matrix A, denoted by $A^\dagger$, is defined by

$$A^\dagger = \overline{(A^T)} \tag{28.7}$$

Example-3: Find the conjugate transpose $A^\dagger$ of the matrix

$$A = \begin{bmatrix} 1 & 1+i \\ 2+3i & 2i \end{bmatrix}$$

Solution: We first find the transpose of A to obtain

$$A^T = \begin{bmatrix} 1 & 2+3i \\ 1+i & 2i \end{bmatrix}$$

so the conjugate transpose of A is given by

$$A^\dagger = \overline{(A^T)} = \begin{bmatrix} \overline{1} & \overline{2+3i} \\ \overline{1+i} & \overline{2i} \end{bmatrix} = \begin{bmatrix} 1 & 2-3i \\ 1-i & -2i \end{bmatrix}$$

Example-4: We can rewrite the Euclidean inner product of any two vectors $\mathbf{u}, \mathbf{v} \in \mathbb{C}^n$ in terms of the conjugate transpose as

$$\langle \mathbf{u}, \mathbf{v} \rangle = \begin{bmatrix} \overline{u_1} & \overline{u_2} & \cdots & \overline{u_n} \end{bmatrix} \begin{bmatrix} v_1 \\ v_2 \\ \vdots \\ v_n \end{bmatrix} = \mathbf{u}^\dagger \mathbf{v} \qquad (28.8$$

where $\mathbf{u}^\dagger$ is the conjugate transpose of $\mathbf{u}$. Recall that $\mathbf{u}$ is a column vector, so $\mathbf{u}^\dagger$ is a *row vector* whose elements are the complex conjugates of the elements of $\mathbf{u}$; that is,

$$\mathbf{u}^\dagger = \overline{\mathbf{u}^T} = \begin{bmatrix} \overline{u_1} & \overline{u_2} & \cdots & \overline{u_n} \end{bmatrix}$$

Definition: *(Norm and Distance in Complex Vector Spaces)*

Let V be a complex vector space equipped with an inner product $\langle\, , \rangle$. Then the **norm** (or **length**) of any vector $\mathbf{v} \in V$ is defined by

$$\|\mathbf{v}\| = \sqrt{\langle \mathbf{v}, \mathbf{v} \rangle} \qquad (28.9$$

and the **distance** $d(\mathbf{u}, \mathbf{v})$ between the vectors $\mathbf{u}$ and $\mathbf{v}$ in V is given by

$$d(\mathbf{u}, \mathbf{v}) = \|\mathbf{u} - \mathbf{v}\| \qquad (28.1$$

Strictly speaking, the norm defined in (28.9) is an **induced norm** on V since its definition follows directly from the specified inner product on V.

Theorem 28.1: *(Properties of the Norm)*

Let V be a complex inner product space with the induced norm $\| \quad \|$. Then the induced norm satisfies the following properties:

(a) $\|\mathbf{v}\| \geq 0$

(b) $\|\mathbf{v}\| = 0$ if and only if $\mathbf{v} = \mathbf{0}$.

(c) $\|c\mathbf{v}\| = |c|\|\mathbf{v}\|$ for any complex scalar c.

(d) For any vectors $\mathbf{u}, \mathbf{v} \in V$,

$$|\langle \mathbf{u}, \mathbf{v} \rangle| \leq \|\mathbf{u}\|\|\mathbf{v}\| \quad \text{(Cauchy-Schwarz Inequality)} \qquad (28.1$$

(e) For any vectors $\mathbf{u}, \mathbf{v} \in V$,

$$\|\mathbf{u} + \mathbf{v}\| \leq \|\mathbf{u}\| + \|\mathbf{v}\| \quad \text{(Triangle Inequality)} \qquad (28.1$$

Proof: The proofs of properties (a) and (b) are obvious. To establish (c), observe that

$$\|c\mathbf{v}\|^2 = \langle c\mathbf{v}, c\mathbf{v}\rangle = \bar{c}c\langle \mathbf{v}, \mathbf{v}\rangle = |c|^2\|\mathbf{v}\|^2$$

from which (c) follows immediately.

We now seek to establish property (d). If $\mathbf{u} = \mathbf{0}$ then there is nothing to show, since we obtain $0 \le 0$, which is certainly true. So, let's assume that $\mathbf{u} \neq \mathbf{0}$. For any *real number* λ, we have

$$\langle \lambda\mathbf{u} + \mathbf{v}, \lambda\mathbf{u} + \mathbf{v}\rangle \ge 0$$

Next, we expand the left-hand side, to obtain

$$\langle \mathbf{u}, \mathbf{u}\rangle\lambda^2 + 2(\text{Re}\langle \mathbf{u}, \mathbf{v}\rangle)\lambda + \langle \mathbf{v}, \mathbf{v}\rangle \ge 0$$

where $\text{Re}\langle \mathbf{u}, \mathbf{v}\rangle$ denotes the real part of $\langle \mathbf{u}, \mathbf{v}\rangle$. But, this expression also implies that

$$\langle \mathbf{u}, \mathbf{u}\rangle\lambda^2 + 2|\langle \mathbf{u}, \mathbf{v}\rangle|\lambda + \langle \mathbf{v}, \mathbf{v}\rangle \ge 0$$

since $|\langle \mathbf{u}, \mathbf{v}\rangle| \ge \text{Re}\langle \mathbf{u}, \mathbf{v}\rangle$. Now, the last expression is a quadratic inequality of the form $ax^2 + bx + c \ge 0$ where $a = \langle \mathbf{u}, \mathbf{u}\rangle$, $b = 2|\langle \mathbf{u}, \mathbf{v}\rangle|$ and $c = \langle \mathbf{v}, \mathbf{v}\rangle$. So, the corresponding quadratic equation $ax^2 + bx + c = 0$ must have either complex conjugate roots, or a repeated real root. Consequently, the discriminant $\Delta = b^2 - 4ac$ must be less than or equal to zero. Thus,

$$\Delta = b^2 - 4ac = 4\left[|\langle \mathbf{u}, \mathbf{v}\rangle|^2 - \langle \mathbf{u}, \mathbf{u}\rangle\langle \mathbf{v}, \mathbf{v}\rangle\right] \le 0$$

which, in turn, implies

$$|\langle \mathbf{u}, \mathbf{v}\rangle|^2 \le \langle \mathbf{u}, \mathbf{u}\rangle\langle \mathbf{v}, \mathbf{v}\rangle$$

from which (d) follows immediately. The proof of (e) is left as an exercise. $\square$.

Exercise Set 28

1. Given the vectors in $\mathbb{C}^3$:

$$\mathbf{u} = \begin{bmatrix} 1 \\ i \\ 0 \end{bmatrix}, \quad \mathbf{v} = \begin{bmatrix} i \\ 0 \\ 1+i \end{bmatrix}, \quad \mathbf{w} = \begin{bmatrix} 1 \\ 0 \\ 1 \end{bmatrix}$$

compute each of the following:
(a) $\langle \mathbf{u}, \mathbf{v}\rangle$

(b) $\langle \mathbf{v}, \mathbf{u} \rangle$
(c) $\langle \mathbf{v}, \mathbf{w} \rangle$
(d) $\| \mathbf{u} + \mathbf{v} \|$
(e) $\| \mathbf{u} \| + \| \mathbf{v} \|$
(f) $\langle \mathbf{u} + \mathbf{v}, \mathbf{u} - \mathbf{v} \rangle$

2. Find the conjugate transpose $A^{\dagger}$ given that

$$\textbf{(a)} \quad A = \begin{bmatrix} 1+i & 3+2i \\ 5i & -2i \end{bmatrix} \qquad \textbf{(b)} \quad A = \begin{bmatrix} 2i & 3+4i & 1 \\ 2+i & 0 & 4i \end{bmatrix}$$

3. Show that the vectors $\mathbf{u}, \mathbf{v}$, and $\mathbf{w}$ of Exercise-1 are *linearly independent.*

4. Given that V is a complex inner product space, show that

$$\langle \mathbf{u}, c_1\mathbf{v} + c_2\mathbf{w} \rangle = c_1\langle \mathbf{u}, \mathbf{v} \rangle + c_2\langle \mathbf{u}, \mathbf{w} \rangle$$

for all vectors $\mathbf{u}, \mathbf{v}, \mathbf{w} \in V$ and for all complex scalars $c_1, c_2 \in \mathbb{C}$.

5. If $\mathbf{u}, \mathbf{v} \in \mathbb{C}^n$, show that
(a)

$$(c\mathbf{u})^{\dagger} = \bar{c}\mathbf{u}^{\dagger} \quad \text{for all complex scalars } c$$

(b)

$$(\mathbf{u} + \mathbf{v})^{\dagger} = \mathbf{u}^{\dagger} + \mathbf{v}^{\dagger}$$

6. Provide a proof of the Triangle Inequality, i.e., relation (28.12). [**Hint**: Expand the right-hand side of the expression $\| \mathbf{u} + \mathbf{v} \|^2 = \langle \mathbf{u} + \mathbf{v}, \mathbf{u} + \mathbf{v} \rangle$.]

7. Given the vectors

$$\mathbf{f}(x) = x + ix^2, \quad \mathbf{g}(x) = 1 + ix, \quad \text{and} \quad \mathbf{h}(x) = \cos x + i \sin x$$

in $\mathbb{C}[0,1]$ with the complex inner product:

$$\langle \mathbf{f}(x), \mathbf{g}(x) \rangle = \int_0^1 \overline{\mathbf{f}(x)}\mathbf{g}(x)dx$$

calculate each of the following:
(a) $\langle \mathbf{f}(x), \mathbf{g}(x) \rangle$
(b) $\langle \mathbf{g}(x), \mathbf{h}(x) \rangle$
(c) $\| \mathbf{g}(x) \|$
(d) $\| \mathbf{h}(x) \|$

8. Let V be a complex inner product space with induced norm $\| \quad \|$. Show that

$$d(\mathbf{x},\mathbf{z}) \leq d(\mathbf{x},\mathbf{y}) + d(\mathbf{y},\mathbf{z})$$

for all vectors $\mathbf{x},\mathbf{y},\mathbf{z} \in V$.

9. *(Pythagorean Theorem)* Let V be a complex inner product space with induced norm $\| \quad \|$. Show that if the vectors $\mathbf{u},\mathbf{v} \in V$ are orthogonal, then

$$\|\mathbf{u}+\mathbf{v}\|^2 = \|\mathbf{u}\|^2 + \|\mathbf{v}\|^2$$

10. *(Generalized Pythagorean Theorem)* Let V be a complex inner product space with induced norm $\| \quad \|$, and let the vectors $\mathbf{v}_1,\mathbf{v}_2,\ldots,\mathbf{v}_n \in V$ be pairwise orthogonal; that is,

$$\langle \mathbf{v}_j,\mathbf{v}_k \rangle = 0 \quad \text{for } j \neq k$$

Show that

$$\left\|\mathbf{v}_1+\mathbf{v}_2+\ldots+\mathbf{v}_n\right\|^2 = \|\mathbf{v}_1\|^2 + \|\mathbf{v}_2\|^2 + \ldots + \|\mathbf{v}_n\|^2$$

11. Let A and B be square matrices of order n with complex elements. Show that:

(a)

$$(A+B)^\dagger = A^\dagger + B^\dagger$$

(b)

$$(AB)^\dagger = B^\dagger A^\dagger$$

(c)

$$(cA)^\dagger = \bar{c}A^\dagger \quad \text{for any complex scalar } c$$

(d)

$$(A^\dagger)^\dagger = A$$

12. If A is a square complex matrix, then show that

(a)

$$A^\dagger = \left(\overline{A}\right)^T$$

(b)

$$\det(A^\dagger) = \overline{\det(A)}$$

29. Unitary and Hermitian Matrices

In this lesson, we introduce two special types of complex matrices called **unitary** and **Hermitian** matrices. Such matrices have many applications in the physical sciences, and also prove useful in the process of diagonalizing complex matrices.

Unitary matrices are the complex analogues of real orthogonal matrices. From a previous lesson, recall that a real matrix A is orthogonal if $A^{-1} = A^T$. On the other hand, a complex matrix A is said to be **unitary** if it satisfies the property $A^{-1} = A^{\dagger}$. This makes sense since operation of conjugate transposition plays a role for complex matrices which is similar to the role that the operation of transposition plays for real matrices.

Definition: *(Unitary Matrix)*

A square complex matrix U is **unitary** if

$$UU^{\dagger} = U^{\dagger}U = I \tag{29.1}$$

or equivalently, if

$$U^{-1} = U^{\dagger} \tag{29.2}$$

This definition implies that every unitary matrix with *real elements* is also an orthogonal matrix as well. Unitary matrices enjoy many useful properties which we now discuss.

Theorem 29.1: *(Properties of Unitary Matrices)*

Let U be a square matrix of order n. Then U is unitary if and only if its column vectors form an orthonormal set with respect to the Euclidean inner product on $\mathbb{C}^n$.

Proof: Our proof will parallel the one given for orthogonal matrices. First, let's assume that U is unitary, so that $U^{\dagger}U = I$. Now, let

$$U = \begin{bmatrix} \mathbf{u}_1 | & \mathbf{u}_2 | & \cdots & |\mathbf{u}_n \end{bmatrix} = \begin{bmatrix} u_{11} & u_{12} & \cdots & u_{1n} \\ u_{21} & u_{22} & \cdots & u_{2n} \\ \vdots & \vdots & & \vdots \\ u_{n1} & u_{n2} & \cdots & u_{nn} \end{bmatrix}$$

so we must have

$$U^\dagger U = \begin{bmatrix} \overline{u_{11}} & \overline{u_{21}} & \cdots & \overline{u_{n1}} \\ \overline{u_{12}} & \overline{u_{22}} & \cdots & \overline{u_{n2}} \\ \vdots & \vdots & & \vdots \\ \overline{u_{1n}} & \overline{u_{2n}} & \cdots & \overline{u_{nn}} \end{bmatrix} \begin{bmatrix} u_{11} & u_{12} & \cdots & u_{1n} \\ u_{21} & u_{22} & \cdots & u_{2n} \\ \vdots & \vdots & & \vdots \\ u_{n1} & u_{n2} & \cdots & u_{nn} \end{bmatrix}$$

$$= \begin{bmatrix} \langle \mathbf{u}_1, \mathbf{u}_1 \rangle & \langle \mathbf{u}_1, \mathbf{u}_2 \rangle & \cdots & \langle \mathbf{u}_1, \mathbf{u}_n \rangle \\ \langle \mathbf{u}_2, \mathbf{u}_1 \rangle & \langle \mathbf{u}_2, \mathbf{u}_2 \rangle & \cdots & \langle \mathbf{u}_2, \mathbf{u}_n \rangle \\ \vdots & \vdots & & \vdots \\ \langle \mathbf{u}_n, \mathbf{u}_1 \rangle & \langle \mathbf{u}_n, \mathbf{u}_2 \rangle & \cdots & \langle \mathbf{u}_n, \mathbf{u}_n \rangle \end{bmatrix} = I \qquad (29.3$$

Since the corresponding elements of equal matrices must themselves be equal, then

$$\langle \mathbf{u}_i, \mathbf{u}_j \rangle = \delta_{ij} \qquad (29.4$$

so column vectors of U form an orthonormal set. On the other hand, if the column vectors of U are orthogonal, then (29.4) is satisfied, and (29.3) is satisfied as well, so $U^\dagger U = I$ and consequently, $U^{-1} = U^\dagger$. □.

Example-1: Show that the matrix

$$U = \frac{1}{\sqrt{3}} \begin{bmatrix} -1 & 1+i \\ 1-i & 1 \end{bmatrix}$$

is unitary.

Solution: Observe that

$$UU^\dagger = \frac{1}{\sqrt{3}} \begin{bmatrix} -1 & 1+i \\ 1-i & 1 \end{bmatrix} \cdot \frac{1}{\sqrt{3}} \begin{bmatrix} -1 & 1+i \\ 1-i & 1 \end{bmatrix} = \begin{bmatrix} 1 & 0 \\ 0 & 1 \end{bmatrix} = I$$

Since $UU^\dagger = I$, then $U^{-1} = U^\dagger$ and U is unitary.

Recall that a real (square) matrix A is said to be symmetric if $A^T = A$. In the complex realm, however, we say that a complex square matrix A is **Hermitian** if it is identical to its conjugate transpose; that is, if $A = A^\dagger$.

Definition: *(Hermitian Matrix)*

A square matrix A is said to be **Hermitian** if

$$A = A^\dagger \qquad (29.5$$

Since Hermitian matrices are the complex cousins of symmetric matrices, we should suspect that Hermitian matrices will share some of the properties of their real counterparts. The next two theorems will verify that our suspicions are correct.

Theorem 29.2: *(Hermitian Matrices Have Real Eigenvalues)*

If A is a Hermitian matrix, then all of its eigenvalues are real.

Proof: Assume that λ is an eigenvalue of A with corresponding eigenvector $\mathbf{x}$, so we have

$$A\mathbf{x} = \lambda\mathbf{x} \tag{29.6}$$

Taking the conjugate transpose of both sides this equation, while keeping in mind that $A^\dagger = A$, we get

$$\mathbf{x}^\dagger A = \bar{\lambda}\mathbf{x}^\dagger \tag{29.7}$$

Next, we multiply both sides of (29.6) on the left by $\mathbf{x}^\dagger$, and both sides of (29.7) on the right by $\mathbf{x}$, to obtain, respectively

$$\mathbf{x}^\dagger A\mathbf{x} = \lambda\mathbf{x}^\dagger\mathbf{x}$$

$$\mathbf{x}^\dagger A\mathbf{x} = \bar{\lambda}\mathbf{x}^\dagger\mathbf{x}$$

Subtracting the last two equations, gives

$$(\lambda - \bar{\lambda})\mathbf{x}^\dagger\mathbf{x} = (\lambda - \bar{\lambda})\|\mathbf{x}\|^2 = 0$$

But since $\mathbf{x}$ is an eigenvector, we must have $\|\mathbf{x}\| \neq 0$, and the last equation implies $\lambda = \bar{\lambda}$, so λ must be a real number. □.

Theorem 29.3: *(Eigenvectors of a Hermitian Matrix)*

If A is a Hermitian matrix, then the eigenvectors that correspond to distinct eigenvalues of A are *orthogonal* with respect to the Euclidean inner product on $\mathbb{C}^n$.

Proof: Let λ_1 and λ_2 be distinct eigenvalues of A with eigenvectors $\mathbf{x}_1$ and $\mathbf{x}_2$ respectively, so that

$$A\mathbf{x}_1 = \lambda_1\mathbf{x}_1 \tag{29.8}$$

$$A\mathbf{x}_2 = \lambda_2\mathbf{x}_2 \tag{29.9}$$

To establish the result of the theorem, we must show that

$$\langle \mathbf{x}_1, \mathbf{x}_2 \rangle = \mathbf{x}_1^\dagger \mathbf{x}_2 = 0$$

where $\langle \mathbf{x}_1, \mathbf{x}_2 \rangle$ denotes the Euclidean inner product. We first take the conjugate transpose of (29.8), remembering that the eigenvalue λ_1 is real, to obtain

$$\mathbf{x}_1^\dagger A = \lambda_1 \mathbf{x}_1^\dagger \qquad (29.1$$

Next, we multiply (29.9) on the left by $\mathbf{x}^\dagger$, and we multiply (29.10) on the right by $\mathbf{x}_2$ to get

$$\mathbf{x}_1^\dagger A \mathbf{x}_2 = \lambda_2 \mathbf{x}_1^\dagger \mathbf{x}_2$$
$$\mathbf{x}_1^\dagger A \mathbf{x}_2 = \lambda_1 \mathbf{x}_1^\dagger \mathbf{x}_2$$

Finally, subtracting the last two equations gives

$$(\lambda_2 - \lambda_1)\mathbf{x}_1^\dagger \mathbf{x}_2 = (\lambda_2 - \lambda_1)\langle \mathbf{x}_1, \mathbf{x}_2 \rangle = 0$$

By hypothesis $(\lambda_2 - \lambda_1) \neq 0$, so we must have $\langle \mathbf{x}_1, \mathbf{x}_2 \rangle = 0$; that is, the eigenvectors $\mathbf{x}_1$ and $\mathbf{x}_2$ are orthogonal. □.

Example-2: Show that the matrix

$$A = \begin{bmatrix} 1 & 4+2i \\ 4-2i & 1 \end{bmatrix}$$

is Hermitian, its eigenvalues are real and distinct, and consequently, its eigenvectors are orthogonal.

Solution: The matrix is Hermitian, since

$$A^\dagger = \overline{(A^T)} = \begin{bmatrix} \overline{1} & \overline{4-2i} \\ \overline{4+2i} & \overline{1} \end{bmatrix} = \begin{bmatrix} 1 & 4+2i \\ 4-2i & 1 \end{bmatrix} = A$$

Next, we find the eigenvalues of A by setting $|A - \lambda I| = 0$ to obtain

$$\begin{vmatrix} 1-\lambda & 4+2i \\ 4-2i & 1-\lambda \end{vmatrix} = \lambda^2 - 2\lambda - 19 = 0$$

whose solution is $\lambda_1 = 1 - 2\sqrt{5}$ and $\lambda_2 = 1 + 2\sqrt{5}$ so the eigenvalues are real. As in the past, we can find the eigenvectors of A by solving the linear system $(A - \lambda I)\mathbf{x} = \mathbf{0}$ for each value of λ. In doing so, we find that the required eigenvectors are

$$\mathbf{x}_1 = \begin{bmatrix} \left(-\frac{2}{5} - \frac{1}{5}i\right)\sqrt{5} \\ 1 \end{bmatrix} \text{ and } \mathbf{x}_2 = \begin{bmatrix} \left(\frac{2}{5} + \frac{1}{5}i\right)\sqrt{5} \\ 1 \end{bmatrix}$$

Observe that the eigenvectors are orthogonal since

$$\langle \mathbf{x}_1, \mathbf{x}_2 \rangle = \mathbf{x}_1^\dagger \mathbf{x}_2 = \begin{bmatrix} \left(-\frac{2}{5} + \frac{1}{5}i\right)\sqrt{5} & 1 \end{bmatrix} \begin{bmatrix} \left(\frac{2}{5} + \frac{1}{5}i\right)\sqrt{5} \\ 1 \end{bmatrix} = 0$$

If A is a Hermitian matrix of order n such that its eigenvalues $\lambda_1, \lambda_2, \ldots, \lambda_n$ are *distinct*, then Theorem 29.3 ensures that A will have n pairwise *orthogonal* eigenvectors, say $\mathbf{x}_1, \mathbf{x}_2, \ldots, \mathbf{x}_n$. Now, we can normalize these eigenvectors by taking

$$\mathbf{u}_1 = \frac{1}{\|\mathbf{x}_1\|}\mathbf{x}_1, \ \mathbf{u}_2 = \frac{1}{\|\mathbf{x}_2\|}\mathbf{x}_2, \ldots, \mathbf{u}_n = \frac{1}{\|\mathbf{x}_n\|}\mathbf{x}_n,$$

so we will have an *orthonormal* set $\{\mathbf{u}_1, \mathbf{u}_2, \ldots, \mathbf{u}_n\}$ of eigenvectors of A that satisfy the relation

$$\langle \mathbf{u}_i, \mathbf{u}_j \rangle = \delta_{ij}$$

We are then guaranteed that if we form the $n \times n$ unitary matrix U whose columns are formed from the normalized eigenvectors $\mathbf{u}_1, \mathbf{u}_2, \ldots, \mathbf{u}_n$, then the matrix $U^{-1}AU$ will be diagonal.

In the event that a Hermitian matrix A of order n does not have n distinct eigenvalues, then at least one of its eigenvalues, say λ, has an algebraic multiplicity of $k > 1$. In this case, in a manner similar to the case of real symmetric matrices, it may be shown that there still exist k *linearly independent* eigenvectors corresponding to the eigenvalue λ.

In other words, if an eigenvalue λ of A is repeated, then we can still use the Gram-Schmidt process to construct an *orthonormal* set of eigenvectors corresponding to λ. So, even if a Hermitian matrix A doesn't have n distinct eigenvalues, we can still diagonalize A with a unitary matrix U.

Theorem 29.4: *(All Hermitian Matrices are Unitarily Diagonalizable)*

Every Hermitian matrix A of order n is **unitarily diagonalizable**; that is there exists a unitary matrix U such that

$$U^{-1}AU = D \tag{29.11}$$

is a diagonal matrix, where the n eigenvalues of A lie along the main diagonal of D.

Example-3: Given the Hermitian matrix

$$A = \begin{bmatrix} 0 & 1+i & 0 \\ 1-i & 0 & 0 \\ 0 & 0 & 1 \end{bmatrix}$$

Determine:
(**a**) the eigenvalues of A.
(**b**) the corresponding eigenvectors.
(**c**) an orthonormal set of eigenvectors of A.
(**d**) a unitary matrix U that diagonalizes A.

Solution:
(**a**) In order to find the eigenvalues of A, we set $\det(A - \lambda I) = 0$. In doing so, we obtain

$$\begin{vmatrix} -\lambda & 1+i & 0 \\ 1-i & -\lambda & 0 \\ 0 & 0 & 1-\lambda \end{vmatrix} = -\lambda^3 + \lambda^2 + 2\lambda - 2 = 0$$

Consequently, $\lambda^3 - \lambda^2 - 2\lambda + 2 = (\lambda - 1)(\lambda^2 - 2) = 0$ and we find the distinct eigenvalues $\lambda_1 = 1$, $\lambda_2 = -\sqrt{2}$ and $\lambda_3 = \sqrt{2}$.
(**b**) We find the eigenvectors of A by solving the linear system $(A - \lambda I)\mathbf{x} = \mathbf{0}$ for each value of λ. After some work, we find that the required eigenvectors are

$$\mathbf{x}_1 = \begin{bmatrix} 0 \\ 0 \\ 1 \end{bmatrix}, \quad \mathbf{x}_2 = \begin{bmatrix} \left(-\frac{1}{2} - \frac{1}{2}i\right)\sqrt{2} \\ 1 \\ 0 \end{bmatrix}, \quad \text{and} \quad \mathbf{x}_3 = \begin{bmatrix} \left(\frac{1}{2} + \frac{1}{2}i\right)\sqrt{2} \\ 1 \\ 0 \end{bmatrix}$$

(**c**) Next, we normalize each of the eigenvectors by setting

$$\mathbf{u}_1 = \frac{\mathbf{x}_1}{\|\mathbf{x}_1\|} = \begin{bmatrix} 0 \\ 0 \\ 1 \end{bmatrix}, \quad \mathbf{u}_2 = \frac{\mathbf{x}_2}{\|\mathbf{x}_2\|} = \begin{bmatrix} \left(-\frac{1}{2} - \frac{1}{2}i\right) \\ 1/\sqrt{2} \\ 0 \end{bmatrix},$$

$$\mathbf{u}_3 = \frac{\mathbf{x}_3}{\|\mathbf{x}_3\|} = \begin{bmatrix} \left(\frac{1}{2} + \frac{1}{2}i\right) \\ 1/\sqrt{2} \\ 0 \end{bmatrix}$$

By Theorem 29.3, we are guaranteed that the vectors $\mathbf{u}_1, \mathbf{u}_2$ and $\mathbf{u}_2$ will form an orthonormal set.
(**d**) The unitary matrix U is the matrix whose columns are formed from our orthonormal set of eigenvectors; that is,

$$U = \begin{bmatrix} \mathbf{u}_1 | & \mathbf{u}_2 | & \mathbf{u}_3 \end{bmatrix} = \begin{bmatrix} 0 & (-\frac{1}{2} - \frac{1}{2}i) & (\frac{1}{2} + \frac{1}{2}i) \\ 0 & 1/\sqrt{2} & 1/\sqrt{2} \\ 1 & 0 & 0 \end{bmatrix}$$

In fact, a direct calculation shows that

$$U^{-1}AU = U^{\dagger}AU = \begin{bmatrix} 1 & 0 & 0 \\ 0 & -\sqrt{2} & 0 \\ 0 & 0 & \sqrt{2} \end{bmatrix}$$

as it should.

We now write down a general procedure for diagonalizing any Hermitian matrix which closely parallels the corresponding procedure for symmetric matrices.

Procedure for Diagonalizing a Hermitian Matrix

Given a Hermitian matrix A of order n, to find a unitary matrix U that diagonalizes A, we proceed as follows:

1. Find the eigenvalues of A and note their respective multiplicities.

2. If λ is a **simple eigenvalue** of A (with multiplicity $k = 1$), and has a corresponding eigenvector $\mathbf{x}$, then simply normalize this eigenvector by taking

$$\mathbf{u} = \frac{\mathbf{x}}{\|\mathbf{x}\|}$$

Repeat this step for each of the remaining simple eigenvalues of A.

3. If λ is a **repeated eigenvalue** of A (with multiplicity $k > 1$) with eigenvectors $\mathbf{x}_1, \mathbf{x}_2, \ldots, \mathbf{x}_k$, then apply the Gram-Schmidt procedure to this set of eigenvectors to obtain an *orthonormal* set of eigenvectors $\mathbf{u}_1, \mathbf{u}_2, \ldots, \mathbf{u}_k$. Repeat this step for each of the remaining repeated eigenvalues of A.

4. Construct a matrix U whose columns are formed from the eigenvectors obtained in Step 2 and Step 3. The matrix U will be a *unitary matrix* that diagonalizes A; that is $U^{-1}AU$ will be a diagonal matrix whose elements along its main diagonal will be the eigenvalues of A.

Exercise Set 29

In Exercises 1-3, find the conjugate transpose $A^\dagger$ of the given matrix A.

1. Find the conjugate transpose $A^\dagger$ of the given matrix A.

$$\text{(a)}\quad A = \begin{bmatrix} 1 & 1+i \\ 2i & 3 \end{bmatrix} \qquad \text{(b)}\quad A = \begin{bmatrix} 0 & 1+i & i \\ i & 0 & -i \\ 2i & 0 & 4i \end{bmatrix}$$

2. Verify that each matrix U is unitary by showing that $UU^\dagger = I$.

$$\text{(a)}\quad U = \frac{1}{2}\begin{bmatrix} 1+i & 1+i \\ 1-i & -1+i \end{bmatrix} \qquad \text{(b)}\quad U = \begin{bmatrix} 0 & (-\frac{1}{2}-\frac{1}{2}i) & (\frac{1}{2}+\frac{1}{2}i) \\ 0 & 1/\sqrt{2} & 1/\sqrt{2} \\ 1 & 0 & 0 \end{bmatrix}$$

3. Verify that each matrix A is Hermitian by showing that $A^\dagger = A$.

$$\text{(a)}\quad A = \begin{bmatrix} 1 & i & 1+i \\ -i & -1 & 3 \\ 1-i & 3 & 0 \end{bmatrix} \qquad \text{(b)}\quad A = \begin{bmatrix} 1 & i \\ -i & 2 \end{bmatrix}$$

4. Find the eigenvalues and eigenvectors of each matrix A. Also, find a unitary matrix U that diagonalizes A.

$$\text{(a)}\quad A = \begin{bmatrix} 1 & i \\ -i & 1 \end{bmatrix} \qquad \text{(b)}\quad A = \begin{bmatrix} 0 & -i & i \\ i & 0 & -i \\ -i & i & 0 \end{bmatrix}$$

$$\text{(c)}\quad A = \begin{bmatrix} 0 & 1-i & 0 \\ 1+i & 0 & 0 \\ 0 & 0 & 1 \end{bmatrix} \qquad \text{(d)}\quad A = \begin{bmatrix} 1 & -2i \\ 2i & 1 \end{bmatrix}$$

5. Let U be a unitary matrix of order n, and consider the linear transformation $T : \mathbb{C}^n \to \mathbb{C}^n$ defined by

$$T(\mathbf{x}) = U\mathbf{x} \quad \text{for all } \mathbf{x} \in \mathbb{C}^n$$

We call T a **unitary transformation**. Show that for all vectors $\mathbf{x}, \mathbf{y} \in \mathbb{C}^n$

$$\langle T\mathbf{x}, T\mathbf{y} \rangle = \langle \mathbf{x}, \mathbf{y} \rangle$$

Thus, the inner product of any two vectors is *invariant* under a unitary transformation.

6. Show that if A and B are Hermitian matrices the same order, and if $AB = BA$, then AB is also Hermitian.

7. Let $A = [a_{ij}]$ be a Hermitian matrix of order n. Show that

$$a_{ii} \in \mathbb{R} \quad \text{for all } i = 1,2,3,\ldots,n$$
$$a_{ij} = \overline{a_{ji}} \quad \text{for all } i \neq j$$

8. Show that the eigenvalues of a unitary matrix all have a complex modulus of unity.

9. Show that if U is a unitary matrix, and H is a Hermitian matrix, then the matrix $U^{-1}HU$ is Hermitian.

10. Given that H is a Hermitian matrix, show that $(H - iI)(H + iI)^{-1}$ is unitary.

11. Show that if U_1 and U_2 are unitary matrices the same order, then U_1U_2 is also a unitary.

12. A square matrix A is said to be **anti-Hermitian** if $A^{\dagger} = -A$. Show that the matrix

$$A = \begin{bmatrix} i & 1+i \\ -1+i & 2i \end{bmatrix}$$

is anti-Hermitian.

13. Prove that the eigenvalues of an anti-Hermitian matrix are either pure imaginary or zero.

14. A square matrix A is said to be **normal** if $AA^{\dagger} = A^{\dagger}A$. Show that the matrix

$$A = \begin{bmatrix} 1 & i & 1+i \\ -i & 1 & 2i \\ 1-i & -2i & 1 \end{bmatrix}$$

is normal.

15. Prove that every unitary matrix is a normal.

Unit VIII: Advanced Topics

Marie Ennemond Camille Jordan (1838-1922)

Camille Jordan was an important French mathematician and engineer who made significant contributions to topology, linear algebra, complex analysis, group theory and measure theory. Born in Lyon, he was later trained as an engineer at the famous Ecole Polytechnique. He was appointed as professor of analysis at the Ecole Polytechnique in 1876, and later served as a professor at the College de France in 1883.

In his important work *Traitie des substitutions*, which first appeared in print in 1870, dealt primarily with permutation groups; however, in Book II of this work, Jordan employed the notions of matrix similarity and the characteristic equation of a matrix to show that any matrix can be transformed into a special, block diagonal form, known today as its Jordan canonical form, or normal form.

Today, several important theorems of modern mathematics bear Jordan's name. These include the Jordan curve theorem in topology, and the Jordan-Holder theorem in group theory.

30. Powers of Matrices

As we shall see, in many applications, it is necessary to compute the powers of a square matrix. Given a square matrix A, and a non-negative integer k, we define

$$A^k = A \cdot A \cdot A \cdots A \quad (k \text{ factors, and } k > 0) \tag{30.1}$$

$$A^0 = I \tag{30.2}$$

If A is non-singular, then we can also define negative powers of A as

$$A^{-k} = (A^{-1})^k = A^{-1} \cdot A^{-1} \cdot A^{-1} \cdots A^{-1} \quad (k \text{ factors, and } k > 0) \tag{30.3}$$

In general, the computation of the powers of a square matrix can be quite tedious, especially if either the given matrix is of large order, or if the exponent k is particularly large. Suppose, however, that A can be diagonalized by a matrix P so that

$$P^{-1}AP = D$$

where D is a diagonal matrix with the eigenvalues of A along its main diagonal. Then we find that

$$\begin{aligned} A^1 &= A = PDP^{-1} \\ A^2 &= PDP^{-1}(PDP^{-1}) = PD^2P^{-1} \\ &\vdots \\ A^k &= PD^kP^{-1} \end{aligned}$$

Since D is a diagonal matrix, D^k is easily calculated by raising each element along its main diagonal to the k-th power. Let's see how this works.

Example-1: Given the matrix

$$A = \begin{bmatrix} 1 & 2 \\ 2 & 1 \end{bmatrix},$$

compute A^5.

Solution: Proceeding as in the past, we find that the eigenvalues of A are $\lambda_1 = 1$, $\lambda_2 = 3$ with the corresponding eigenvectors

$$\mathbf{x}_1 = \begin{bmatrix} -1 \\ 1 \end{bmatrix} \quad \text{and} \quad \mathbf{x}_2 = \begin{bmatrix} 1 \\ 1 \end{bmatrix}$$

Consequently, a matrix that diagonalizes A is

$$P = \begin{bmatrix} -1 & 1 \\ 1 & 1 \end{bmatrix}$$

Thus,

$$A^5 = P\begin{bmatrix} 1 & 0 \\ 0 & 3 \end{bmatrix}^5 P^{-1} = \begin{bmatrix} -1 & 1 \\ 1 & 1 \end{bmatrix}\begin{bmatrix} 1^5 & 0 \\ 0 & 3^5 \end{bmatrix}\begin{bmatrix} -\frac{1}{2} & \frac{1}{2} \\ \frac{1}{2} & \frac{1}{2} \end{bmatrix}$$

$$= \begin{bmatrix} 122 & 121 \\ 121 & 122 \end{bmatrix}$$

We can now summarize our work in the form of theorem.

Theorem 30.1: *(Powers of a Matrix)*

Let A be a square matrix of order n which can be diagonalized by the matrix P so that $P^{-1}AP = D$ is a diagonal matrix whose diagonal elements are the eigenvalues $\lambda_1, \lambda_2, \ldots, \lambda_n$ of A. Then for any integer $k \geq 0$,

$$A^k = PD^kP^{-1} = P\begin{bmatrix} \lambda_1^k & 0 & \cdots & 0 \\ 0 & \lambda_2^k & \cdots & 0 \\ \vdots & \vdots & \cdots & \vdots \\ 0 & 0 & \cdots & \lambda_n^k \end{bmatrix}P^{-1} \quad (30.4$$

It is clear that the procedure given in the previous theorem breaks down if A is not diagonalizable. An alternative method, however, can be used that involves the Cayley-Hamilton theorem, which we now discuss.

Theorem 30.2: *(Cayley-Hamilton)*

If A is a square matrix of order n whose characteristic equation is

$$\lambda^n + a_{n-1}\lambda^{n-1} + a_{n-2}\lambda^{n-2} + \ldots + a_1\lambda + a_0 = 0$$

then the corresponding polynomial equation in A is satisfied as well; i.e.,

$$A^n + a_{n-1}A^{n-1} + a_{n-2}A^{n-2} + \ldots + a_1A + a_0I = 0$$

In other words, every square matrix satisfies its own characteristic equation.

Proof: Let A be a square matrix of order n whose characteristic polynomial is

$$\Delta(\lambda) = \det(A - \lambda I) = (-1)^n(\lambda^n + a_{n-1}\lambda^{n-1} + a_{n-2}\lambda^{n-2} + \ldots + a_1\lambda + a_0)$$

Now, if C^T is the transpose of the matrix of cofactors of $A - \lambda I$, then each element of C^T will be a polynomial in λ whose degree does not exceed $n - 1$. So, we can write

$$C^T = C^T(\lambda) = C_{n-1}\lambda^{n-1} + C_{n-2}\lambda^{n-2} + C_{n-3}\lambda^{n-3} + \ldots + C_1\lambda + C_0$$

where $C_{n-1}, \ldots, C_0$ are themselves square matrices of order n.

Recalling (8.2), we have

$$(A - \lambda I)C^T(\lambda) = \det(A - \lambda I) \cdot I$$

where I is the identity matrix of order n. Consequently,

$$(A - \lambda I)\left(C_{n-1}\lambda^{n-1} + \ldots + C_1\lambda + C_0\right) = (-1)^n(\lambda^n + \ldots + a_1\lambda + a_0) \cdot I$$

If we equate the coefficients of corresponding powers of λ, we obtain

$$\begin{aligned} -C_{n-1} &= (-1)^n I \\ -C_{n-2} + AC_{n-1} &= (-1)^n a_{n-1} I \\ -C_{n-3} + AC_{n-2} &= (-1)^n a_{n-2} I \\ &\vdots \\ AC_0 &= (-1)^n a_0 I \end{aligned}$$

Next, we multiply the above equations (on the left) $A^n, A^{n-1}, \ldots, A, I$, respectively to get

$$\begin{aligned} -A^n C_{n-1} &= (-1)^n A^n \\ -A^{n-1}C_{n-2} + A^n C_{n-1} &= (-1)^n a_{n-1} A^{n-1} \\ -A^{n-2}C_{n-3} + A^{n-1}C_{n-2} &= (-1)^n a_{n-2} A^{n-2} \\ &\vdots \\ AC_0 &= (-1)^n a_0 I \end{aligned}$$

and then add, we finally obtain

$$A^n + a_{n-1}A^{n-1} + a_{n-2}A^{n-2} + \ldots + a_1 A + a_0 I = 0$$

as claimed. □.

Example-2: Verify the Cayley-Hamilton theorem for the matrix

$$A = \begin{bmatrix} 1 & 0 \\ 1 & 2 \end{bmatrix}$$

Solution: The characteristic equation of A is $\lambda^2 - 3\lambda + 2 = 0$. So, since A must satisfy its own characteristic equation, we should have

$$A^2 - 3A + 2I = 0$$

Let's check this by direct computation. We find that

$$A^2 - 3A + 2I = \begin{bmatrix} 1 & 0 \\ 1 & 2 \end{bmatrix}^2 - 3\begin{bmatrix} 1 & 0 \\ 1 & 2 \end{bmatrix} + 2\begin{bmatrix} 1 & 0 \\ 0 & 1 \end{bmatrix}$$

$$= \begin{bmatrix} 1 & 0 \\ 3 & 4 \end{bmatrix} - \begin{bmatrix} 3 & 0 \\ 3 & 6 \end{bmatrix} + \begin{bmatrix} 2 & 0 \\ 0 & 2 \end{bmatrix}$$

$$= \begin{bmatrix} 0 & 0 \\ 0 & 0 \end{bmatrix}$$

as it should.

In order to see how the Cayley-Hamilton theorem can be used to find any power of a square matrix, take a look at the next example.

Example-3: Given the matrix

$$A = \begin{bmatrix} 1 & 0 \\ 1 & 2 \end{bmatrix}$$

from Example-2, use the Cayley-Hamilton theorem to compute A^4.

Solution: From the previous example, we have $A^2 - 3A + 2I = 0$. Solving for A^2 we get

$$A^2 = 3A - 2I$$

Consequently,

$$A^4 = A^2A^2 = (3A - 2I)(3A - 2I) = 9A^2 - 12A + 4I$$

$$= 9\begin{bmatrix} 1 & 0 \\ 3 & 4 \end{bmatrix} - 12\begin{bmatrix} 1 & 0 \\ 1 & 2 \end{bmatrix} + 4\begin{bmatrix} 1 & 0 \\ 0 & 1 \end{bmatrix}$$

$$= \begin{bmatrix} 1 & 0 \\ 15 & 16 \end{bmatrix}$$

If A is a non-singular matrix, then the Cayley-Hamilton theorem can also be used to find A^{-1}. To see this, assume that the characteristic polynomial of A is given by

$$\Delta(\lambda) = \det(A - \lambda I) = (-1)^n\left(\lambda^n + a_{n-1}\lambda^{n-1} + a_{n-2}\lambda^{n-2} + \ldots + a_1\lambda + a_0\right)$$

If we set $\lambda = 0$, we find that

$$\Delta(0) = \det(A) = (-1)^n a_0$$

so we must have $a_0 \neq 0$ whenever A is non-singular. Now, according to the Cayley-Hamilton theorem, A must satisfy its own characteristic equation; that is,

$$A^n + a_{n-1}A^{n-1} + a_{n-2}A^{n-2} + \ldots + a_1 A + a_0 I = 0$$

which implies that

$$I = -\frac{1}{a_0}\left(A^n + a_{n-1}A^{n-1} + a_{n-2}A^{n-2} + \ldots + a_1 A\right)$$

Multiplying both sides of the last equation by A^{-1} gives

$$A^{-1} = -\frac{1}{a_0}\left(A^{n-1} + a_{n-1}A^{n-2} + a_{n-2}A^{n-3} + \ldots + a_1 I\right) \tag{30.5}$$

Observe that this equation allows us to express the inverse of any non-singular matrix in terms of its integral powers. Additionally, if A is of order n then the highest power of A we must compute is A^{n-1}.

Example-5: Use the Cayley-Hamilton theorem to find the inverse of

$$A = \begin{bmatrix} 1 & 0 & 0 \\ 0 & 2 & 1 \\ -1 & 0 & 1 \end{bmatrix}$$

Solution: We first set $\det(A - \lambda I) = 0$ to find the characteristic equation of A:

$$\begin{vmatrix} 1-\lambda & 0 & 0 \\ 0 & 2-\lambda & 1 \\ -1 & 0 & 1-\lambda \end{vmatrix} = -\lambda^3 + 4\lambda^2 - 5\lambda + 2 = 0$$

Consequently, the characteristic equation is

$$\lambda^3 - 4\lambda^2 + 5\lambda - 2 = 0$$

By the Cayley-Hamilton theorem, we must have

$$A^3 - 4A^2 + 5A - 2I = 0$$

from which we obtain

$$I = \frac{1}{2}(A^3 - 4A^2 + 5A)$$

Multiplying both sides of the last equation by A^{-1} gives

$$A^{-1} = \frac{1}{2}(A^2 - 4A + 5I)$$

Now, in order to use the last formula, we need A^2; we find

$$A^2 = \begin{bmatrix} 1 & 0 & 0 \\ 0 & 2 & 1 \\ -1 & 0 & 1 \end{bmatrix} \begin{bmatrix} 1 & 0 & 0 \\ 0 & 2 & 1 \\ -1 & 0 & 1 \end{bmatrix} = \begin{bmatrix} 1 & 0 & 0 \\ -1 & 4 & 3 \\ -2 & 0 & 1 \end{bmatrix}$$

Consequently,

$$A^{-1} = \frac{1}{2}\left(\begin{bmatrix} 1 & 0 & 0 \\ -1 & 4 & 3 \\ -2 & 0 & 1 \end{bmatrix} - 4\begin{bmatrix} 1 & 0 & 0 \\ 0 & 2 & 1 \\ -1 & 0 & 1 \end{bmatrix} + 5\begin{bmatrix} 1 & 0 & 0 \\ 0 & 1 & 0 \\ 0 & 0 & 1 \end{bmatrix} \right)$$

$$= \begin{bmatrix} 1 & 0 & 0 \\ -\frac{1}{2} & \frac{1}{2} & -\frac{1}{2} \\ 1 & 0 & 1 \end{bmatrix}$$

Exercise Set 30

1. For each matrix A, use (30.4) to find A^4.

(a) $A = \begin{bmatrix} 2 & 1 \\ 0 & 1 \end{bmatrix}$ **(b)** $A = \begin{bmatrix} 1 & 2 & 2 \\ 2 & 1 & 2 \\ 2 & 2 & 1 \end{bmatrix}$

2. Verify the Cayley-Hamilton theorem for each of the matrices in Exercise-1

3. For each matrix A, use the Cayley-Hamilton theorem to find A^{-1}.

(a) $A = \begin{bmatrix} 2 & 1 \\ 1 & 2 \end{bmatrix}$ **(b)** $A = \begin{bmatrix} 3 & 2 \\ 1 & 2 \end{bmatrix}$ **(c)** $A = \begin{bmatrix} 1 & 0 & -1 \\ 0 & 1 & 0 \\ 1 & 1 & 0 \end{bmatrix}$

4. If a square matrix A is *diagonalizable*, then use Theorem 30.1 to show that it satisfies its own characteristic equation, thus providing an alternative proof of the Cayley-Hamilton theorem.

31. Functions of a Square Matrix

We now generalize our work in the previous lesson. If we are given any polynomial in a scalar variable x, such as

$$p(x) = a_n x^n + a_{n-1}x^{n-1} + \ldots + a_1 x + a_0$$

then for any square matrix A, we can construct the corresponding matrix function of A by defining

$$p(A) = a_n A^n + a_{n-1}A^{n-1} + \ldots + a_1 A + A_0 I$$

We call $p(A)$ a **matrix polynomial**, and it can be easily computed whenever A can be diagonalized.

Theorem 31.1: *(Evaluating a Matrix Polynomial)*

Let A be a square matrix of order n which can be diagonalized by a matrix P so that $P^{-1}AP = D$ is a diagonal matrix whose diagonal elements are the eigenvalues $\lambda_1, \lambda_2, \ldots, \lambda_n$ of A. Then for any polynomial,

$$p(x) = a_n x^n + a_{n-1}x^{n-1} + \ldots + a_1 x + a_0 \tag{31.1}$$

the corresponding matrix polynomial $p(A)$ is given by

$$p(A) = P[p(D)]P^{-1} = P\begin{bmatrix} p(\lambda_1) & 0 & \cdots & 0 \\ 0 & p(\lambda_2) & \cdots & 0 \\ \vdots & \vdots & \cdots & \vdots \\ 0 & 0 & \cdots & p(\lambda_n) \end{bmatrix}P^{-1} \tag{31.2}$$

Proof: The proof follows immediately from Theorem 30.1. For any integer $k \geq 0$, we have

$$a_k A^k = P(a_k D^k)P^{-1}$$

Adding all such expressions for $k = 0, 1, 2, \ldots, n$, we obtain

$$\begin{aligned} p(A) &= a_n A^n + a_{n-1}A^{n-1} + \ldots + a_1 A + A_0 I \\ &= P(a_n D^n)P^{-1} + P(a_{n-1}D^{n-1})P^{-1} + \ldots + P(a_1 D)P^{-1} + P(a_0 D^0)P^{-1} \\ &= P(a_n D^n + a_{n-1}D^{n-1} + a_1 D + a_0 I)P^{-1} \\ &= P[p(D)]P^{-1} \end{aligned}$$

Consequently,

$$p(A) = P\begin{bmatrix} p(\lambda_1) & 0 & \cdots & 0 \\ 0 & p(\lambda_2) & \cdots & 0 \\ \vdots & \vdots & \cdots & \vdots \\ 0 & 0 & \cdots & p(\lambda_n) \end{bmatrix}P^{-1}$$

as claimed. □.

Example-1: If $g(x) = x^2 - 5x + 6$ and

$$A = \begin{bmatrix} 2 & 1 \\ 1 & 2 \end{bmatrix}$$

then find $g(A)$.

Solution: We find that the eigenvalues of A are $\lambda_1 = 1$ and $\lambda_2 = 3$, with corresponding eigenvectors

$$\mathbf{x}_1 = \begin{bmatrix} -1 \\ 1 \end{bmatrix} \quad \text{and} \quad \mathbf{x}_2 = \begin{bmatrix} 1 \\ 1 \end{bmatrix}$$

Consequently,

$$P = \begin{bmatrix} -1 & 1 \\ 1 & 1 \end{bmatrix} \quad \text{and} \quad P^{-1} = \begin{bmatrix} -\frac{1}{2} & \frac{1}{2} \\ \frac{1}{2} & \frac{1}{2} \end{bmatrix}$$

Using (31.2), we obtain

$$\begin{aligned} g(A) &= P\begin{bmatrix} g(1) & 0 \\ 0 & g(3) \end{bmatrix}P^{-1} \\ &= \begin{bmatrix} -1 & 1 \\ 1 & 1 \end{bmatrix}\begin{bmatrix} 2 & 0 \\ 0 & 0 \end{bmatrix}\begin{bmatrix} -\frac{1}{2} & \frac{1}{2} \\ \frac{1}{2} & \frac{1}{2} \end{bmatrix} \\ &= \begin{bmatrix} 1 & -1 \\ -1 & 1 \end{bmatrix} \end{aligned}$$

Now, we can go even further and generalize (31.2) handle any function $g(x)$ of a scalar x. For example, even if we are presented with functions that are not polynomials, such as $g(x) = e^x$ or $h(x) = \cos x$, we can still, in many cases, compute the corresponding **matrix functions** $g(A) = e^A$ and $h(A) = \cos A$. The basic idea is contained in the following definition.

Definition: *(Matrix Function)*

Let A be a square matrix of order n which can be diagonalized by the matrix P so that $P^{-1}AP = D$ is a diagonal matrix whose diagonal elements are the eigenvalues $\lambda_1, \lambda_2, \ldots, \lambda_n$ of A. Then for any function $f(x)$ of a scalar variable x, we define the corresponding **matrix function** $f(A)$ by

$$f(A) = Pf(D)P^{-1} = P\begin{bmatrix} f(\lambda_1) & 0 & \cdots & 0 \\ 0 & f(\lambda_2) & \cdots & 0 \\ \vdots & \vdots & \cdots & \vdots \\ 0 & 0 & \cdots & f(\lambda_n) \end{bmatrix}P^{-1} \tag{31.3}$$

provided that each of the quantities $f(\lambda_1), f(\lambda_2), \ldots, f(\lambda_n)$ is defined.

Example-2: Given the matrix

$$A = \begin{bmatrix} 1 & 2 \\ 2 & 1 \end{bmatrix}$$

from Example-1, and the function $f(x) = \sqrt{x}$, use (31.3) to compute the square root of A.

Solution: From Example-1, the eigenvalues of A are $\lambda_1 = 1$ and $\lambda_2 = 3$, while

$$P = \begin{bmatrix} -1 & 1 \\ 1 & 1 \end{bmatrix} \quad \text{and} \quad P^{-1} = \begin{bmatrix} -\frac{1}{2} & \frac{1}{2} \\ \frac{1}{2} & \frac{1}{2} \end{bmatrix}$$

Using $f(x) = \sqrt{x}$, and employing (31.3), we can write

$$f(A) = \sqrt{A} = P\sqrt{D}\,P^{-1} = \begin{bmatrix} -1 & 1 \\ 1 & 1 \end{bmatrix}\begin{bmatrix} \sqrt{1} & 0 \\ 0 & \sqrt{3} \end{bmatrix}\begin{bmatrix} -\frac{1}{2} & \frac{1}{2} \\ \frac{1}{2} & \frac{1}{2} \end{bmatrix}$$

$$= \frac{1}{2}\begin{bmatrix} 1+\sqrt{3} & -1+\sqrt{3} \\ -1+\sqrt{3} & 1+\sqrt{3} \end{bmatrix}$$

Remark: Since we used the *principal values* of $\sqrt{1}$ and $\sqrt{3}$ in the above computation of $\sqrt{A}$, we shall call our result the **principal square root** of A.

Example-3: Given the matrix

$$A = \begin{bmatrix} 1 & 0 & 1 \\ 0 & 1 & 0 \\ 1 & 0 & 1 \end{bmatrix}$$

compute e^A.

Solution: We find that the eigenvalues of A are $\lambda_1 = 0$, $\lambda_2 = 1$, and $\lambda_3 = 2$ with the corresponding eigenvectors

$$\mathbf{x}_1 = \begin{bmatrix} -1 \\ 0 \\ 1 \end{bmatrix}, \quad \mathbf{x}_2 = \begin{bmatrix} 0 \\ 1 \\ 0 \end{bmatrix}, \quad \text{and} \quad \mathbf{x}_3 = \begin{bmatrix} 1 \\ 0 \\ 1 \end{bmatrix}$$

so that

$$P = \begin{bmatrix} -1 & 0 & 1 \\ 0 & 1 & 0 \\ 1 & 0 & 1 \end{bmatrix} \quad \text{and} \quad P^{-1} = \begin{bmatrix} -\frac{1}{2} & 0 & \frac{1}{2} \\ 0 & 1 & 0 \\ \frac{1}{2} & 0 & \frac{1}{2} \end{bmatrix}$$

Setting $f(x) = e^x$ and using (31.3), we obtain

$$\begin{aligned} f(A) &= e^A = Pe^D P^{-1} \\ &= \begin{bmatrix} -1 & 0 & 1 \\ 0 & 1 & 0 \\ 1 & 0 & 1 \end{bmatrix} \begin{bmatrix} e^0 & 0 & 0 \\ 0 & e^1 & 0 \\ 0 & 0 & e^2 \end{bmatrix} \begin{bmatrix} -\frac{1}{2} & 0 & \frac{1}{2} \\ 0 & 1 & 0 \\ \frac{1}{2} & 0 & \frac{1}{2} \end{bmatrix} \\ &= \frac{1}{2} \begin{bmatrix} e^2+1 & 0 & e^2-1 \\ 0 & e & 0 \\ e^2-1 & 0 & e^2+1 \end{bmatrix} \end{aligned}$$

Exercise Set 31

---

1. Given the matrix

$$A = \begin{bmatrix} 3 & 2 \\ 2 & 3 \end{bmatrix}$$

compute $p(A) = A^2 - 6A + 9I$.

2. Find $\sqrt{B}$ (the principal square root of B) for the matrix

$$B = \begin{bmatrix} 4 & 1 \\ 1 & 4 \end{bmatrix}$$

3. Calculate 2^A for the matrix

$$A = \begin{bmatrix} -1 & 3 \\ 1 & 1 \end{bmatrix}$$

4. Using the same matrix A in Exercise-1, show that

$$A^2 - 6A + 5I = 0$$

in two different ways.

5. For the matrix B given in Exercise-2, calculate
(a) $\cos B$
(b) $\sin B$

6. Use the result of Exercise-5 to show that

$$\cos^2 B + \sin^2 B = I$$

7. Given the matrix

$$B = \begin{bmatrix} 1 & 0 & 0 \\ 1 & 2 & 0 \\ 2 & 1 & 3 \end{bmatrix}$$

find an expression for B^k where k is any positive integer.

8. Using the matrix A in Exercise-1, evaluate $\ln A$.

32. Matrix Power Series

Here, we extend the results of previous lesson to handle scalar-valued functions, which may be complex-valued, that are defined in terms of power series. For example, if a function $f(x)$ is defined by the power series

$$f(x) = \sum_{k=0}^{\infty} a_k x^k \tag{32.1}$$

where the series converges for $|x| < R$, then we are naturally led to consider the corresponding **matrix power series**

$$f(A) = \sum_{k=0}^{\infty} a_k A^k \tag{32.2}$$

for all square matrices A, provided that (32.2) converges in some sense.

Consequently, we need to extend the concept of convergence to an infinite series of matrices. As we shall see, this extension is straightforward and parallels the development of the same concept for any infinite series of scalar-valued functions. We begin with the notion of convergence of an infinite sequence of square matrices.

Definition: *(Convergence of a Sequence of Matrices)*

Given an infinite sequence of square matrices, say $\{A^{(k)}\}_{k=1}^{\infty}$, where each matrix $A^{(k)} = \left[a_{ij}^{(k)}\right]$ is a square matrix of order n, we say that the sequence **converges** to the **limit** $A = [a_{ij}]$, and write

$$\lim_{k\to\infty} A^{(k)} = A \tag{32.3}$$

if and only if

$$\lim_{k\to\infty} a_{ij}^{(k)} = a_{ij} \quad \text{for all } i,j = 1,2,3,\ldots,n \tag{32.4}$$

Example-1: Show that the infinite sequence of matrices $\{A^{(k)}\}_{k=1}^{\infty}$ where

$$A^{(k)} = \begin{bmatrix} (2+\frac{1}{k}) & 1/k^2 \\ -1/k^3 & (1-\frac{1}{k}) \end{bmatrix}$$

converges.

Solution: We must show that $\lim_{k\to\infty} a_{ij}^{(k)} = a_{ij}$ for all $i,j = 1,2$. We find that

$$\lim_{k\to\infty} a_{11}^{(k)} = \lim_{k\to\infty}(2 + \frac{1}{k}) = 2, \quad \lim_{k\to\infty} a_{12}^{(k)} = \lim_{k\to\infty}(1/k^2) = 0$$

$$\lim_{k\to\infty} a_{21}^{(k)} = \lim_{k\to\infty}(-1/k^3) = 0, \quad \lim_{k\to\infty} a_{22}^{(k)} = \lim_{k\to\infty}(1 - \frac{1}{k}) = 1$$

Consequently,

$$\lim_{k\to\infty} A^{(k)} = \begin{bmatrix} 2 & 0 \\ 0 & 1 \end{bmatrix} = A$$

Definition: *(Convergence of an Infinite Series of Matrices)*

Given an infinite series of matrices

$$\sum_{k=1}^{\infty} A^{(k)}$$

where each matrix $A^{(k)} = \left[a_{ij}^{(k)}\right]$ is a square matrix of order n, we say that the series **converges** to the **sum** S, and write

$$\sum_{k=1}^{\infty} A^{(k)} = S \tag{32.5}$$

if the limit of the infinite sequence $\{S^{(k)}\}_{k=1}^{\infty}$ of **partial sums**, where

$$S^{(k)} = \sum_{i=1}^{k} A^{(i)} = A^{(1)} + A^{(2)} + \ldots + A^{(k)} \tag{32.6}$$

exists, and

$$\lim_{k\to\infty} S^{(k)} = S \tag{32.7}$$

Example-2: Consider the (scalar) exponential series

$$e^x = \sum_{k=0}^{\infty} \frac{x^k}{k!}$$

which is known to converge for all real values of x. Let

$$A = \begin{bmatrix} 2 & 1 \\ 1 & 2 \end{bmatrix}$$

and show that the corresponding matrix series e^A converges.

Solution: We first diagonalize the matrix A. Here, A has the eigenvalues $\lambda_1 = 1$ and $\lambda_2 = 3$, with corresponding eigenvectors

$$x_1 = \begin{bmatrix} -1 \\ 1 \end{bmatrix}, \text{ and } x_2 = \begin{bmatrix} 1 \\ 1 \end{bmatrix}$$

where

$$P = \begin{bmatrix} -1 & 1 \\ 1 & 1 \end{bmatrix} \text{ and } P^{-1} = \begin{bmatrix} -\frac{1}{2} & \frac{1}{2} \\ \frac{1}{2} & \frac{1}{2} \end{bmatrix}$$

so that

$$A = PDP^{-1} = \begin{bmatrix} -1 & 1 \\ 1 & 1 \end{bmatrix} \begin{bmatrix} 1 & 0 \\ 0 & 3 \end{bmatrix} \begin{bmatrix} -\frac{1}{2} & \frac{1}{2} \\ \frac{1}{2} & \frac{1}{2} \end{bmatrix}$$

We now form the sequence of partial sums $S^{(k)}$ by writing

$$S^{(k)} = \sum_{i=1}^{k} \frac{A^{(i)}}{i!} = \begin{bmatrix} -1 & 1 \\ 1 & 1 \end{bmatrix} \begin{bmatrix} \sum_{i=1}^{k} 1^i/i! & 0 \\ 0 & \sum_{i=1}^{k} 3^i/i! \end{bmatrix} \begin{bmatrix} -\frac{1}{2} & \frac{1}{2} \\ \frac{1}{2} & \frac{1}{2} \end{bmatrix}$$

and taking the limit as $k \to \infty$, to obtain

$$S = \begin{bmatrix} -1 & 1 \\ 1 & 1 \end{bmatrix} \begin{bmatrix} \lim_{k\to\infty} \sum_{i=1}^{k} 1^i/i! & 0 \\ 0 & \lim_{k\to\infty} \sum_{i=1}^{k} 3^i/i! \end{bmatrix} \begin{bmatrix} -\frac{1}{2} & \frac{1}{2} \\ \frac{1}{2} & \frac{1}{2} \end{bmatrix}$$

$$= \begin{bmatrix} -1 & 1 \\ 1 & 1 \end{bmatrix} \begin{bmatrix} e & 0 \\ 0 & e^3 \end{bmatrix} \begin{bmatrix} -\frac{1}{2} & \frac{1}{2} \\ \frac{1}{2} & \frac{1}{2} \end{bmatrix}$$

$$= \frac{1}{2} \begin{bmatrix} e + e^3 & e^3 - e \\ e^3 - e & e + e^3 \end{bmatrix}$$

We conclude that

$$e^A = \sum_{k=0}^{\infty} \frac{A^k}{k!} = \frac{1}{2} \begin{bmatrix} e + e^3 & e^3 - e \\ e^3 - e & e + e^3 \end{bmatrix}$$

so, the corresponding matrix series converges. We note, in passing, that we would have obtained the same result by employing (31.3), so the result is consistent with our previous work.

Now, assume that we are given a matrix power series

$$f(A) = \sum_{k=0}^{\infty} a_k A^k$$

where the matrix A can be *diagonalized*, and the corresponding (scalar) series

$$f(x) = \sum_{k=0}^{\infty} a_k x^k$$

converges for $|x| < R$, then it is obvious that the given matrix power series will converge provided that $|\lambda_i| < R$ for each eigenvalue λ_i of A. Now, if A cannot be diagonalized, then our analysis breaks down. Consequently, we are led to consider how we can analyze more general situations. One way of resolving this dilemma is to introduce the concept of **matrix norm**.

Definition: *(Matrix Norm)*

If A is an $m \times n$ matrix with complex elements, then the **norm** of A, denoted by $\|A\|$, is defined as

$$\|A\| = \sup_{\|\mathbf{u}\|=1} \|A\mathbf{u}\| \qquad (32.8$$

where $\mathbf{u} \in \mathbb{C}^n$.

In this definition, the symbol sup denotes the **supremum**, or least upper bound; for this reason, $\|A\|$ is often called the "sup norm." In other words,

$$\|A\| = \text{least upper bound of } \left\{\|A\mathbf{u}\| \ : \ \text{where } \mathbf{u} \in \mathbb{C}^n \text{ and } \|\mathbf{u}\| = 1\right\} \qquad (32.9$$

Stated differently, $\|A\|$ is the least upper bound of all possible values of $\|A\mathbf{u}\|$ for all *unit vectors* $\mathbf{u} \in \mathbb{C}^n$. Consequently,

$$\|A\mathbf{u}\| \le \|A\| \qquad (32.1$$

for any unit vector $\mathbf{u}$.

Now, if $\mathbf{v}$ any non-zero vector, then $\mathbf{u} = (1/\|\mathbf{v}\|)\mathbf{v}$ is a unit vector that has the same direction as $\mathbf{v}$, so $\mathbf{v} = \|\mathbf{v}\|\mathbf{u}$, and we can write

$$\|A\mathbf{v}\| = \|A(\|\mathbf{v}\|\mathbf{u})\| = \|\mathbf{v}\| \cdot \|A\mathbf{u}\| \le \|\mathbf{v}\| \cdot \|A\|$$

so that

$$\|A\mathbf{v}\| \le \|A\| \cdot \|\mathbf{v}\| \qquad (32.1$$

for any non-zero vector $\mathbf{v} \in \mathbb{C}^n$.

Since A is an $m \times n$ matrix, then the matrix transformation

$$T(\mathbf{v}) = A\mathbf{v}$$

defines a linear transformation $T : \mathbb{C}^n \to \mathbb{C}^m$, such that T maps each vector $\mathbf{v} \in \mathbb{C}^n$ to a unique image vector $A\mathbf{v} \in \mathbb{C}^m$. Relation (32.11) tells us that $\|A\|$ is simply the *maximum magnification* that $\mathbf{v}$ undergoes under the linear transformation T.

Theorem 32.1: *(Properties of the Matrix Norm)*

If A and B are square matrices of order n, with complex elements, then:

(a)

$$\|cA\| = |c| \cdot \|A\| \quad \text{for all complex scalars } c \tag{32.12}$$

(b)

$$\|A + B\| \le \|A\| + \|B\| \tag{32.13}$$

(c)

$$\|AB\| \le \|A\| \cdot \|B\| \tag{32.14}$$

(d)

$$|a_{ij}| \le \|A\| \quad \text{for all } i,j = 1,2,\ldots,n \tag{32.15}$$

Proof: Given any unit vector $\mathbf{u} \in \mathbb{C}^n$ we can write

$$\|(cA)\mathbf{u}\| = \|c(A\mathbf{u})\| = |c| \cdot \|A\mathbf{u}\|$$

so that

$$\|cA\| = \sup_{\|\mathbf{u}\|=1} \|(cA)\mathbf{u}\| = |c| \cdot \sup_{\|\mathbf{u}\|=1} \|A\mathbf{u}\| = |c| \cdot \|A\|$$

and then, property (a) follows immediately.

To estabish property (b), observe that

$$\begin{aligned} \|A + B\| &= \sup_{\|\mathbf{u}\|=1} \|(A + B)\mathbf{u}\| = \sup_{\|\mathbf{u}\|=1} \|A\mathbf{u} + B\mathbf{u}\| \\ &\le \sup_{\|\mathbf{u}\|=1} (\|A\mathbf{u}\| + \|B\mathbf{u}\|) \\ &\le \sup_{\|\mathbf{u}\|=1} \|A\mathbf{u}\| + \sup_{\|\mathbf{u}\|=1} \|B\mathbf{u}\| \\ &\le \|A\| + \|B\| \end{aligned}$$

The proof of property (c) is left as an exercise. To prove (d), we first introduce the standard (orthonormal) basis $\mathbf{e}_1, \mathbf{e}_2, \ldots, \mathbf{e}_n$ for $\mathbb{C}^n$. Now, we can write

$$\langle \mathbf{e}_i, A\mathbf{e}_j \rangle = \mathbf{e}_i^\dagger A \mathbf{e}_j = \mathbf{e}_i^\dagger \begin{bmatrix} a_{1j} \\ a_{2j} \\ \vdots \\ a_{nj} \end{bmatrix} = a_{ij}$$

Consequently, we have

$$\begin{aligned} |a_{ij}| = |\langle \mathbf{e}_i, A\mathbf{e}_j \rangle| &\leq \|\mathbf{e}_i\| \cdot \|A\mathbf{e}_j\| \quad \text{[by the Cauchy-Schwarz inequality]} \\ &\leq 1 \cdot \|A\| \quad \text{[by (32.10)]} \end{aligned}$$

as required. □.

Theorem 32.2: *(Relationship Between Eigenvalues and Matrix Norm)*

If A is any $m \times n$ matrix with complex elements, then
(a) The $n \times n$ matrix $H = A^\dagger A$ is Hermitian.
(b) All of the eigenvalues of H are positive real numbers.
(c) The norm of A is given by

$$\|A\| = \sqrt{\lambda_{\max}} \tag{32.1}$$

where $\lambda_{\max}$ is the largest eigenvalue of H.

Proof: Clearly, H is Hermitian since

$$H^\dagger = (A^\dagger A)^\dagger = A^\dagger A = H$$

Furthermore, since H is Hermitian, its eigenvalues are real. Let λ be any eigenvalue of H with corresponding eigenvector $\mathbf{v}$, so that $H\mathbf{v} = \lambda\mathbf{v}$. Since λ is real, we can write

$$\begin{aligned} \lambda \cdot \|\mathbf{v}\|^2 &= \langle \lambda\mathbf{v}, \mathbf{v} \rangle = \langle H\mathbf{v}, \mathbf{v} \rangle \\ &= \langle (A^\dagger A)\mathbf{v}, \mathbf{v} \rangle = \mathbf{v}^\dagger A^\dagger A \mathbf{v} \\ &= \langle A\mathbf{v}, A\mathbf{v} \rangle \\ &= \|A\mathbf{v}\|^2 \end{aligned}$$

So, λ is a positive real number.

Finally, we establish property (c). Since H is Hermitian, there exists an orthonormal basis of eigenvectors $\mathbf{u}_1, \mathbf{u}_2, \ldots, \mathbf{u}_n$ of H such that

$$H\mathbf{u}_i = \lambda_i \mathbf{u}_i \quad \text{for } i = 1, 2, 3, \ldots, n$$

Next, we re-index the eigenvectors, if necessary, to ensure that the corresponding eigenvalues form a non-decreasing sequence of positive numbers:

$$0 < \lambda_1 \le \lambda_2 \le \lambda_3 \le \cdots \le \lambda_n$$

Let $\mathbf{v} \in \mathbb{C}^n$ be any *unit* vector written in terms of this orthonomal basis as

$$\mathbf{v} = \sum_{k=1}^{n} c_k \mathbf{u}_k \quad \text{where } \|\mathbf{v}\|^2 = \sum_{k=1}^{n} |c_k|^2 = 1$$

then

$$\begin{aligned}
\|A\mathbf{v}\|^2 &= \langle A\mathbf{v}, A\mathbf{v}\rangle = \mathbf{v}^\dagger H\mathbf{v} = \mathbf{v}^\dagger \sum_{k=1}^{n\dagger} Hc_k\mathbf{u}_k \\
&= \mathbf{v}^\dagger \sum_{k=1}^{n} c_k(A\mathbf{u}_k) = \mathbf{v}^\dagger \sum_{k=1}^{n} c_k\lambda_k\mathbf{e}_k \\
&= \sum_{k=1}^{n} c_k\lambda_k(\mathbf{v}^\dagger\mathbf{e}_k) = \sum_{k=1}^{n} c_k\lambda_k\overline{\langle \mathbf{v}, \mathbf{e}_k\rangle} \\
&= \sum_{k=1}^{n} |c_k|^2\lambda_k \le \lambda_n\left(\sum_{k=1}^{n}|c_k|^2\right) \\
&\le \lambda_n
\end{aligned}$$

Consequently,

$$\|A\| = \sup_{\|\mathbf{v}\|=1} \|A\mathbf{v}\| \le \sqrt{\lambda_n}$$

Thus, $\sqrt{\lambda_n}$ is an upper bound of $\|A\|$. But, $\|A\|$ actually *achieves* this upper bound for the choice $\mathbf{v} = \mathbf{u}_n$ since

$$\|A\mathbf{u}_n\| = \sqrt{\|A\mathbf{u}_n\|^2} = \sqrt{\mathbf{u}_n^\dagger H\mathbf{u}_n} = \sqrt{\lambda_n}$$

so we must have $\|A\| = \sqrt{\lambda_n} = \sqrt{\lambda_{\max}}$ as claimed. □.

Example-2: Determine the norm of the matrix

$$A = \begin{bmatrix} 1+i & 2 \\ 2 & 1-i \end{bmatrix}$$

Solution: By the previous theorem, we must first compute the matrix $H = A^\dagger A$. We find that

$$H = A^\dagger A = \begin{bmatrix} 1-i & 2 \\ 2 & 1+i \end{bmatrix}\begin{bmatrix} 1+i & 2 \\ 2 & 1-i \end{bmatrix} = \begin{bmatrix} 6 & 4-4i \\ 4+4i & 6 \end{bmatrix}$$

and consequently, the characteristic equation of H is

$$\begin{vmatrix} 6-\lambda & 4-4i \\ 4+4i & 6-\lambda \end{vmatrix} = \lambda^2 - 12\lambda + 4 = 0$$

Solving for λ, we find that the matrix H has the two eigenvalues $\lambda = 6 \pm 4\sqrt{2}$. Clearly, the largest eigenvalue is $\lambda_{\text{max}} = 6 + 4\sqrt{2}$; therefore

$$\|A\| = \sqrt{\left(6+4\sqrt{2}\right)} \approx 3.41$$

Lemma 32.3: *(An Upper Bound on Absolute Values of Eigenvalues)*

Let A be a square matrix of order n with complex elements. Then any eigenvalue λ of A satisfies the inequality

$$|\lambda| \le \|A\| \qquad (32.1$$

In other words, $\|A\|$ is an upper bound for the absolute values of the eigenvalues of A.

Proof: Assume that λ is an eigenvalue of A with corresponding eigenvector $\mathbf{x}$, so $A\mathbf{x} = \lambda\mathbf{x}$, and consequently,

$$\|A\mathbf{x}\| = |\lambda| \cdot \|\mathbf{x}\|$$

Since $\mathbf{x}$ is an eigenvector of A, then $\|\mathbf{x}\| \ne 0$, so

$$|\lambda| = \frac{\|A\mathbf{x}\|}{\|\mathbf{x}\|} \le \frac{\|A\| \cdot \|\mathbf{x}\|}{\|\mathbf{x}\|} = \|A\| \quad \text{[by (32.11)]}$$

as claimed. □.

Theorem 32.4: *(Norm of a Hermitian Matrix)*

If A is a Hermitian matrix of order n with eigenvalues $\lambda_1, \lambda_2, \lambda_3, \ldots, \lambda_n$, then the norm of A is equal to the eigenvalue of A which has the greatest absolute value; that is,

$$\|A\| = \max_k |\lambda_k| \qquad (32.1$$

and each eigenvalue λ_k of A satisfies the inequality

$$-\|A\| \le \lambda_k \le \|A\| \quad (\text{for } k = 1,2,3,\ldots,n) \qquad (32.1$$

Proof: Since A is Hermitian, then $A^2 = A^\dagger A$ and the eigenvalues of A^2 must be $\lambda_1^2, \lambda_2^2, \lambda_3^2, \ldots, \lambda_n^2$. By Theorem 32.2, however,

$$\|A\| = \sqrt{\max_k \lambda_k^2} = \max_k |\lambda_k|$$

which establishes (32.18). Now, by Lemma 32.3, we have

$$|\lambda_k| \leq \|A\| \quad (\text{for } k = 1,2,3,\ldots,n)$$

Since the eigenvalues of A are real numbers, this inequality implies that

$$-\|A\| \leq \lambda_k \leq \|A\| \quad (\text{for } k = 1,2,3,\ldots,n)$$

as claimed. □.

Example-3: Use (32.16) to verify the result of the previous theorem for the Hermitian matrix

$$A = \begin{bmatrix} 1 & 1+i \\ 1-i & 1 \end{bmatrix}$$

Solution: In order to use (32.16), we must first compute the matrix $H = A^{\dagger}A$. We find that

$$H = A^{\dagger}A = \begin{bmatrix} 1 & 1+i \\ 1-i & 1 \end{bmatrix}\begin{bmatrix} 1 & 1+i \\ 1-i & 1 \end{bmatrix} = \begin{bmatrix} 3 & 2+2i \\ 2-2i & 3 \end{bmatrix}$$

and consequently, the characteristic equation of H is

$$\begin{vmatrix} 3-\lambda & 2+2i \\ 2-2i & 3-\lambda \end{vmatrix} = \lambda^2 - 6\lambda + 1 = 0$$

Solving for λ, we find that the matrix H has the two eigenvalues $\lambda = 3 \pm 2\sqrt{2}$. Clearly, the largest eigenvalue is $\lambda_{\max} = 3 + 2\sqrt{2}$; therefore

$$\|A\| = \sqrt{\left(3+2\sqrt{2}\right)} = \sqrt{\left(1+\sqrt{2}\right)^2} = 1+\sqrt{2}$$

On the other hand, we find that the eigenvalues of the original matrix A are $\lambda_1 = 1 - \sqrt{2}$ and $\lambda_2 = 1 + \sqrt{2}$, so that by Theorem (32.4),

$$\|A\| = \max\left\{\left|1-\sqrt{2}\right|, \left|1+\sqrt{2}\right|\right\} = 1+\sqrt{2}$$

as we obtained previously.

Now that we have explored the basic properties of the matrix norm, we can provide a simple criterion for the convergence of matrix any power series that corresponds to a convergent power series of scalar-valued functions. This is the content of the next theorem.

Theorem 32.5: *(Convergence of a Matrix Power Series)*

If the power series of scalar-valued functions

$$f(x) = \sum_{k=0}^{\infty} a_k x^k \tag{32.2}$$

converges for $|x| < R$, then the corresponding matrix power series, where A is a square matrix of order n, with complex elements,

$$f(A) = \sum_{k=0}^{\infty} a_k A^k \tag{32.2}$$

converges for $\|A\| < R$. Furthermore, if $\lambda_1, \lambda_2, \lambda_3, \ldots \lambda_n$ are the eigenvalues of A, then

$$f(\lambda_1), f(\lambda_2), f(\lambda_3), \ldots, f(\lambda_n)$$

are the eigenvalues of $f(A)$.

Proof: Let A be a square matrix of order n where $\|A\| < R$. For any integer $k \geq 0$, let

$$a_k A^k = \left[(a_k A^k)_{ij} \right] \quad \text{(for all } i,j = 1,2,3,\ldots,n)$$

and consider the sequence $\{S^{(N)}\}_{N=1}^{\infty}$ of partial sums. We have

$$\left| S_{ij}^{(N)} \right| = \left| \sum_{k=1}^{N} (a_k A^k)_{ij} \right| \leq \sum_{k=1}^{N} \left| (a_k A^k)_{ij} \right| \leq \sum_{k=1}^{N} |a_k| \cdot \left| (A^k)_{ij} \right|$$

$$\leq \sum_{k=1}^{N} |a_k| \cdot \|A_{ij}\|^k$$

In other words,

$$\left| S_{ij}^{(N)} \right| = \left| \sum_{k=1}^{N} (a_k A^k)_{ij} \right| \leq \sum_{k=1}^{N} |a_k| \cdot \|A_{ij}\|^k$$

Now, let $N \to \infty$; since $\|A\| < R$, the power series on the right converges absolutely. So, by the Comparison Test, we conclude that the matrix power series (32.21) converges as well.

To prove the last part of the theorem, let λ be an arbitrary eigenvalue of A with corresponding eigenvector $\mathbf{x}$. Observe that for $\|A\| < R$,

$$f(A)\mathbf{x} = \left(\sum_{k=0}^{\infty} a_k A^k \right) \mathbf{x} = \sum_{k=0}^{\infty} a_k (A^k \mathbf{x}) = \left(\sum_{k=0}^{\infty} a_k \lambda^k \right) \mathbf{x}$$

But the series of scalars (on the right) converges to $f(\lambda)$ since $|\lambda| \leq \|A\| < R$.

Thus,

$$f(A)\mathbf{x} = f(\lambda)\mathbf{x}$$

and $f(\lambda)$ is an eigenvalue of $f(A)$. This completes the proof. □.

Example-4: Consider the power series

$$f(x) = \frac{1}{1-x} = 1 + x + x^2 + x^3 + \ldots$$

which is known to converge for $|x| < 1$. By the previous theorem, we know that the corresponding matrix series

$$f(A) = I + A + A^2 + A^3 + \ldots$$

will converge for $\|A\| < 1$. In fact, we claim that

$$f(A) = (I - A)^{-1} = I + A + A^2 + A^3 + \ldots \quad \text{(for } \|A\| < 1\text{)} \tag{32.22}$$

To see this, observe that if $\mathbf{x}$ is any eigenvector of A with corresponding eigenvalue λ, then we have both

$$I\mathbf{x} = \mathbf{x} \quad \text{and} \quad A\mathbf{x} = \lambda\mathbf{x}$$

Subtracting, we get

$$(I - A)\mathbf{x} = (1 - \lambda)\mathbf{x}$$

so $(1 - \lambda)$ is an eigenvalue of $(I - A)$. But $|\lambda| \leq \|A\| < 1$ so we must have $(1 - \lambda) \neq 0$ and consequently, $(I - A)^{-1}$ exists. Finally, observe that

$$\begin{aligned}(I - A)(I + A + A^2 + A^3 + \ldots) &= I + (A + A^2 + A^3 + \ldots) - (A + A^2 + A^3 + \ldots) \\ &= I\end{aligned}$$

Consequently, (32.22) follows immediately.

We now turn our attention to the differentiation and integration of matrices. If $A(t) = [a_{ij}(t)]$ is an $m \times n$ matrix, where each of its elements depend upon a parameter t, then we define

$$\frac{dA}{dt} = \left[\frac{d}{dt}a_{ij}(t)\right] \tag{32.23a}$$

$$\int_{t_0}^{t} A(t)dt = \left[\int_{t_0}^{t} a_{ij}(t)dt\right] \tag{32.23b}$$

provided that each of the quantities on the right-hand side of (32.23) exist.

From real analysis, it well known that if the power series

$$f(x) = \sum_{k=0}^{\infty} a_k x^k$$

converges for $|x| < R$, then we can differentiate the series term by term, and the resulting series will converge for $|x| < R$. Similarly, the given series can be integrated term by term over any interval $[a, b]$ which is contained within the interval of convergence, and the resulting series will converge for $|x| < R$. The next theorem extends these properties to matrix power series.

Theorem 32.5: *(Differentiation and Integration of a Matrix Power Series)*

Suppose that the scalar power series

$$f(x) = \sum_{k=0}^{\infty} a_k x^k$$

converges for $|x| < R$, and let A be any square matrix of order n (possibly with complex elements) such that $\|A\| \neq 0$. Then:

(a) The matrix power series, where A is a square matrix of order n, with complex elements, given by

$$f(At) = \sum_{k=0}^{\infty} a_k (At)^k \tag{32.}$$

converges for all $|t| < R/\|A\|$.

(b) The matrix function $f(At)$ is differentiable, and the matrix series

$$\frac{d}{dt} f(At) = \sum_{k=0}^{\infty} a_k \frac{d}{dt}(At)^k \tag{32.}$$

converges for all $|t| < R/\|A\|$.

(c) The matrix function $f(At)$ is integrable, and the matrix series

$$\int_a^b f(At)dt = \sum_{k=0}^{\infty} a_k \left(\int_a^b (At)^k dt \right) \tag{32.2}$$

converges for all $|t| < R/\|A\|$, provided that $[a, b]$ is contained within its interval of convergence.

Proof: Follows immediately from Theorem 32.4, and the properties of scalar power series.

Corollary 35.6: *(Derivative of the Matrix Exponential)*

If A is a square matrix, then for all values of the parameter t, we have

$$\frac{d}{dt}(e^{At}) = e^{tA} A \tag{32.2}$$

Proof: Left as an exercise for the reader.

Exercise Set 32

1. Compute $\|A\|$ for each of the matrices:

$$\textbf{(a)}\ A = \begin{bmatrix} 2 & 1 \\ 1 & 2 \end{bmatrix} \quad \textbf{(b)}\ A = \begin{bmatrix} -1 & 3 \\ 5 & 1 \end{bmatrix} \quad \textbf{(c)}\ A = \begin{bmatrix} 1 & 0 & 2 \\ 0 & 2 & 0 \\ 2 & 0 & 1 \end{bmatrix}$$

2. Given the matrix

$$H = \begin{bmatrix} 1 & i \\ -i & 1 \end{bmatrix}$$

(a) Show that H is Hermitian.
(b) Compute $\|H\|$.

3. If A is a square matrix such that $\|A\| < 1$, then show that

$$\lim_{k \to \infty} A^k = 0$$

4. Show that if U is any unitary matrix, then $\|U\| = 1$.

5. For any square matrix A, show that e^A is invertible, and

$$(e^A)^{-1} = e^{-A} \tag{32.28}$$

6. For any square matrix A, show that

$$\det(e^A) = e^{Tr(A)} \tag{32.29}$$

7. Given that A and B are square matrices of the same order that *commute*; i.e., $AB = BA$, show that

$$e^{A+B} = e^A \cdot e^B \tag{32.30}$$

8. Given that the series

$$\sin x = \sum_{k=0}^{\infty} (-1)^k \frac{x^{2k+1}}{(2k+1)!} = x - \frac{x^3}{3!} + \frac{x^5}{5!} - \ldots$$

converges for all scalars x,

(a) Write down the corresponding matrix power series for $\sin A$.
(b) Where does the matrix series converge?
(c) Given the matrix

$$A = \begin{bmatrix} -1 & 1 \\ 3 & 1 \end{bmatrix}$$

compute $\sin A$.

9. Prove Corollary 35.6.

10. Let $A(t) = [a_{ij}(t)]$ and $B(t) = [b_{ij}(t)]$ be square matrices of the same order whose respective elements are differentiable over the same set of values for t. Show that

(a)

$$\frac{d}{dt}\{cA(t)\} = c\frac{dA(t)}{dt} \quad \text{for any scalar } c \qquad (32.:$$

(b)

$$\frac{d}{dt}\{A(t) + B(t)\} = \frac{dA(t)}{dt} + \frac{dB(t)}{dt} \qquad (32.:$$

(c)

$$\frac{d}{dt}\{A(t)B(t)\} = A(t)\frac{dB(t)}{dt} + \frac{dA(t)}{dt}B(t) \qquad (32.:$$

(d) If $A(t)$ is invertible, then

$$\frac{d}{dt}[A(t)]^{-1} = -\left\{[A(t)]^{-1}\frac{dA(t)}{dt}[A(t)]^{-1}\right\} \qquad (32.:$$

11. Let $A(t) = [a_{ij}(t)]$ be an $m \times n$ matrix. We say that $A(t)$ is **differentiable** at $t = t_0$ and write

$$A'(t_0) = \lim_{h\to 0}\left\{\frac{1}{h}[A(t_0 + h) - A(t_0)]\right\} \qquad (32.3$$

if and only if the limit exists. Show that if $A(t)$ is differentiable at $t = t_0$ then

$$A'(t_0) = \left[a'_{ij}(t_0)\right] \qquad (32.3$$

(This result provides the motivation for the definition given in (32.23a)).

33. Minimal Polynomials

If A is a square matrix of order n, with distinct eigenvalues $\lambda_1, \lambda_2, \lambda_3, \ldots, \lambda_k$ of multiplicities $r_1, r_2, r_3, \ldots, r_k$ respectively, then (by repeated application of the Fundamental Theorem of Algebra) its characteristic polynomial $\Delta(\lambda)$ can be written in the factored form

$$\Delta(x) = (-1)^n \det(A - xI) = \det(xI - A) = (x - \lambda_1)^{r_1}(x - \lambda_2)^{r_2} \cdots (x - \lambda_k)^{r_k} \tag{33.1}$$

As this is a polynomial of degree n, we have

$$r_1 + r_2 + r_3 + \ldots + r_k = n \tag{33.2}$$

Recall that from the Cayley-Hamilton Theorem, we have $\Delta(A) = 0$, since any square matrix must satisfy its own characteristic equation. It may happen that there exists a polynomial, say $m(x)$ of degree less than n such that $m(A) = 0$ as well. This leads to the concept of a minimal polynomial.

Definition: *(Minimal Polynomial)*

Let A be a square matrix of order n. The **minimal polynomial** $\mu(x)$ of A is the monic polynomial of *lowest degree* such that

$$\mu(A) = 0 \tag{33.3}$$

From Algebra, a polynomial $p(x) \neq 0$ is said to be **monic** if its leading coefficient is equal to one. It turns out that the minimum polynomial of a square matrix has many interesting properties, some of which we now explore.

Theorem 33.1: *(The Minimal Polynomial $\mu(x)$ Divides $\Delta(x)$)*

Let A be a square matrix, and let $p(x)$ be any polynomial such that $p(A) = 0$. Then the minimum polynomial $\mu(x)$ of A divides $p(x)$. In particular, $\mu(x)$ also divides the characteristic polynomial $\Delta(x)$ of A.

Proof: Assume that $p(x)$ is any polynomial such that $p(A) = 0$. By the division algorithm, there exist polynomials $q(x)$ and $r(x)$ such that

$$p(x) = \mu(x)q(x) + r(x)$$

where the degree of $r(x)$ is less than the degree of $\mu(x)$. Consequently,

$$p(A) = \mu(A)q(A) + r(A)$$

Since $p(A) = 0$ and $\mu(A) = 0$, then we must have $r(A) = 0$ as well.

Now, if $r(x) \neq 0$, then $r(x)$ would be a polynomial of degree less than $\mu(x)$ such that $r(A) = 0$. But this would contradict the fact the $\mu(x)$ is the minimal polynomial of A; we must have $r(x) = 0$ and therefore, $\mu(x)$ divides $p(x)$. Finally, $\mu(x)$ must divide $\Delta(x)$ since $\Delta(A) = 0$ by the Cayley-Hamilton theorem. □.

Theorem 33.2: *(Any Linear Factor of $\Delta(x)$ is a Factor of $\mu(x)$)*

Let A be a square matrix with characteristic polynomial $\Delta(x)$ and minimal polynomial $\mu(x)$. If $(x - \lambda_0)$ is any linear factor of $\Delta(x)$, then $(x - \lambda_0)$ is also a factor of $\mu(x)$.

Proof: By the division algorithm, there exist a polynomial $q(x)$ and a scalar r such that

$$\mu(x) = (x - \lambda_0)q(x) + r$$

Replacing x by A, we obtain

$$\mu(A) = (A - \lambda_0 I)q(A) + rI$$

But since $\mu(x)$ is the minimal polynomial of A, then $\mu(A) = 0$, and the last equation implies that

$$(A - \lambda_0 I)q(A) = -rI$$

Taking the determinant of both sides of the last equation gives

$$\det(A - \lambda_0 I) \cdot \det q(A) = \det(-rI) = (-r)^n$$

But since $(x - \lambda_0)$ is a factor of A, then $\lambda = \lambda_0$ is an eigenvalue of A, and consequently, $\det(A - \lambda_0 I) = 0$ and $r = 0$ as well. Thus, $(x - \lambda_0)$ divides $\mu(x)$, and therefore, it is a factor of $\mu(x)$. □.

Theorem 33.3: *(General Form of a Minimal Polynomial)*

Let A be a square matrix of order n. If

$$\Delta(x) = (x - \lambda_1)^{r_1}(x - \lambda_2)^{r_2}\cdots(x - \lambda_k)^{r_k}$$

is the characteristic polynomial of A, then the minimal polynomial $\mu(x)$ of A must have the form

$$\mu(x) = (x - \lambda_1)^{s_1}(x - \lambda_2)^{s_2}\cdots(x - \lambda_k)^{s_k}$$

where

$$1 \leq s_i \leq r_i \quad (\text{for all } i = 1,2,3,\ldots,k)$$

and $(s_1 + s_2 + s_3 + \ldots + s_k) \leq n$.

Proof: The proof follows immediately from Theorem 33.2 and is left as an exercise.

The previous theorem is extremely useful in helping us find the minimal polynomial of any square matrix. Let's see how this works.

Example-1: Find the minimal polynomial $\mu(x)$ of the matrix

$$A = \begin{bmatrix} -3 & -2 & 1 \\ -3 & -8 & 3 \\ -3 & -6 & 1 \end{bmatrix}$$

Solution: We first find the characteristic polynomial $\Delta(x)$ of A by setting

$$\det(xI - A) = \begin{vmatrix} x+3 & 2 & -1 \\ 3 & x+8 & -3 \\ 3 & 6 & x-1 \end{vmatrix} = x^3 + 10x^2 + 28x + 24$$

Upon factoring, we get $\Delta(x) = (x+6)(x+2)^2$. So, by the previous theorem, either $p(x) = (x+6)(x+2)^2$ or $q(x) = (x+6)(x+2)$ is the miminal polynomial of A. Now, $p(A) = 0$ by the Cayley-Hamilton theorem, so we only need to check if $q(A) = 0$. In doing so, we find

$$q(A) = (A+6I)(A+2I) = \begin{bmatrix} 3 & -2 & 1 \\ -3 & -2 & 3 \\ -3 & -6 & 7 \end{bmatrix}\begin{bmatrix} -1 & -2 & 1 \\ -3 & -6 & 3 \\ -3 & -6 & 3 \end{bmatrix}$$

$$= \begin{bmatrix} 0 & 0 & 0 \\ 0 & 0 & 0 \\ 0 & 0 & 0 \end{bmatrix}$$

So, we conclude that $\mu(x) = (x+6)(x+2)$ is the minimal polynomial of A.

Given any square matrix A, it turns out that its characteristic polynomial and minimum polynomial have the same roots. This is the content of the next theorem.

Theorem 33.4: *($\Delta(x)$ and $\mu(x)$ Have the Same Roots)*

For any square matrix A, its characteristic polynomial $\Delta(x)$ and minimal polynomial $\mu(x)$ have the same roots.

Proof: Let $x = \lambda_0$ be any root of $\Delta(x)$ so that $\Delta(\lambda_0) = 0$. By the factor theorem, $(x - \lambda_0)$ must be a factor of $\Delta(x)$. But then, by Theorem 33.2, $(x - \lambda_0)$ must be a factor of $\mu(x)$ so that $\mu(\lambda_0) = 0$ as well. Conversely, if $\mu(\lambda_0) = 0$ then $(x - \lambda_0)$ is a factor of $\mu(x)$. Since $\mu(x)$ divides $\Delta(x)$, then $(x - \lambda_0)$ is also a factor of $\Delta(x)$, so $\Delta(\lambda_0) = 0$. □.

Before we leave this lesson, it's important to see how many of the concepts we have already discussed, which have been expressed in terms of square matrices, can be extended to *linear operators* defined in terms of such matrices.

Suppose that T is a linear operator on a finite dimensional space V. We already know that relative to any chosen basis, we can always find a square matrix A such that

$$T(\mathbf{v}) = A\mathbf{v}$$

Now, recall that a non-zero vector $\mathbf{v}$ is an eigenvector of A if $A\mathbf{v} = \lambda\mathbf{v}$ for some scalar λ. It is clear that $\mathbf{v}$ is an eigenvector of A if and only if

$$T(\mathbf{v}) = \lambda\mathbf{v} \tag{33.4}$$

Thus, we say that a non-zero vector $\mathbf{v}$ is an **eigenvector** of a *linear operator* T with corresponding **eigenvalue** λ if $A\mathbf{v} = \lambda\mathbf{v}$ for some scalar λ. This means that T and A have the same eigenvectors and eigenvalues. It should be clear that the "matrix point of view" and the "linear operator" point of view are logically equivalent. Consequently, each of the definitions and theorems of this lesson can be rephrased in terms of linear operators.

For example, if $T(\mathbf{v}) = A\mathbf{v}$ is any linear operator on V, then we define the **characteristic polynomial** $\Delta(x)$ of T to be the same as the characteristic polynomial of A. This definition makes sense since the characteristic polynomial of the linear operator is *independent* of our choice of basis. To see this, recall that if B is any other matrix representation of T relative to a new basis, then we must have $B = P^{-1}AP$, where P is the transition matrix which maps the the "new basis" to the "old basis" But then the matrices A and B are *similar* and consequently, they must have the same characteristic equation.

Furthermore, if $p(x)$ is any polynomial, then from (33.4), it is clear that $p(A) = 0$ if and only if $p(T) = 0$. So, the Cayley-Hamilton theorem for linear operators becomes $\Delta(T) = 0$ where the 0 on the right-hand side of this equation is the *zero operator*. In other words, *any linear operator is always a root of its characteristic equation.* Similarly, we define the **minimal polynomial** $\mu(x)$ of T as the monic polynomial (in the operator T) of *lowest degree* such that $\mu(T) = 0$.

Example-2: Find the characteristic and minimal polynomials of the linear operator $T(\mathbf{v}) = A\mathbf{v}$ where A is the same matrix as in Example-1.

Solution: In Example-1, we found that the characteristic and minimal polynomials of A were $\Delta(x) = (x+6)(x+2)^2$ and $\mu(x) = (x+6)(x+2)$. Consequently, by definition, these are also characteristic and minimal polynomials of T. Furthermore, observe that for an arbitrary vector $\mathbf{v}$

$$\Delta(T)\mathbf{v} = (T+6I)(T+2I)^2\mathbf{v} = (A+6I)(A+2I)^2\mathbf{v} = \mathbf{0}, \text{ and}$$
$$\mu(T)\mathbf{v} = (T+6I)(T+2I)\mathbf{v} = (A+6I)(A+2I)\mathbf{v} = \mathbf{0}$$

so both $\Delta(T)$ and $\mu(T)$ are *zero operators*; that is, $\Delta(T) = 0$ and $\mu(T) = 0$ as they should be.

Before leaving this lesson, we restate its theorems in the language of linear operators for easy reference. The proofs of the following theorems are virtually the same as given above:

Theorem 33.5: *(The Minimal Polynomial $\mu(x)$ of a Linear Operator)*

Let T be a linear operator on a finite-dimensional vector space V, and let $p(x)$ be any polynomial such that $p(T) = 0$. Then the minimum polynomial $\mu(x)$ of T divides $p(x)$. In particular, $\mu(x)$ also divides the characteristic polynomial $\Delta(x)$ of T.

Theorem 33.6: *(Minimal Polynomial of a Linear Operator)*

If T is a linear operator on a finite-dimensional vector space V and

$$\Delta(x) = (x-\lambda_1)^{r_1}(x-\lambda_2)^{r_2}\cdots(x-\lambda_k)^{r_k}$$

is the characteristic polynomial of T, then the minimal polynomial $\mu(x)$ of T must have the form

$$\mu(x) = (x-\lambda_1)^{s_1}(x-\lambda_2)^{s_2}\cdots(x-\lambda_k)^{s_k}$$

where

$$1 \leq s_i \leq r_i \quad (\text{for all } i = 1,2,3,\ldots,k)$$

and $(s_1 + s_2 + s_3 + \ldots + s_k) \leq n$.

Theorem 33.7: *($\Delta(x)$ and $\mu(x)$ Have the Same Roots)*

Let T be a linear operator on a finite-dimensional vector space V. Then its characteristic polynomial $\Delta(x)$ and minimal polynomial $\mu(x)$ have the same roots.

Exercise Set 33

1. Find the characteristic and minimal polynomials of each matrix:

(a) $\begin{bmatrix} 1 & 3 \\ -1 & 2 \end{bmatrix}$ **(b)** $\begin{bmatrix} -2 & -6 & -9 \\ 3 & 7 & 9 \\ -1 & -2 & -2 \end{bmatrix}$ **(c)** $\begin{bmatrix} -2 & 1 & 1 \\ -2 & 1 & 1 \\ -2 & 1 & 1 \end{bmatrix}$

(d) $\begin{bmatrix} 1 & 0 & 1 \\ 2 & -1 & 0 \\ 3 & 1 & 1 \end{bmatrix}$

2. For any square matrix A show that both A and A^T have the same minimal polynomial.

3. Show that if A is a square matrix such that $A = \lambda I$ for some scalar λ, then $\mu(x) = (x - \lambda)$.

4. Let A be a square matrix, and let $\mu_1(x)$ and $\mu_2(x)$ be monic polynomials such that $\mu_1(A) = 0$ and $\mu_2(A) = 0$. Show that $\mu_1(x) = \mu_2(x)$; that is, *the minimum polynomial of any square matrix is unique.*

5. Show if a square matrix A of order n has n *distinct* eigenvalues, then its characteristic polynomial and minimal polynomial are identical.

6. Let

$$\mu(x) = x^s + a_{s-1}x^{s-1} + a_{s-2}x^{s-2} + \ldots + a_0$$

be the minimal polynomial of a square matrix A where $a_0 \neq 0$. Show that
(a) A^{-1} exists
(b) A^{-1} is a polynomial in A of degree $s - 1$.

7. Let $T : V \to V$ be a linear operator. If $p(T)$ is any polynomial in T; that is,

$$p(T) = a_n T^n + a_{n-1} T^{n-1} + \ldots + a_1 T + a_0 I$$

for some non-negative integer n, then show that for any vector $\mathbf{v} \in V$,

$$T[p(T)]\mathbf{v} = p(T)[T(\mathbf{v})]$$

so the linear operators $P(T)$ and T *commute*.

34. Direct Sum Decompositions

In this lesson, we discuss how the minimal polynomial of a linear operator on a vector space V can be used to decompose V and write it as the direct sum of an appropriate collection of subspaces of V. Before doing so, however, we need to introduce several new concepts, including the notion of an **invariant subspace.**

Definition: *(Invariant Subspace)*

Let $T : V \to V$ be a linear operator, and assume that U is any subspace of V. Then we say that **U is invariant under T**, or **U is T-invariant** if and only if $\mathbf{v} \in V$ implies that $T(\mathbf{v}) \in V$ as well.

Now, if U is invariant under T, and we restrict the domain of T to U, then T is also a linear operator on U. We denote this new operator by $T|_U$, and call it the **restriction of T to U**. In other words, if U is any subspace of V which is invariant under T, then we define

$$T|_U(\mathbf{v}) = T(\mathbf{v})$$

for all $\mathbf{v} \in U$.

Example-1: If $T : V \to V$ is a linear operator, then show that ker(T) is an invariant subspace of V.

Solution: By definition, if any vector $\mathbf{v} \in \ker(T)$, then $T(\mathbf{v}) = \mathbf{0}$. But since $\mathbf{0} \in \ker(T)$, then $T(\mathbf{v}) \in \ker(T)$ as well. Similarly, it's easy to show that *range*(T) is also an invariant subspace of T.

Example-2: *(Rotation About the z-Axis)* Consider the linear operator T on $\mathbb{R}^3$ given by

$$T(\mathbf{v}) = \begin{bmatrix} \cos\alpha & -\sin\alpha & 0 \\ \sin\alpha & \cos\alpha & 0 \\ 0 & 0 & 1 \end{bmatrix} \begin{bmatrix} v_1 \\ v_2 \\ v_3 \end{bmatrix}$$

This rotates any vector $\mathbf{v} = [v_1, v_2, v_3]^T$ counterclockwise about the z-axis by an angle α. Consider the subspaces of $\mathbb{R}^3$ given by

$$U_1 = \left\{[0,0,c]^T : c \in \mathbb{R}\right\} \quad \text{and} \quad U_2 = \left\{[a,b,0]^T : a,b \in \mathbb{R}\right\}$$

Here, U_1 is the subspace which consists of all vectors lying along the z-axis, while U_2 consists of all vectors in the xy-plane. It turns out that both U_1 and

U_2 are invariant subspaces of T. To see this, let $\mathbf{v} = [0, 0, v_3]^T \in U_1$, then $T(\mathbf{v}) = [0, 0, v_3]^T = \mathbf{v}$, so $T(v) \in U_1$ and U_1 is an invariant subspace of T. Similarly, for any vector $\mathbf{v} = [v_1, v_2, 0]^T \in U_2$ we have

$$T(\mathbf{v}) = [v_1 \cos\alpha - v_2 \sin\alpha, v_2 \cos\alpha + v_1 \sin\alpha, 0]^T$$

so $T(\mathbf{v}) \in U_2$ as well. Consequently, U_2 is also an invariant subspace of T.

Theorem 34.1: *(Kernel of Polynomial in T)*

Let $T : V \rightarrow V$ be a linear operator. If $p(T)$ is any polynomial in T; that is,

$$p(T) = a_n T^n + a_{n-1} T^{n-1} + \ldots + a_1 T + a_0 I$$

for some non-negative integer n, then $\ker p(T)$ is invariant under T.

Proof: Since $p(T) : V \rightarrow V$ is a linear operator on V, then by Theorem 22.1, $\ker p(T)$ is a subspace of V. For any vector $\mathbf{v} \in \ker p(T)$. we must show that $T(\mathbf{v}) \in \ker p(T)$. By hypothesis, $p(T)(\mathbf{v}) = \mathbf{0}$, so that $T[p(T)(\mathbf{v})] = \mathbf{0}$ as well. Now, observe that

$$\begin{aligned} T[p(T)(\mathbf{v})] &= \left[T\left(a_n T^n + a_{n-1} T^{n-1} + \ldots + a_1 T + a_0 I\right) \right](\mathbf{v}) \\ &= \left[\left(a_n T^{n+1} + a_{n-1} T^n + \ldots + a_1 T^2 + a_0 T\right) \right](\mathbf{v}) \\ &= \left[\left(a_n T^n + a_{n-1} T^{n-1} + \ldots + a_1 T + a_0 I\right) T \right](\mathbf{v}) \\ &= [p(T)](T\mathbf{v}) = \mathbf{0} \end{aligned}$$

so $T(\mathbf{v}) \in \ker p(T)$ as required. □.

Before we can proceed in our journey, we need to discuss the greatest common divisor of a collection of nonzero polynomials. In the Theory of Numbers, we define the greatest common divisor of any two natural numbers as the largest natural number which divides each of the given numbers. In a similar fashion, we define the **greatest common divisor** of a finite set of nonzero polynomials to be the polynomial $d(x)$ of *greatest degree* which divides each element of the set. The next lemma provides a proof of both the existence and uniqueness of $d(x)$.

Lemma 34.2: *(Greatest Common Divisor)*

If $P = \{p_1(x), p_2(x), \ldots, p_n(x)\}$ is a collection of *nonzero* but otherwise arbitrary polynomials, then

(a) The elements of P have a **greatest common divisor** $d(x)$

(b) The greatest common divisor $d(x)$ is *unique*, and there exist polynomials $a_1(x), a_2(x), \ldots, a_n(x)$ such that $d(x)$ can be written in the form

$$d(x) = a_1(x)p_1(x) + a_2(x)p_2(x) + \ldots + a_n(x)p_n(x) \tag{34.1}$$

Proof: Our proof is an adaptation of the proof given in Curtiss [4], and we shall employ the **Well-Ordering Axiom** for the set of natural numbers, $\mathbb{N}$. Recall that this axiom states that if S is any non-empty subset of $\mathbb{N}$, then S has a *least element* s_0; that is, $s_0 \leq s$ for all $s \in S$.

To establish (a), let $a_1(x), a_2(x), \ldots, a_n(x)$ be arbitrary polynomials, and define the set S as the collection of all polynomials $s(x)$ of the form

$$s(x) = \sum_{k=1}^{n} a_k(x)p_k(x)$$

Clearly, $P \subseteq S$, and if $s_1(x), s_2(x) \in S$ then $s_1(x) \cdot s_2(x) \in S$ as well. Now, let D be the collection of the *degrees* of all *nonzero* polynomials in S. Since $D \subseteq \mathbb{N}$, then by the Well-Ordering Axiom, there exists a non-zero polynomial $d(x) \in S$ such that

$$d(x) = a_1(x)p_1(x) + a_2(x)p_2(x) + \ldots + a_n(x)p_n(x)$$

and $\deg d(x) \leq \deg s(x)$ for all nonzero polynomials $s(x) \in S$. We now show that $d(x)$ divides each $p_k(x)$. For each k where $1 \leq k \leq n$, the Division Algorithm ensures there exist polynomials $q_k(x)$ and $r_k(x)$ such that

$$p_k(x) = q_k(x)d(x) + r_k(x)$$

and where either $r_k(x) = 0$ or $\deg r_k(x) < \deg d(x)$. Now since

$$r_k(x) = p_k(x) - q_k(x)d(x) \in S$$

and by our construction, $d(x)$ is the polynomial of *least* degree in S, we must have $r_k(x) = 0$. So, $d(x)$ is a common divisor of the elements in P.

To show that $d(x)$ is a *greatest* common divisor, assume that $d'(x)$ is any other common divisor of the elements in P. Then, there must exist polynomials, say $q'_1(x), q'_2(x), \ldots, q'_n(x)$ such that $p_k(x) = d'(x)q'_k(x)$ for $1 \leq k \leq n$. Consequently,

$$\begin{aligned} d(x) &= \sum_{k=1}^{n} a_k(x)p_k(x) = \sum_{k=1}^{n} a_k(x)[d'(x)q'_k(x)] \\ &= d'(x)\left(\sum_{k=1}^{n} a_k(x)q'_k(x)\right) \end{aligned}$$

So $d'(x)$ must divide $d(x)$ and consequently, $\deg d'(x) \leq \deg d(x)$.

Finally, we show that $d(x)$ is *unique*. Let $d''(x)$ be any common divisor of the elements of P. Then $d(x)$ divides $d''(x)$, and conversely $d''(x)$ must divide $d(x)$. Thus, there exist polynomials, say $q(x)$ and $q''(x)$ such that $d''(x) = q''(x)d(x)$ and $d(x) = q(x)d''(x)$. But, these equations imply that $q(x)q''(x) = 1$. Consequently, $q(x) = 1$ and $q''(x) = 1$. $\square$.

Theorem 34.3: *(Special Decomposition Theorem)*

Let $T : V \to V$ be a linear operator. If there exist polynomials $p_1(x)$ and $p_2(x)$ such that

$$p(x) = p_1(x)p_2(x)$$

where $p_1(x)$ and $p_2(x)$ are relatively prime, and $p(T) = 0$, then:
(a) The vector spaces $K_1 = \ker[p_1(T)]$ and $K_2 = \ker[p_2(T)]$ are invariant under T.
(b) The vector space V has the direct sum decomposition

$$V = K_1 \oplus K_2$$

Proof: To establish (a), note that both $p_1(T)$ and $p_2(T)$ are polynomials in T; hence, by the previous theorem, both $\ker[p_1(T)]$ and $\ker[p_2(T)]$ are invariant under T.

To establish (b), note that since $p_1(x)$ and $p_2(x)$ are relatively prime, their greatest common divisor $d(x) = 1$. So by the previous lemma, there must exist polynomials $q_1(x)$ and $q_2(x)$ such that

$$p_1(x)q_1(x) + p_2(x)q_2(x) = 1$$

Consequently, we can write

$$p_1(T)q_1(T) + p_2(T)q_2(T) = I$$

so that for any $\mathbf{v} \in V$, we have

$$[p_1(T)q_1(T) + p_2(T)q_2(T)](\mathbf{v}) = \mathbf{v} = [p_1(T)q_1(T)](\mathbf{v}) + [p_2(T)q_2(T)](\mathbf{v}) \qquad (\bigstar)$$

Now, for any $\mathbf{v} \in V$, we must have $[p_1(T)q_1(T)](\mathbf{v}) \in K_2$ since

$$\begin{aligned} p_2(T)[p_1(T)q_1(T)](\mathbf{v}) &= q_1(T)[p_1(T)p_2(T)](\mathbf{v}) \\ &= q_1(T)[p(T)(\mathbf{v})] \\ &= q_1(T) \cdot \mathbf{0} = \mathbf{0} \end{aligned}$$

Similarly, we can show $[p_2(T)q_2(T)](\mathbf{v}) \in K_1$. Consequently, by ($\bigstar$) we have $V = K_1 + K_2$. We now show that $K_1 \cap K_2 = \{\mathbf{0}\}$. To this end, assume that $\mathbf{x} \in K_1 \cap K_2$, so that we have $[p_1(T)](\mathbf{x}) = \mathbf{0}$ and $[p_2(T)](\mathbf{x}) = \mathbf{0}$. However, by ($\bigstar$), we have

$$\begin{aligned} \mathbf{x} &= [p_1(T)q_1(T)](\mathbf{x}) + [p_2(T)q_2(T)](\mathbf{x}) = q_1(T)[p_1(T)](\mathbf{x}) + q_2(T)[p_2(T)](\mathbf{x}) \\ &= q_1(T) \cdot \mathbf{0} + q_2(T) \cdot \mathbf{0} = \mathbf{0} \end{aligned}$$

so $K_1 \cap K_2 = \{\mathbf{0}\}$. By Theorem 15.3, we have

$$V = \ker[p_1(T)] \oplus \ker[p_2(T)]$$

as claimed. $\Box$.

Corollary 34.4: *(Minimal Polynomial)*

Let $T : V \to V$ be a linear operator, and let $\mu(T)$ be the minimal polynomial of T. If there exist monic polynomials $\mu_1(x)$ and $\mu_2(x)$ such that

$$\mu(x) = \mu_1(x)\mu_2(x)$$

where $\mu_1(x)$ and $\mu_2(x)$ are relatively prime, then $\mu_1(x)$ and $\mu_2(x)$ are the minimal polynomials of the restrictions $T|_{K_1}$ and $T|_{K_2}$ of T to K_1 and K_2, respectively, where

$$K_1 = \ker[\mu_1(T)] \quad \text{and} \quad K_2 = \ker[\mu_2(T)].$$

Proof: In order to simplify notation, let $T_1 = T|_{K_1}$ be the restriction of T to K_1. Now, for any vector $\mathbf{x} \in K_1$, we have $\mu_1(T)\mathbf{x} = \mu_1(T_1)\mathbf{x} = \mathbf{0}$, so $\mu_1(T_1) = 0$. Similarly, if let $T_2 = T|_{K_2}$ be the restriction of T to K_2, then $\mu_2(T) = 0$ as well. Clearly, if $p_1(x)$ is the minimal polynomial of $\mu_1(T)$, then since $\mu_1(T_1) = 0$, by a previous theorem $p_1(x)$ must divide $\mu_1(x)$. At the same time, if $p_2(x)$ is the minimal polynomial of $\mu_2(T)$, then $\mu_2(T_1) = 0$ implies that $p_2(x)$ divides $\mu_2(x)$. But since, by hypothesis, the polynomial factors $\mu_1(x)$ and $\mu_2(x)$ are both monic and relatively prime, and $\mu(x)$ is the *minimal* polynomial of T, we must have $\mu_1(x) = p_1(x)$ and $\mu_2(x) = p_2(x)$. $\Box$.

We now generalize the definition of direct sum, which was first introduced in Lesson 15.

Definition: *(Direct Sum)*

If $V_1, V_2, \ldots, V_n$ are subspaces of a vector space V, then we shall say that V is the **direct sum** of the spaces $V_1, V_2, \ldots, V_n$, and write

$$V = V_1 \oplus V_2 \oplus \cdots \oplus V_n$$

if and only if for each vector $\mathbf{v} \in V$, there exist vectors $\mathbf{v}_k \in V_k$ for $1 \le k \le n$ such that

$$\mathbf{v} = \mathbf{v}_1 + \mathbf{v}_2 + \ldots + \mathbf{v}_n \tag{34.2}$$

where the vectors on the right-hand side of (34.2) are *unique*.

Lemma 34.5: *(Criterion for Expressing V as a Direct Sum)*

If $V_1, V_2, \ldots, V_n$ are subspaces of a vector space V, then

$$V = V_1 \oplus V_2 \oplus \cdots \oplus V_n$$

if and only if:
(a) For each vector $\mathbf{v} \in V$, there exist vectors $\mathbf{v}_k \in V_k$ for $1 \le k \le n$ such that

$$\mathbf{v} = \mathbf{v}_1 + \mathbf{v}_2 + \ldots + \mathbf{v}_n$$

so that $V = V_1 + V_2 + \ldots + V_n$.
(b) If the vectors $\mathbf{v}_k \in V_k$ for $1 \le k \le n$ are such that

$$\mathbf{v}_1 + \mathbf{v}_2 + \ldots + \mathbf{v}_n = \mathbf{0}$$

then $\mathbf{v}_1 = \mathbf{v}_2 = \ldots = \mathbf{v}_n = \mathbf{0}$.

Proof: First, assume that $V = V_1 \oplus V_2 \oplus \cdots \oplus V_n$. By definition, property (a) is immediately satisfied. If the vectors $\mathbf{v}_k \in V_k$ for $1 \le k \le n$ are such that

$$\mathbf{v}_1 + \mathbf{v}_2 + \ldots + \mathbf{v}_n = \mathbf{0}$$

then we can write

$$\mathbf{v}_1 + \mathbf{v}_2 + \ldots + \mathbf{v}_n = \mathbf{0} + \mathbf{0} + \ldots + \mathbf{0}$$

which implies $\mathbf{v}_1 = \mathbf{v}_2 = \ldots = \mathbf{v}_n = \mathbf{0}$ by the uniqueness of the decomposition of a vector space into a direct sum; so (b) holds as well.

Now, assume that both (a) and (b) hold. We then have immediately that $V = V_1 + V_2 + \ldots + V_n$, so it suffices to prove the *uniqueness* of any decomposition. So suppose we have both

$$\mathbf{v} = \mathbf{v}_1 + \mathbf{v}_2 + \ldots + \mathbf{v}_n \quad \text{and} \quad \mathbf{v} = \mathbf{v}_1' + \mathbf{v}_2' + \ldots + \mathbf{v}_n'$$

Upon subtracting, we obtain

$$(\mathbf{v}_1 - \mathbf{v}_1') + (\mathbf{v}_2 - \mathbf{v}_1') + \ldots + (\mathbf{v}_n - \mathbf{v}_n') = \mathbf{0}$$

By property (b), we then have $\mathbf{v}_k - \mathbf{v}_k' = \mathbf{0}$ for all $1 \le k \le n$, and so, $\mathbf{v}_k = \mathbf{v}_k'$ for all k. $\square$.

Finally, we are in a position to show that if the minimal polynomial $\mu(x)$ of a linear operator T on a vector space V can be written as the product of relatively prime polynomials, then it induces a decomposition of V as the direct sum of subspaces of V, each of which is the kernel of a factor of the minimal polynomial, and has the special property of being invariant under T.

Theorem 34.6: *(General Decomposition Theorem)*

Let $T : V \to V$ be a linear operator on a finite-dimensional vector space V. If the minimal polynomial $\mu(x)$ of T has the form

$$\mu(x) = [p_1(x)]^{r_1} \cdot [p_2(x)]^{r_2} \cdots [p_n(x)]^{r_n}$$

where the monic polynomials $p_1(x), p_2(x), \ldots, p_n(x)$ are relatively prime, then
(a) The space V has the direct sum decomposition

$$V = V_1 \oplus V_2 \oplus \cdots \oplus V_n$$

where each subspace V_k is invariant under T and

$$V_k = \ker[p_k(T)]^{r_k} \quad \text{for } 1 \le k \le n$$

(b) Each polynomial $[p_k(T)]^{r_k}$ for $1 \le k \le n$, is the minimal polynomial of the restriction of T to V_k.

Proof: We prove (a) and leave (b) as an exercise. Assume that $1 \le k \le n$, and define the polynomials

$$q_k(x) = \frac{\mu(x)}{[p_k(T)]^{r_k}}$$

By construction, the set of polynomials $\{q_1(x), q_2(x), \ldots, q_n(x)\}$ are also monic and relatively prime. Consequently, by Lemma 34.2, there exist polynomials $a_1(x), a_2(x), \ldots, a_n(x)$ such that

$$a_1(x)q_1(x) + a_2(x)q_2(x) + \ldots + a_n(x)q_n(x) = 1$$

Replacing x by T, we obtain

$$a_1(T)q_1(T) + a_2(T)q_2(T) + \ldots + a_n(T)q_n(T) = I$$

and for any $\mathbf{v} \in V$, we can write

$$\mathbf{v} = I\mathbf{v} = a_1(T)q_1(T)\mathbf{v} + a_2(T)q_2(T)\mathbf{v} + \ldots + a_n(T)q_n(T)\mathbf{v}$$

Now, for each k, observe that

$$[p_k(T)]^{r_k}(a_k(T)q_k(T)\mathbf{v}) = a_k(T)\{[p_k(T)]^{r_k}q_k(T)\}\mathbf{v} = a_k(T)[\mu(T)\mathbf{v}] = \mathbf{0}$$

so that $a_k(T)q_k(T)\mathbf{v} \in \ker[p_k(T)]^{r_k}$ for each k, and

$$\begin{aligned} V &= \ker[p_1(T)]^{r_1} + \ker[p_2(T)]^{r_2} + \ldots + \ker[p_n(T)]^{r_n} \\ &= V_1 + V_2 + \ldots + V_n \end{aligned}$$

where we set $V_k = \ker[p_k(T)]^{r_k}$. It remains to show that the decomposition is unique. So suppose that for $\mathbf{v}_k \in V_k$ we have

$$\mathbf{v}_1 + \mathbf{v}_2 + \ldots + \mathbf{v}_n = \mathbf{0}$$

By the previous lemma, it suffices to show that $\mathbf{v}_1 = \mathbf{v}_2 = \cdots = \mathbf{v}_n = \mathbf{0}$. By Lemma 34.2, however, there exist polynomials $a_1(x)$ and $a_2(x)$ such that

$$a_1(x)[p_1(x)]^{r_1} + a_2(x)\left\{[p_2(x)]^{r_2}\cdots[p_n(x)]^{r_n}\right\} = 1$$

so that

$$a_1(T)[p_1(T)]^{r_1} + a_2(T)\left\{[p_2(T)]^{r_2}\cdots[p_n(T)]^{r_n}\right\} = I$$

But then

$$\begin{aligned}\mathbf{v}_1 &= I\mathbf{v}_1 = a_1(T)[p_1(T)]^{r_1}\mathbf{v}_1 + a_2(T)\left\{[p_2(T)]^{r_2}\cdots[p_n(T)]^{r_n}\right\}\left(-\mathbf{v}_2 - \ldots - \mathbf{v}_n\right)\\ &= a_1(T)[p_1(T)]^{r_1}\mathbf{v}_1 - \left\{[p_3(T)]^{r_3}\cdots[p_n(T)]^{r_n}\right\}a_2(T)[p_2(T)]^{r_2}\mathbf{v}_2 - \cdots\\ &\quad - \left\{[p_2(T)]^{r_2}\cdots[p_{n-1}(T)]^{r_{n-1}}\right\} a_n(T)\,[p_n(T)]^{r_n}\mathbf{v}_n\\ &= \mathbf{0}\end{aligned}$$

and consequently, $\mathbf{v}_1 = \mathbf{0}$. Similarly, we can show that $\mathbf{v}_2 = \mathbf{v}_3 = \ldots = \mathbf{v}_n = \mathbf{0}$ as well. □.

Example-3: Verify the previous theorem for the linear operator on $\mathbb{R}^4$ given by $T(\mathbf{v}) = A\mathbf{v}$ where

$$A = \begin{bmatrix} -2 & 0 & 0 & 0 \\ 0 & -3 & -2 & 1 \\ 0 & -3 & -8 & 3 \\ 0 & -3 & -6 & 1 \end{bmatrix}$$

Solution: It can be shown that the characteristic and minimum polynomials of T are

$$\Delta(x) = (x+6)(x+2)^3 \text{ and}$$
$$\mu(x) = (x+6)(x+2)$$

respectively. Let's find $V_1 = \ker(T+6I)$ and $V_2 = \ker(T+2I)$. To find V_1 we must solve the homogenous system

$$(T+6I)\mathbf{v} = \begin{bmatrix} 4 & 0 & 0 & 0 \\ 0 & 3 & -2 & 1 \\ 0 & -3 & -2 & 3 \\ 0 & -3 & -6 & 7 \end{bmatrix}\begin{bmatrix} v_1 \\ v_2 \\ v_3 \\ v_4 \end{bmatrix} = \begin{bmatrix} 0 \\ 0 \\ 0 \\ 0 \end{bmatrix}$$

Using the Gauss-Jordan method, we find that the general solution is

$$\mathbf{v} = s_1 \begin{bmatrix} 0 \\ \frac{1}{3} \\ 1 \\ 1 \end{bmatrix}$$

where s_1 is a real parameter. Consequently, $V_1 = span\{[0, 1/3, 1, 1]^T\}$. Similarly, to find V_2 we must solve the homogenous system

$$(T + 2I)\mathbf{v} = \begin{bmatrix} 0 & 0 & 0 & 0 \\ 0 & -1 & -2 & 1 \\ 0 & -3 & -6 & 3 \\ 0 & -3 & -6 & 3 \end{bmatrix} \begin{bmatrix} v_1 \\ v_2 \\ v_3 \\ v_4 \end{bmatrix} = \begin{bmatrix} 0 \\ 0 \\ 0 \\ 0 \end{bmatrix}$$

We find that the general solution is

$$\mathbf{v} = t_1 \begin{bmatrix} 1 \\ 0 \\ 0 \\ 0 \end{bmatrix} + t_2 \begin{bmatrix} 0 \\ -2 \\ 1 \\ 0 \end{bmatrix} + t_3 \begin{bmatrix} 0 \\ 1 \\ 0 \\ 1 \end{bmatrix}$$

where t_1, t_2, t_3 are real parameters. Consequently,

$$V_2 = span\{[1, 0, 0, 0]^T, [0. -2, 1, 0]^T, [0, 1, 0, 1]^T\}$$

The previous theorem guarantees that $\mathbb{R}^4 = V_1 \oplus V_2$. This is easy to check since the set of vectors

$$B = \{[0, 1/3, 1, 1]^T, [1, 0, 0, 0]^T, [0. -2, 1, 0]^T, [0, 1, 0, 1]^T\}$$

is linearly independent, and hence, forms a basis for $\mathbb{R}^4$.

Exercise Set 34

---

1. Let T be a linear operator on a vector space V. Show that each of the following subspaces is invariant under T:
(a) $\{\mathbf{0}\}$
(b) V
(c) *range* T

2. Let $\mathbf{v}$ be an eigenvector of the linear operator $T : \mathbb{C}^n \to \mathbb{C}^n$ and consider the one dimensional subspace S of $\mathbb{C}^n$ given by

$$S = \{c\mathbf{v} : c \in \mathbb{C}\}$$

Show that S is invariant under T.

3. Let $T(\mathbf{x}) = A\mathbf{x}$ be a linear operator on $\mathbb{R}^2$ where

$$A = \begin{bmatrix} 1 & 2 \\ 2 & 1 \end{bmatrix}$$

Find all invariant subspaces of T.

4. Let T be a linear operator on V. Show that if each of the subspaces S_1 and S_2 is invariant under T, then $S_1 \cap S_2$ is invariant under T.

5. Consider the linear operator on $\mathbb{R}^3$ given by $T(\mathbf{v}) = A\mathbf{v}$ where

$$A = \begin{bmatrix} 1 & 0 & 0 \\ 0 & 1 & 1 \\ 2 & 0 & 2 \end{bmatrix}$$

(**a**) Show that the minimal polynomial of A is $\mu(x) = (x-1)^2(x-2)$.
(**b**) Find a bases B_1 and B_2 for the subspaces of $\mathbb{R}^3$ given by

$$V_1 = \ker(A - I)^2 \text{ and } V_2 = \ker(A - 2I).$$

(**c**) Show that $B = B_1 \cup B_2$ is a basis for $\mathbb{R}^3$ so that $\mathbb{R}^3 = V_1 \oplus V_2$.
(**d**) Find a matrix representation of T in terms of the basis B.

6. Provide a proof of property (b) of Theorem 34.6.

35. Jordan Canonical Form

In our past work, we have seen that any square matrix A of order n, which has n-linearly independent eigenvectors, can be reduced to diagonal form by an appropriate similarity transformation. We have also seen that this procedure is not possible if A does not have n-linearly independent eigenvectors. We will show in this lesson that any non-diagonalizable matrix can still be brought into a simpler form, called its **Jordan canonical form**, that is almost as easy to work with as a diagonal matrix. But first, we need to establish some preliminary results.

Lemma 35.1: *(An Important Property of Direct Sums)*

Suppose that $V_1, V_2, \ldots, V_k$ are subspaces of a vector space V, such that

$$V = V_1 \oplus V_2 \oplus \cdots \oplus V_k$$

and

$$B_1 = \{b_{11}, b_{12}, \ldots, b_{1n_1}\}, \ldots, B_k = \{b_{k1}, b_{k2}, \ldots, b_{kn_k}\}$$

are bases of $V_1, V_2, \ldots, V_k$ respectively. Then $B = B_1 \cup B_2 \cup \ldots \cup B_k$ is a basis for V.

Proof: Since V is the direct sum of the V_i, then for any vector $\mathbf{v} \in V$, there exist vectors $\mathbf{v}_i \in V_i$ such that

$$\mathbf{v} = \mathbf{v}_1 + \mathbf{v}_2 + \ldots + \mathbf{v}_k$$

Each set B_i is a basis for V_i, so each vector $\mathbf{v}_i$ can be written as a linear combination of the vectors in B_i, and consequently, B spans V. We now show that B is linearly independent. So assume that

$$(c_{11}\mathbf{b}_{11} + c_{12}\mathbf{b}_{12} + \ldots, c_{1n_1}\mathbf{b}_{1n_1}) + \ldots + (c_{k1}\mathbf{b}_{11} + c_{k2}\mathbf{b}_{12} + \ldots, c_{kn_k}\mathbf{b}_{kn_k}) = \mathbf{0}$$

But we also have $\mathbf{0} + \mathbf{0} + \ldots + \mathbf{0} = \mathbf{0}$ and this sum must be unique. Thus,

$$(c_{11}\mathbf{b}_{11} + c_{12}\mathbf{b}_{12} + \ldots, c_{1n_1}\mathbf{b}_{1n_1}) = \mathbf{0}, \ldots, (c_{k1}\mathbf{b}_{11} + c_{k2}\mathbf{b}_{12} + \ldots, c_{kn_k}\mathbf{b}_{kn_k}) = \mathbf{0}$$

and we find that all of c's are zero owing to the linear independence of each set of basis vectors. □.

Assume that we have a linear operator $T : V \to V$ where V can be written as the direct sum of the subspaces $V_1,,, V_k$ of V. Assume also, that each subspace V_i is *invariant* under T, so we have

$$V = V_1 \oplus V_2 \oplus \cdots \oplus V_k \quad \text{where } T(V_i) \subseteq V_i \quad \text{for } i = 1, 2, \ldots, k \tag{35.1}$$

Let $T_i = T|_{V_i}$ be the restriction of T to each subspace V_i. Then we say that T is the **direct sum** of the T_i, and we agree to write formally

$$T = T_1 \oplus T_2 \oplus \cdots \oplus T_k \tag{35.2}$$

Alternatively, we say that the subspaces $V_1, \ldots, V_k$ form a T-**invariant decomposition** of V. In such cases, the matrix representation of T can be made to assume a particularly simple form.

To see this, let's suppose that $V = V_1 \oplus V_2$ where $\dim V_1 = 2$, $\dim V_2 = 2$ where the sets of vectors given by

$$B_1 = \{\mathbf{u}_1, \mathbf{u}_2\} \text{ and } B_2 = \{\mathbf{w}_1, \mathbf{w}_2\}$$

are bases for V_1 and V_2 respectively. Assume further that

$$T_1(\mathbf{u}_1) = a_{11}\mathbf{u}_1 + a_{21}\mathbf{u}_2 \quad \text{and} \quad T_1(\mathbf{u}_2) = a_{12}\mathbf{u}_1 + a_{22}\mathbf{u}_2$$
$$T_2(\mathbf{w}_1) = b_{11}\mathbf{w}_1 + b_{21}\mathbf{w}_2 \quad \text{and} \quad T_2(\mathbf{w}_2) = b_{12}\mathbf{w}_1 + b_{22}\mathbf{w}_2$$

Then the matrix representations of T_1 and T_2 are respectively

$$A_1 = \begin{bmatrix} a_{11} & a_{12} \\ a_{21} & a_{22} \end{bmatrix} \quad \text{and} \quad A_2 = \begin{bmatrix} b_{11} & b_{12} \\ b_{21} & b_{22} \end{bmatrix}$$

Now, by Lemma 35.1, the set $B = \{\mathbf{u}_1, \mathbf{u}_2, \mathbf{w}_1, \mathbf{w}_2\}$ is a basis for V. So for any vector $\mathbf{v} \in V$ there exist unique scalars $v_1, \ldots, v_4$ such that

$$\mathbf{v} = v_1\mathbf{u}_1 + v_2\mathbf{u}_2 + v_3\mathbf{w}_1 + v_4\mathbf{w}_2$$

Consequently,

$$\begin{aligned} T(\mathbf{v}) &= v_1T_1(\mathbf{u}_1) + v_2T_1(\mathbf{u}_2) + v_3T_2(\mathbf{w}_1) + v_4T_2(\mathbf{w}_2) \\ &= v_1(a_{11}\mathbf{u}_1 + a_{21}\mathbf{u}_2) + v_2(a_{12}\mathbf{u}_1 + a_{22}\mathbf{u}_2) \\ &\quad + v_3(b_{11}\mathbf{w}_1 + b_{21}\mathbf{w}_2) + v_4(b_{12}\mathbf{w}_1 + b_{22}\mathbf{w}_2) \end{aligned}$$

Upon rearranging terms, the last equation becomes

$$\begin{aligned} T(\mathbf{v}) &= (a_{11}v_1 + a_{12}v_2)\mathbf{u}_1 + (a_{21}v_1 + a_{22}v_2)\mathbf{u}_2 \\ &\quad + (b_{11}v_3 + b_{12}v_4)\mathbf{w}_1 + (b_{21}v_3 + b_{22}v_4)\mathbf{w}_2 \end{aligned}$$

We can write this in matrix form as

$$T\left(\begin{bmatrix} v_1 \\ v_2 \\ v_3 \\ v_4 \end{bmatrix}\right) = \begin{bmatrix} a_{11} & a_{12} & 0 & 0 \\ a_{21} & a_{22} & 0 & 0 \\ 0 & 0 & b_{11} & b_{12} \\ 0 & 0 & b_{21} & b_{22} \end{bmatrix}\begin{bmatrix} v_1 \\ v_2 \\ v_3 \\ v_4 \end{bmatrix}$$

Evidently, the matrix

$$A = \begin{bmatrix} a_{11} & a_{12} & 0 & 0 \\ a_{21} & a_{22} & 0 & 0 \\ 0 & 0 & b_{11} & b_{12} \\ 0 & 0 & b_{21} & b_{22} \end{bmatrix} = \begin{bmatrix} A_1 & 0 \\ 0 & A_2 \end{bmatrix} = diag\,(A_1, A_2)$$

is the matrix of T relative to the basis B. Each of the 2×2 matrices A_1 and A_2 are called **blocks** and A is called a **block-diagonal matrix**. We can easily extend the above argument to a general theorem:

Theorem 35.2: *(Block Diagonal Representation)*

Let T be a linear operator on a finite-dimensional space V. If V can be written as the direct sum of T-invariant subspaces $V_1, V_2, \ldots, V_k$; that is, if

$$V = V_1 \oplus V_2 \oplus \cdots \oplus V_k \quad \text{where } T(V_i) \subseteq V_i \quad \text{for } i = 1,2,\ldots,k$$

and each A_i is the matrix representation of the restriction $T_i = T|_{V_i}$ to V_i, then there exists a block-diagonal matrix representation A of T in the form

$$A = diag(A_1, A_2, \ldots, A_k) = \begin{bmatrix} A_1 & & & \\ & A_2 & & \\ & & \ddots & \\ & & & A_k \end{bmatrix} \tag{35.3}$$

Furthermore, each block A_i is a square matrix of order $\dim(V_i)$.

Proof: Left as an exercise for the reader.

Definition: *(Nilpotent Operator and Nilpotent Matrix)*

Let T be a linear operator on a vector space V, and let k be a natural number. Then T is said to be **nilpotent operator of index** k if $T^k = 0$ but $T^{k-1} \neq 0$. Similarly, a square matrix A is said to be a **nilpotent matrix of index** k if $A^k = 0$ but $A^{k-1} \neq 0$. In each case, we call k the **index of nilpotency**.

Now, if T is a nilpotent operator of order k and if λ is any eigenvalue of T with corresponding eigenvector $\mathbf{x}$, then $A\mathbf{x} = \lambda\mathbf{x}$ implies that

$$A^k\mathbf{x} = \lambda^k\mathbf{x} = \mathbf{0}$$

so $\lambda = 0$. Hence, $\lambda = 0$ is the only eigenvalue of T. Furthermore, the minimal polynomial of T must simply be $\mu(x) = x^k$.

Example-1: It may be easily verified that the matrix

$$A = \begin{bmatrix} 0 & 1 & 0 & 1 \\ 0 & 0 & 1 & 1 \\ 0 & 0 & 0 & 0 \\ 0 & 0 & 0 & 0 \end{bmatrix}$$

is a nilpotent matrix of index 3 since $A^3 = 0$ but $A^2 \neq 0$.

There are two special types of square matrices, called **Jordan nilpotent blocks** and **Jordan blocks**, that we will be working with for the remainder of this lesson. Later on, we will show that the matrix representation of any linear operator can be written as a Jordan block matrix such that each of its diagonal blocks are the sum of a scalar multiple of the identity matrix and a Jordan nilpotent block matrix.

Definition: *(Jordan Nilpotent Block and Jordan Block)*

A **Jordan nilpotent block** N_k is a square matrix of order k which is of the form

$$N_k = \begin{bmatrix} 0 & 1 & 0 & \cdots & 0 \\ 0 & 0 & 1 & \cdots & 0 \\ 0 & 0 & 0 & \ddots & 0 \\ \vdots & \vdots & \vdots & \ddots & 1 \\ 0 & 0 & 0 & \cdots & 0 \end{bmatrix} \qquad (35.4$$

such that N_k has ones all along the superdiagonal (the diagonal just above the main diagonal), and zeros elsewhere. Given any scalar λ, a **Jordan block** $J(\lambda)$ is a square matrix of order k which has the form

$$J_k(\lambda) = \begin{bmatrix} \lambda & 1 & 0 & \cdots & 0 \\ 0 & \lambda & 1 & \cdots & 0 \\ 0 & 0 & \lambda & \ddots & 0 \\ \vdots & \vdots & \vdots & \ddots & 1 \\ 0 & 0 & 0 & \cdots & \lambda \end{bmatrix} = \lambda I + N_k \qquad (35.5$$

It's easy to show that a Jordan nilpotent block N_k is a *nilpotent matrix* with index k. Later on, we will show that given any linear operator $T(\mathbf{x}) = A\mathbf{x}$ on a vector space V, we can always find a basis for V such that the matrix representation of T can be expressed in terms of one or more Jordan block matrices. This is the simplest representation of T that is available when A cannot be diagonalized.

Lemma 35.3: *(Matrix Representation of Nilpotent Operators)*

Let $T : V \to V$ be a nilpotent operator of index k on a finite-dimensional space V. Then:
(**a**) There exists a vector $\mathbf{v} \in V$ such that the set

$$C = \{\mathbf{v}, T(\mathbf{v}), T^2(\mathbf{v}), \ldots, T^{k-1}(\mathbf{v})\} \tag{35.6}$$

is linearly independent.
(**b**) The subspace $S = span\{\mathbf{v}, T(\mathbf{v}), \ldots, T^{k-1}(\mathbf{v})\}$ is invariant under T.
(**c**) Relative to the basis $\{T^{k-1}(\mathbf{v}), \ldots, T(\mathbf{v}), \mathbf{v}\}$, the matrix representation of the restriction $T|_S$ of T to S is the Jordan nilpotent block N_k.

Before proving this theorem, we should mention some terminology. Any set of linearly independent vectors which has the same form as the set C of vectors in (35.6) is called a **Jordan chain of length** k. Furthermore, if there exists a collection of Jordan chains $C_1, C_2, \ldots, C_m$ whose union forms a basis for a vector space V, then that collection of Jordan chains is called a **Jordan basis** for V.

Proof: Since $T : V \to V$ is a nilpotent operator of index k, then $T^k = 0$ but $T^{k-1} \neq 0$, so there must exist a *non-zero* vector $\mathbf{v} \in V$ such that $T^{k-1}(\mathbf{v}) \neq \mathbf{0}$. Assume there exist scalars $c_1, \ldots, c_k$ such that

$$c_1\mathbf{v} + c_2T(\mathbf{v}) + c_3T^2(\mathbf{v}) + \ldots + c_kT^{k-1}(\mathbf{v}) = \mathbf{0}$$

By T^{k-1} to this equation, we obtain $c_1T^{k-1}(\mathbf{v}) = \mathbf{0}$ so that $c_1 = 0$, and the last equation simplifies to

$$c_2T(\mathbf{v}) + c_3T^2(\mathbf{v}) + \ldots + c_kT^{k-1}(\mathbf{v}) = \mathbf{0}$$

Now apply T^{k-2} to this equation to get $c_2T^{k-1}(\mathbf{v}) = \mathbf{0}$, so $c_2 = 0$. Clearly, we can continue in this fashion to show that all of the remaining scalar coefficients are zero. So, the set S is linearly independent.

We now show that S is invariant under T. Since C is a basis for S, for any $\mathbf{x} \in S$ there must exist unique scalars $x_1, x_2, \ldots, x_k$ such that

$$\mathbf{x} = x_1\mathbf{v} + x_2T(\mathbf{v}) + \ldots + x_kT^{k-1}(\mathbf{v})$$

But then

$$T(\mathbf{x}) = x_1T(\mathbf{v}) + x_2T^2(\mathbf{v}) + \ldots + x_{k-1}T^{k-2}(\mathbf{v})$$

so $T(\mathbf{x}) \in S$, and S is invariant under T.

To establish (c), we relabel the vectors in C as

$$\mathbf{v}_1 = T^{k-1}(\mathbf{v}),\ \mathbf{v}_2 = T^{k-2}(\mathbf{v}), \ldots,\ \mathbf{v}_{k-1} = T(\mathbf{v}),\ \mathbf{v}_k = \mathbf{v}$$

so that

$$\begin{aligned}
T_S(\mathbf{v}_1) &= T_S[T^{k-1}(\mathbf{v})] = T^k(\mathbf{v}) = \mathbf{0} \\
T_S(\mathbf{v}_2) &= T_S[T^{k-2}(\mathbf{v})] = T^{k-1}(\mathbf{v}) = \mathbf{v}_1 \\
T_S(\mathbf{v}_3) &= T_S[T^{k-3}(\mathbf{v})] = T^{k-2}(\mathbf{v}) = \mathbf{v}_2 \\
&\vdots \\
T_S(\mathbf{v}_k) &= T(\mathbf{v}) = \mathbf{v}_{k-1}
\end{aligned}$$

Thus, given any vector $\mathbf{y} = y_1\mathbf{v}_1 + y_2\mathbf{v}_2 + \ldots + y_k\mathbf{v}_k \in S$ we have

$$T_S(\mathbf{y}) = y_1(\mathbf{0}) + y_2\mathbf{v}_1 + y_3\mathbf{v}_2 + \ldots + y_k\mathbf{v}_{k-1}$$

Now, we write the last equation in matrix form to obtain

$$T_S\begin{bmatrix} y_1 \\ y_2 \\ y_3 \\ \vdots \\ y_k \end{bmatrix} = \begin{bmatrix} 0 & 1 & 0 & \cdots & 0 \\ 0 & 0 & 1 & & \\ 0 & 0 & \ddots & \ddots & \\ \vdots & \vdots & \vdots & 0 & 1 \\ 0 & 0 & 0 & 0 & 0 \end{bmatrix}\begin{bmatrix} y_1 \\ y_2 \\ y_3 \\ \vdots \\ y_k \end{bmatrix} = \begin{bmatrix} y_2 \\ y_3 \\ \vdots \\ y_k \\ 0 \end{bmatrix}$$

Thus, the matrix representation of the restriction $T|_S$ of T to S is the Jordan nilpotent block N_k. □.

Now, we generalize the previous lemma by showing that a Jordan basis always exists for any nilpotent operator on a finite-dimensional vector space.

Theorem 35.4: *(Existence of Jordan Basis for Nilpotent Operators)*

Let $T : V \to V$ be a nilpotent operator of index k. Then there exists a Jordan basis for V such that the matrix N of T, relative to this basis, assumes the block diagonal form

$$N = \begin{bmatrix} N_1 & & & \\ & N_2 & & \\ & & \ddots & \\ & & & N_m \end{bmatrix} \qquad (35.$$

where each diagonal entry N_i is a Jordan nilpotent block, and at least one of the blocks N_i has order k while all other blocks have orders $\leq k$. Furthermore, the total number blocks is equal to the nullity of T.

Proof: Assume that $\dim V = n$. We shall construct a proof by induction on the dimension of V. Assume that $n = 1$. Since the index of T is k, then there must exist a vector $\mathbf{x} \in V$ such that $T^{k-1}\mathbf{x} \neq \mathbf{0}$. But then, by a previous lemma, the set of vectors $\{\mathbf{x}, T\mathbf{x}, T^2\mathbf{x}, \ldots, T^{k-1}\mathbf{x}\}$ must be linearly independent. This is only possible if $k = 1$. Thus, $T(\mathbf{x}) = 0$ for all $\mathbf{x} \in V$, and $\dim \ker T = 1$. In this special case, $N = [0]$ (a 1×1 matrix) and the theorem holds for $n = 1$.

Now, assume the theorem holds for all vector spaces of dimension less than n. Since T is nilpotent with index k, then there must exist a *non-zero* vector $T^{k-1}(\mathbf{x})$ for some $\mathbf{x} \in V$ such that

$$T^k(\mathbf{x}) = T(T^{k-1}\mathbf{x}) = \mathbf{0}$$

So the vector $T^{k-1}(\mathbf{x}) \in \ker V$ and consequently, $\dim \ker T \geq 1$. But then $\dim$ *range* $T < n$, and the hypothesis must apply to the range of T. Consequently, there must exist a basis B of Jordan chains C_i for the *range* T of the form

$$B = \bigcup_{i=1}^{p} C_i \text{ where } C_i = \{\mathbf{v}_i, T(\mathbf{v}_i), T^2(\mathbf{v}_i), \ldots, T^{l_i-1}(\mathbf{v}_i)\} \quad i = 1,2,3,\ldots,p \tag{A}$$

such that $l_i \leq k$ and at least one chain has length k.

Clearly, the p-linearly independent vectors

$$T^{l_1-1}(\mathbf{v}_1), T^{l_2-1}(\mathbf{v}_2), T^{l_3-1}(\mathbf{v}_3), \ldots, T^{l_p-1}(\mathbf{v}_p) \in \ker T$$

so we must have

$$\dim \textit{range } T = r_1 + p$$

where r_1 is the number of linearly independent vectors in B but not in the $\ker T$. Thus, if $\dim \ker T = m$, then there must exist $(m - p)$ linearly independent vectors such that

$$\mathbf{k}_1, \mathbf{k}_2, \ldots, \mathbf{k}_{m-p} \in \ker T$$

but $\mathbf{k}_s \notin B$ for $1 \leq s \leq m - p$. Now, observe that

$$n = \dim V = r_1 + m = (r_1 + p) + (m - p)$$

so the vectors in $B \cup \{\mathbf{k}_1, \mathbf{k}_2, \ldots, \mathbf{k}_{m-p}\}$ will form a basis for V if we can show that this set is linearly independent.

To this end, assume that there exist sets of scalars c_{ij} and d_j such that

$$\sum_{i=1}^{p}\sum_{j=0}^{l_i-1} c_{ij}T^j(\mathbf{v}_i) + \sum_{j=1}^{m-p} d_j\mathbf{k}_j = \mathbf{0} \tag{B}$$

Applying T to this equation, we find that

$$\sum_{i=1}^{p}\sum_{j=0}^{l_i-1} c_{ij}T^{j+1}(\mathbf{v}_i) = \mathbf{0} \tag{C}$$

where we used the fact $T(\mathbf{k}_j) = \mathbf{0}$ for $j = 1,,,m-p$. By hypothesis, however, all of the scalars $c_{ij} = 0$ for $1 \le i \le p$ and $1 \le j \le l_i - 2$. Substituting this result back into (C), we find that

$$\sum_{i=1}^{p} c_{il_i-1}T^{l_i-1}(\mathbf{v}_i) = \mathbf{0} \tag{D}$$

So, the remaining c's are zero as well. Finally, substituting the results of (C) and (D) back into (B), we obtain

$$\sum_{j=1}^{m-p} d_j\mathbf{k}_j = \mathbf{0}$$

so that all of the d's are zero as well. Thus, the m Jordan chains

$$C_1, C_2, C_3, \cdots, C_p, \mathbf{k}_1, \mathbf{k}_2, \ldots, \mathbf{k}_{m-p}$$

form a basis for V, where $\mathbf{k}_1, \mathbf{k}_2, \ldots \mathbf{k}_{m-p} \in \ker T$ are chains of unit length. This completes the proof. □.

Example-2: Find a Jordan basis for the nilpotent matrix A (of Example-1):

$$A = \begin{bmatrix} 0 & 1 & 0 & 1 \\ 0 & 0 & 1 & 1 \\ 0 & 0 & 0 & 0 \\ 0 & 0 & 0 & 0 \end{bmatrix}$$

with respect to which, A assumes the block diagonal form given in (35.7).

Solution: Since the dim $V = 4$, and the dim *range* $T = 2$, then the dim $\ker T = 2$. Thus, by the previous theorem, there must exist exactly 2 Jordan nilpotent blocks in the block diagonal form of A. Now, the index of A is 3, so we first find a Jordan chain of length 3. Thus, we must find a vector $\mathbf{v}$ such that $A^2\mathbf{v} \neq 0$. We find that

$$A^2\mathbf{v} = \begin{bmatrix} 0 & 0 & 1 & 1 \\ 0 & 0 & 0 & 0 \\ 0 & 0 & 0 & 0 \\ 0 & 0 & 0 & 0 \end{bmatrix}\begin{bmatrix} v_1 \\ v_2 \\ v_3 \\ v_4 \end{bmatrix} = \begin{bmatrix} v_3 + v_4 \\ 0 \\ 0 \\ 0 \end{bmatrix}$$

so if we take $\mathbf{v} = [0,0,1,1]^T$ then $A^2\mathbf{v} = [2,0,0,0]^T \neq 0$. We now find the remaining element $A\mathbf{v}$ of the chain:

$$Av = \begin{bmatrix} 0 & 1 & 0 & 1 \\ 0 & 0 & 1 & 1 \\ 0 & 0 & 0 & 0 \\ 0 & 0 & 0 & 0 \end{bmatrix} \begin{bmatrix} 0 \\ 0 \\ 1 \\ 1 \end{bmatrix} = \begin{bmatrix} 1 \\ 2 \\ 0 \\ 0 \end{bmatrix}$$

Thus, a Jordan chain is

$$C = \{A^2\mathbf{v}, A\mathbf{v}, \mathbf{v}\} = \left\{[2,0,0,0]^T, [1,2,0,0]^T, [0,0,1,1]^T\right\}$$

Next, to complete a basis for $\mathbb{R}^4$ we must add a vector $\mathbf{k}_1$ in the $\ker T$ such that the set $\{A^2\mathbf{v}, A\mathbf{v}, \mathbf{v}, \mathbf{k}_1\}$ is linearly independent. One such vector is

$$\mathbf{k}_1 = \begin{bmatrix} 0 \\ -1 \\ -1 \\ 1 \end{bmatrix} \in \ker T$$

which is an *eigenvector* of A. So, an appropriate transition matrix P and its inverse P^{-1} are

$$P = \begin{bmatrix} 2 & 1 & 0 & 0 \\ 0 & 2 & 0 & -1 \\ 0 & 0 & 1 & -1 \\ 0 & 0 & 1 & 1 \end{bmatrix} \text{ and } P^{-1} = \begin{bmatrix} \frac{1}{2} & -\frac{1}{4} & \frac{1}{8} & -\frac{1}{8} \\ 0 & \frac{1}{2} & -\frac{1}{4} & \frac{1}{4} \\ 0 & 0 & \frac{1}{2} & \frac{1}{2} \\ 0 & 0 & -\frac{1}{2} & \frac{1}{2} \end{bmatrix}$$

Consequently, we find that

$$P^{-1}AP = \begin{bmatrix} 0 & 1 & 0 & 0 \\ 0 & 0 & 1 & 0 \\ 0 & 0 & 0 & 0 \\ 0 & 0 & 0 & 0 \end{bmatrix} \equiv \begin{bmatrix} N_1 & \\ & N_2 \end{bmatrix}$$

where

$$N_1 = \begin{bmatrix} 0 & 1 & 0 \\ 0 & 0 & 1 \\ 0 & 0 & 0 \end{bmatrix} \text{ and } N_2 = [0]$$

Note that the total number of nilpotent blocks = 2 = $\dim \ker T$ as guaranteed by the previous theorem.

Theorem 35.5: *(Jordan Canonical Form)*

Let T be a linear operator on a complex vector space V. Suppose further, that the characteristic and minimal polynomials of T are

$$\Delta(x) = (x - \lambda_1)^{n_1}(x - \lambda_2)^{n_2} \cdots (x - \lambda_r)^{n_r}$$
$$\mu(x) = (x - \lambda_1)^{m_1}(x - \lambda_2)^{m_2} \cdots (x - \lambda_r)^{m_r}$$

respectively, where the eigenvalues $\lambda_1, \lambda_2, \ldots, \lambda_r$ of T are distinct. Then there exists a Jordan basis for V such that the matrix J of T, relative to this basis, assumes the block diagonal form

$$J = \begin{bmatrix} J(\lambda_1) & & & \\ & J(\lambda_2) & & \\ & & \ddots & \\ & & & J(\lambda_r) \end{bmatrix} \qquad (35.8)$$

where each diagonal entry $J(\lambda_i)$ consists of one or more Jordan blocks J_{ij} corresponding to λ_i. Furthermore, for each i, the Jordan blocks J_{ij} satisfy the following properties:

(a) The sum of the orders of the blocks J_{ij} in $J(\lambda_i)$ is n_i.
(b) The number of Jordan blocks J_{ij} corrresponding to λ_i is equal to the geometric multiplicity of λ_i.
(c) There is at least one block J_{ij} of order m_i and all of the remaining blocks corresponding to λ_i have orders $\leq m_i$.
(d) The number of Jordan blocks J_{ij} corresponding to λ_i is uniquely determined by T.

Proof: By the General Decomposition Theorem, T has the decomposition

$$T = T_1 \oplus T_2 \oplus \ldots \oplus T_r$$

where $T_i = T|_{K_i}$ and where each subspace $K_i = \ker(T - \lambda_i I)^{m_i}$ of V is invariant under T. Furthermore, for each i, $\mu_i(x) = (x - \lambda_i)^{m_i}$ is the minimal polynomial of T_i. Consequently, we must have

$$(T - \lambda_1 I)^{m_1} = \mathbf{0}, \ (T - \lambda_2 I)^{m_2} = \mathbf{0}, \ \ldots, \ (T - \lambda_r I)^{m_r} = \mathbf{0}$$

Thus, each operator $(T - \lambda_i I)^{m_i}$ is a nilpotent operator of index m_i. Now let $N_i = T - \lambda_i I$ so that for $1 \leq i \leq r$, we have

$$T_i = N_i + \lambda_i I \ \text{ and } \ N_i^{m_i} = 0$$

By the previous theorem, we can find a Jordan basis so that N_i assumes the block diagonal form shown in (35.7). But then, in this basis, each $T_i = N_i + \lambda_i I$ has a matrix representation $J(\lambda_i)$ consisting of Jordan blocks J_{ij}. Finally, by Theorem 35.2, the collection of matrices $J(\lambda_i)$ in J must form the matrix representation of T.

Property (a) follows immediately from the fact that T and J are similar matrices, and therefore, they must have the same characteristic polynomial. Property (b) follows from the previous theorem, and the observation that for each i,

$$\dim \ker N_i = \dim \ker(T - \lambda_i I) = g_i$$

where g_i is the *geometric multiplicity* of the eigenvalue λ_i.

Property (c) follows immediately from the previous theorem, and the fact that each N_i has an index of nilpotency of m_i. Finally, the last property follows from the fact that each T_i (and therefore each N_i) is uniquely determined by T. □.

Given a linear operator $T : \mathbb{C}^n \to \mathbb{C}^n$ such that $T(\mathbf{x}) = A\mathbf{x}$ for all $x \in \mathbb{C}^n$, the previous theorem guarantees the existence of a Jordan basis, and hence, a transition matrix P of order n such that

$$P^{-1}AP = J \tag{35.9}$$

Here, of course, J is a Jordan matrix which may consist of one or more Jordan block matrices, and whose diagonal elements are the eigenvalues of A.

An efficient way of finding such a Jordan basis is to use the concept of a generalized eigenvector. A **generalized eigenvector v** of **degree** k of A corresponding to the eigenvalue λ is any *non-zero* vector $\mathbf{v}$ such that

$$(A - \lambda I)^k \mathbf{v} = \mathbf{0} \tag{35.10}$$

but $(A - \lambda I)^{k-1}\mathbf{v} \neq \mathbf{0}$. Observe that if $k = 1$, then the definition reduces to $(A - \lambda I)\mathbf{v} = \mathbf{0}$ and $\mathbf{v} \neq \mathbf{0}$; this is just the definition of an eigenvector. Thus, the term "generalized eigenvector" is certainly appropriate.

Now, consider the set of generalized eigenvectors corresponding to λ

$$C = \{\mathbf{v}, (A - \lambda I)\mathbf{v}, (A - \lambda I)^2\mathbf{v}, \ldots, (A - \lambda I)^{k-1}\mathbf{v}\} \tag{35.11}$$

If the factor $(x - \lambda)^k$ appears in the minimal polynomial of A, then as discussed in the proof of the previous theorem, $N = (A - \lambda I)^k$ is the minimal polynomial of the restriction of T to the $\ker(A - \lambda I)^k$, and consequently, $(A - \lambda I)$ must be a *nilpotent matrix* of index k. Hence, it follows from Lemma 35.3, that the vectors in C will be linearly independent and thus, they will form a Jordan chain of length k.

We can repeat this analysis for each eigenvalue of A to generate as many Jordan chains as necessary. These chains, together with the chains of unit length that correspond to simple eigenvalues, will form a Jordan basis for C^n.

This simple procedure gives us a practical method of finding a Jordan basis which is illustrated in the following examples.

Example-3: Reduce the following matrix to Jordan canonical form:

$$A = \begin{bmatrix} 1 & 0 \\ 2 & 1 \end{bmatrix}$$

Solution: It turns out that the characteristic and minimal polynomials of A are identical, and

$$\Delta(x) = (x-1)^2$$

From $\Delta(x)$ we see that $\lambda_1 = -1$ is an eigenvalue of multiplicity 2, and

$$\mathbf{v}_1 = \begin{bmatrix} 0 \\ 1 \end{bmatrix}$$

is the corresponding eigenvector. We now construct a Jordan basis. We seek a vector $\mathbf{v}_2$ such that

$$(A-I)\mathbf{v}_2 = \mathbf{v}_1$$

In doing so, we find

$$\begin{bmatrix} 0 & 0 \\ 2 & 0 \end{bmatrix}\begin{bmatrix} x_1 \\ x_2 \end{bmatrix} = \begin{bmatrix} 0 \\ 1 \end{bmatrix}$$

whose solution is $\mathbf{v}_2 = [1/2, t]^T$ where t is a real parameter. So, take $t = 0$ so $\mathbf{v}_2 = [1/2, 0]^T$. Now observe that the matrix $N = A - I$ is a nilpotent matrix of index 2, since

$$(A-I)^2\mathbf{v}_2 = (A-I)\mathbf{v}_1 = 0$$
$$(A-I)\mathbf{v}_2 = \mathbf{v}_1$$

Thus, we have constructed a Jordan chain $C = \{\mathbf{v}_1, \mathbf{v}_2\}$ of length 2, which will serve as a Jordan basis. The vector $\mathbf{v}_2$ is the **generalized eigenvector** of A corresponding to $\lambda_1 = 1$. Our transition matrix P is the matrix whose columns are the vectors $\mathbf{v}_1$ and $\mathbf{v}_2$:

$$P = [\mathbf{v}_1|\mathbf{v}_2] = \begin{bmatrix} 0 & 1/2 \\ 1 & 0 \end{bmatrix} \text{ so } P^{-1} = \begin{bmatrix} 0 & 1 \\ 2 & 0 \end{bmatrix}$$

Consequently,

$$J = P^{-1}AP = \begin{bmatrix} 1 & 1 \\ 0 & 1 \end{bmatrix}$$

Observe that J consists of a single Jordan block of order 2; this is consistent with property (c) of the previous theorem.

Example-4: Find the Jordan canonical form of the matrix

$$A = \begin{bmatrix} 4 & 2 & -1 \\ 2 & 1 & -2 \\ 3 & 2 & 0 \end{bmatrix}$$

Solution: We find that the characteristic and minimal polynomials of A are identical:

$$\Delta(x) = \mu(x) = (x-1)^2(x-3)$$

The eigenvalues of A are $\lambda_1 = 1$ and $\lambda_2 = 3$ with corresponding eigenvectors:

$$\mathbf{v}_1 = \begin{bmatrix} 1 \\ -1 \\ 1 \end{bmatrix} \text{ and } \mathbf{y}_2 = \begin{bmatrix} 1 \\ 0 \\ 1 \end{bmatrix}$$

Since the multiple eigenvalue $\lambda_1 = 1$ corresponds to $\mathbf{v}_1$, we seek a generalized eigenvector $\mathbf{v}_2 = [x_1, x_2, x_3]^T$ such that $(A - I)\mathbf{v}_2 = \mathbf{v}_1$. Consequently,

$$(A - I)\mathbf{v}_2 = \begin{bmatrix} 3 & 2 & -1 \\ 2 & 0 & -2 \\ 3 & 2 & -1 \end{bmatrix} \begin{bmatrix} x_1 \\ x_2 \\ x_3 \end{bmatrix} = \begin{bmatrix} 1 \\ -1 \\ 1 \end{bmatrix}$$

whose solution is $\mathbf{v}_2 = \left[t - \frac{1}{2}, \frac{5}{4} - t, t\right]^T$ where t is a real parameter. We take $t = 1$, so that $\mathbf{v}_2 = \left[\frac{1}{2}, \frac{1}{4}, 1\right]^T$. Observe that we now have two Jordan chains

$$C_1 = \{\mathbf{v}_1, \mathbf{v}_2\} \text{ and } C_2 = \{\mathbf{y}_2\}$$

of lengths $l_1 = 2$ and $l_2 = 1$ whose union forms a basis for $\mathbb{R}^3$. Now, the transition matrix must be

$$P = [\mathbf{v}_1 | \mathbf{v}_2 | \mathbf{y}_2] = \begin{bmatrix} 1 & 1/2 & 1 \\ -1 & 1/4 & 0 \\ 1 & 1 & 1 \end{bmatrix}, \text{ so } P^{-1} = \begin{bmatrix} -\frac{1}{2} & -1 & \frac{1}{2} \\ -2 & 0 & 2 \\ \frac{5}{2} & 1 & -\frac{3}{2} \end{bmatrix}$$

and consequently,

$$J = P^{-1}AP = \begin{bmatrix} 1 & 1 & 0 \\ 0 & 1 & 0 \\ 0 & 0 & 3 \end{bmatrix}$$

Example-5: Find a Jordan basis for the matrix

$$A = \begin{bmatrix} 1 & 2 & 0 & 0 \\ 0 & 1 & 0 & 0 \\ 0 & 0 & 2 & 0 \\ 0 & 0 & 3 & 3 \end{bmatrix}$$

Solution: We find that the characteristic and minimal polynomials of A are identical:

$$\Delta(x) = \mu(x) = (x-1)^2(x-2)(x-3)$$

The eigenvalues of A are $\lambda_1 = 1$, $\lambda_2 = 2$ and $\lambda_3 = 3$ with corresponding eigenvectors:

$$\mathbf{v}_1 = \begin{bmatrix} 1 \\ 0 \\ 0 \\ 0 \end{bmatrix}, \ \mathbf{y}_1 = \begin{bmatrix} 0 \\ 0 \\ -\frac{1}{3} \\ 1 \end{bmatrix}, \ \text{and } \mathbf{y}_2 = \begin{bmatrix} 0 \\ 0 \\ 0 \\ 1 \end{bmatrix}$$

Since the multiple eigenvalue corresponds to the eigenvector $\mathbf{v}_1$, we seek a generalized eigenvector $\mathbf{v}_2 = [x_1, x_2, x_3, x_4]^T$ such that $(A - I)\mathbf{v}_2 = \mathbf{v}_1$. Consequently,

$$(A - I)\mathbf{v}_2 = \begin{bmatrix} 0 & 2 & 0 & 0 \\ 0 & 0 & 0 & 0 \\ 0 & 0 & 1 & 0 \\ 0 & 0 & 3 & 2 \end{bmatrix} \begin{bmatrix} x_1 \\ x_2 \\ x_3 \\ x_4 \end{bmatrix} = \begin{bmatrix} 1 \\ 0 \\ 0 \\ 0 \end{bmatrix}$$

whose solution is $\mathbf{v}_2 = \left[t, \frac{1}{2}, 0, 0\right]^T$ where t is a real parameter. We take $t = 0$, so that $\mathbf{v}_2 = \left[0, \frac{1}{2}, 0, 0\right]^T$. Observe that we now have three Jordan chains

$$C_1 = \{\mathbf{v}_1, \mathbf{v}_2\},\ C_2 = \{\mathbf{y}_1\},\ C_3 = \{\mathbf{y}_2\}$$

of lengths $l_1 = 2$, $l_2 = 1$, and $l_3 = 1$, whose union forms a basis for $\mathbb{R}^4$. Now, the transition matrix must be

$$P = [\mathbf{v}_1 | \mathbf{v}_2 | \mathbf{y}_1 | \mathbf{y}_2] = \begin{bmatrix} 1 & 0 & 0 & 0 \\ 0 & \frac{1}{2} & 0 & 0 \\ 0 & 0 & -\frac{1}{3} & 0 \\ 0 & 0 & 1 & 1 \end{bmatrix} \ \text{and } P^{-1} = \begin{bmatrix} 1 & 0 & 0 & 0 \\ 0 & 2 & 0 & 0 \\ 0 & 0 & -3 & 0 \\ 0 & 0 & 3 & 1 \end{bmatrix}$$

Consequently,

$$J = P^{-1}AP = \begin{bmatrix} 1 & 1 & 0 & 0 \\ 0 & 1 & 0 & 0 \\ 0 & 0 & 2 & 0 \\ 0 & 0 & 0 & 3 \end{bmatrix}$$

Exercise Set 35

1. Provide a proof of Theorem 35.2.

2. By direct calculation, show that the matrix

$$N = \begin{bmatrix} 0 & 1 & 1 & 0 \\ 0 & 0 & 1 & 0 \\ 0 & 0 & 0 & 1 \\ 0 & 0 & 0 & 0 \end{bmatrix}$$

is a nilpotent matrix of index 4.

In Exercises 3-5, find a Jordan basis for each matrix, and write each matrix in Jordan canonical form:

3. $A = \begin{bmatrix} 2 & 3 \\ 0 & 2 \end{bmatrix}$

4. $A = \begin{bmatrix} 1 & 0 & 0 \\ 1 & 1 & 0 \\ 0 & 1 & 1 \end{bmatrix}$

5. $A = \begin{bmatrix} 1 & 0 & 0 \\ 0 & 2 & 0 \\ 1 & 1 & 2 \end{bmatrix}$

6. Show that the matrix N_k given in (35.4) is a nilpotent matrix of index k.

Unit IX: Applications

George Boole (1815-1864)

George Boole was an English mathematician who made important contributions to the fields of mathematical logic and the theory of differential equations. Born in Lincoln, Lincolnshire, England, to the modest family of a shoemaker, John Boole, and Mary Ann Joyce, Boole did not enjoy the benefits of a formal education, and was largely self-taught.

In spite of his humble upbringing, his mathematical talent was soon recognized and in 1849, he was appointed as professor of mathematics at Queen's College in Ireland. Boole went on to publish over 50 journal articles on his research.

Although best known for his *Laws of Thought* (1854), which contained the important notions of Boolean algebra and Boolean logic, he also made significant contributions to the study of differential equations and difference equations. He was one of the first mathematicians to introduce operator methods in the solution of differential equations with constant coefficients.

Boole received many honors for his work. In 1857, he was appointed as a fellow of the Royal Society, and received an honorary doctorate from Oxford in 1859. Unfortunately, he died from pneumonia at the relatively young age of 49. Today, he is generally acknowledged as the principal founder of the Information Age.

36. Systems of First Order Differential Equations

A **system of n first order differential equations** is one of the form

$$\begin{aligned} \dot{x}_1 &= a_{11}x_1 + a_{12}x_2 + \ldots + a_{1n}x_n + f_1(t) \\ \dot{x}_2 &= a_{21}x_1 + a_{22}x_2 + \ldots + a_{2n}x_n + f_2(t) \\ &\vdots \\ \dot{x}_n &= a_{n1}x_1 + a_{n2}x_2 + \ldots + a_{nn}x_n + f_n(t) \end{aligned} \tag{36.1}$$

where the unknown functions $x_1, x_2, \ldots, x_n$ are functions of t, and the dots denote differentiation with respect to t. This system can be written in matrix form as

$$\dot{\mathbf{x}}(t) = A\mathbf{x}(t) + \mathbf{f}(t) \tag{36.2}$$

where the $n \times n$ matrix A has constant elements and is given by

$$A = \begin{bmatrix} a_{11} & a_{12} & \cdots & a_{1n} \\ a_{21} & a_{22} & \cdots & a_{2n} \\ \vdots & \vdots & & \vdots \\ a_{n1} & a_{n2} & \cdots & a_{nn} \end{bmatrix}$$

and the $1 \times n$ matrices $\mathbf{x}(t)$ and $\mathbf{f}(t)$ are

$$\mathbf{x}(t) = \begin{bmatrix} x_1(t) \\ x_2(t) \\ \vdots \\ x_n(t) \end{bmatrix}, \quad \mathbf{f}(t) = \begin{bmatrix} f_1(t) \\ f_2(t) \\ \vdots \\ f_n(t) \end{bmatrix}$$

Now, the system (36.2) is said to be **homogeneous** if $\mathbf{f}(t) = \mathbf{0}$, and **non-homogeneous** if $\mathbf{f}(t) \neq \mathbf{0}$. It is well known that the system (36.2), together with a given *initial condition*, say

$$\mathbf{x}_0 \equiv \mathbf{x}(0) = \begin{bmatrix} x_1(0) \\ x_2(0) \\ \vdots \\ x_n(0) \end{bmatrix} \tag{36.3}$$

has a *unique* solution.

Consider the non-homogeneous *initial value problem*

$$\dot{\mathbf{x}}(t) = A\mathbf{x}(t) + \mathbf{f}(t)$$
$$\mathbf{x}(0) = \mathbf{x}_0 \qquad (36.4)$$

In order to solve (36.4), we first rearrange terms to obtain

$$\dot{\mathbf{x}} - A\mathbf{x} = \mathbf{f}(t)$$

and multiply both sides (on the left) by e^{-At} to obtain

$$e^{-At}\dot{\mathbf{x}} - Ae^{-At}\mathbf{x} = e^{-At}\mathbf{f}(t)$$

so that

$$\frac{d}{dt}[e^{-At}\mathbf{x}(\mathrm{t})] = e^{-At}\mathbf{f}(t)$$

Integrating from 0 to t, we obtain

$$e^{-At}\mathbf{x}(t) - \mathbf{x}(0) = \int_0^t e^{-Au}\mathbf{f}(u)du$$

Thus, the solution of the non-homogeneous initial value problem is

$$\mathbf{x}(t) = \mathbf{e}^{At}\mathbf{x}_0 + \int_0^t e^{A(t-u)}\mathbf{f}(u)du \qquad (36.5)$$

Observe that the solution (36.5) consists of two terms. The first term is the solution of the corresponding homogeneous problem, and the second term is a *particular solution* of the non-homogeneous problem (which satisfies the initial condition $\mathbf{x}_0 = \mathbf{0}$). We have proved the following theorem:

Theorem 36.1: *(Explicit Solution of lst Order System)*

The solution of the non-homogeneous system of first order differential equations

$$\dot{\mathbf{x}}(t) = A\mathbf{x}(t) + \mathbf{f}(t)$$
$$\mathbf{x}(0) = \mathbf{x}_0 \qquad (36.4)$$

is given by

$$\mathbf{x}(t) = \mathbf{e}^{At}\mathbf{x}_0 + \int_0^t e^{A(t-u)}\mathbf{f}(u)du \qquad (36.5)$$

The solution of the corresponding homogeneous problem:

$$\dot{\mathbf{x}}(t) = A\mathbf{x}(t)$$
$$\mathbf{x}(0) = \mathbf{x}_0 \qquad (36.4)$$

is

$$\mathbf{x}(t) = \mathbf{e}^{At}\mathbf{x}_0 \qquad (36.5)$$

Our success in implementing the above solutions hinges upon our ability to evaluate the matrix exponentials involved. From Lesson 31, we already know how to do this if A is diagonalizable. If A is not diagonalizable, then as we shall see presently, we can still carry out our scheme by first expressing A in Jordan canonical form. We first consider the simplest case where A can be diagonalized by the usual procedure.

Example-1: Find the solution of the homogeneous system

$$\dot{x}_1(t) = x_1(t) + 2x_2(t)$$
$$\dot{x}_2(t) = 2x_1(t) + x_2(t)$$
$$x_1(0) = 1,\ x_2(0) = 0$$

Solution: We first write the system in matrix form as

$$\begin{bmatrix} \dot{x}_1(t) \\ \dot{x}_2(t) \end{bmatrix} = \begin{bmatrix} 1 & 2 \\ 2 & 1 \end{bmatrix} \begin{bmatrix} x_1(t) \\ x_2(t) \end{bmatrix}$$
$$\begin{bmatrix} x_1(0) \\ x_2(0) \end{bmatrix} = \begin{bmatrix} 1 \\ 0 \end{bmatrix}$$

Upon comparing this result with (35.4), we see that

$$A = \begin{bmatrix} 1 & 2 \\ 2 & 1 \end{bmatrix}, \text{ and } \mathbf{x}_0 = \begin{bmatrix} 1 \\ 0 \end{bmatrix}$$

so our solution will be given by $\mathbf{x}(t) = e^{At}\mathbf{x}_0$. The eigenvalues of A are $\lambda_1 = -1, \lambda_2 = 3$ with the corresponding eigenvectors $\mathbf{u}_1 = [-1,1]^T$ and $\mathbf{u}_2 = [1,1]^T$. By equation (31.3), we obtain

$$e^{At} = \begin{bmatrix} -1 & 1 \\ 1 & 1 \end{bmatrix} \begin{bmatrix} e^{-t} & 0 \\ 0 & e^{3t} \end{bmatrix} \begin{bmatrix} -\frac{1}{2} & \frac{1}{2} \\ \frac{1}{2} & \frac{1}{2} \end{bmatrix}$$
$$= \begin{bmatrix} \frac{1}{2}e^{-t} + \frac{1}{2}e^{3t} & \frac{1}{2}e^{3t} - \frac{1}{2}e^{-t} \\ \frac{1}{2}e^{3t} - \frac{1}{2}e^{-t} & \frac{1}{2}e^{-t} + \frac{1}{2}e^{3t} \end{bmatrix}$$

So the solution is

$$\mathbf{x}(t) = e^{At}\mathbf{x}_0 = \begin{bmatrix} \frac{1}{2}e^{-t} + \frac{1}{2}e^{3t} \\ \frac{1}{2}e^{3t} - \frac{1}{2}e^{-t} \end{bmatrix}$$

Example-2: Find the solution of the non-homogeneous system:

$$\dot{x}_1(t) = -8x_1(t) - 12x_2(t) + t$$
$$\dot{x}_2(t) = 6x_1(t) + 9x_2(t) + 1$$
$$x_1(0) = 0,\ x_2(0) = 1$$

Once again, we first write the system in matrix form as

$$\begin{bmatrix} \dot{x}_1(t) \\ \dot{x}_2(t) \end{bmatrix} = \begin{bmatrix} -8 & -12 \\ 6 & 9 \end{bmatrix}\begin{bmatrix} x_1(t) \\ x_2(t) \end{bmatrix} + \begin{bmatrix} t \\ 1 \end{bmatrix}$$
$$\begin{bmatrix} x_1(0) \\ x_2(0) \end{bmatrix} = \begin{bmatrix} 0 \\ 1 \end{bmatrix}$$

Upon comparing the given system with (36.4), have

$$A = \begin{bmatrix} -8 & -12 \\ 6 & 9 \end{bmatrix},\ \mathbf{x}_0 = \begin{bmatrix} 0 \\ 1 \end{bmatrix},\ \mathbf{f}(t) = \begin{bmatrix} t \\ 1 \end{bmatrix}$$

The eigenvalues of A are $\lambda_1 = 0, \lambda_2 = 1$ with the corresponding eigenvectors, $\mathbf{u}_1 = [-3,2]^T$ and $\mathbf{u}_2 = [-4,3]^T$. We find that

$$e^{At} = \begin{bmatrix} -3 & -4 \\ 2 & 3 \end{bmatrix}\begin{bmatrix} 1 & 0 \\ 0 & e^t \end{bmatrix}\begin{bmatrix} -3 & -4 \\ 2 & 3 \end{bmatrix}$$
$$= \begin{bmatrix} 9 - 8e^t & 12 - 12e^t \\ 6e^t - 6 & 9e^t - 8 \end{bmatrix}$$

Also,

$$e^{A(t-u)}\mathbf{f}(u) = \begin{bmatrix} 9 - 8e^{t-u} & 12 - 12e^{t-u} \\ 6e^{t-u} - 6 & 9e^{t-u} - 8 \end{bmatrix}\begin{bmatrix} u \\ 1 \end{bmatrix}$$

After some simple but tedious integrations, we find

$$\int_0^t e^{A(t-u)}\mathbf{f}(u)du = \begin{bmatrix} 20t - 20e^t + \frac{9}{2}t^2 + 20 \\ 15e^t - 14t + 3t^2 - 15 \end{bmatrix}$$

Thus,

$$\mathbf{x}(t) = \mathbf{e}^{At}\mathbf{x}_0 + \int_0^t e^{A(t-u)}\mathbf{f}(u)du = \begin{bmatrix} 20t - 32e^t + \frac{9}{2}t^2 + 32 \\ 24e^t - 14t + 3t^2 - 23 \end{bmatrix}$$

We now turn our attention to the case where the matrix A in (36.4) is not diagonalizable. We first prove the following useful lemma:

Lemma 36.2: *(Exponential of a Jordan Block)*

Let J_k denote the $k \times k$ Jordan block matrix

$$J_k = \begin{bmatrix} \lambda_k & 1 & 0 & \cdots & 0 \\ 0 & \lambda_k & 1 & \cdots & 0 \\ 0 & 0 & \lambda_k & \ddots & 0 \\ \vdots & \vdots & \vdots & \ddots & 1 \\ 0 & 0 & 0 & \cdots & \lambda_k \end{bmatrix} = \lambda_k I + N_k \tag{36.6}$$

where N_k is a $k \times k$ nilpotent matrix of index k. Then for any parameter t,

$$e^{J_k t} = e^{\lambda_k t} \begin{bmatrix} 1 & t & \frac{t^2}{2!} & \cdots & \frac{t^{n-1}}{(k-1)!} \\ 0 & 1 & t & \cdots & \frac{t^{n-2}}{(k-2)!} \\ 0 & 0 & 1 & \ddots & \vdots \\ \vdots & \vdots & \vdots & \ddots & t \\ 0 & 0 & 0 & \cdots & 1 \end{bmatrix} \tag{36.7}$$

Proof: By the power series expansion for the matrix exponential, we have

$$e^{J_k t} = \sum_{n=0}^{\infty} \frac{1}{n!} (\lambda_k t I + N_k t)^n$$

By the Binomial Theorem, we can write

$$e^{J_k t} = \sum_{n=0}^{\infty} \frac{1}{n!} \left\{ (\lambda_k t)^n I + \binom{n}{1} (\lambda_k t)^{n-1} N_k t + \ldots + \binom{n}{k-1} \lambda_k^{n-k-1} (N_k t)^{k-1} \right\}$$

where we have used the fact that $N_k^m = 0$ for $m \geq k$.

Now, observe that we can rewrite the last expression as

$$e^{J_k t} = \sum_{n=0}^{\infty} \frac{(\lambda_k t)^n I}{n!} + \frac{N_k t}{1!} \sum_{n=1}^{\infty} \frac{(\lambda_k t)^{n-1}}{(n-1)!} + \frac{(N_k t)^2}{2!} \sum_{n=2}^{\infty} \frac{(\lambda_k t)^{n-2}}{(n-2)!} + \ldots$$
$$\ldots + \frac{(Nt)^{k-1}}{(k-1)!} \sum_{n=k-1}^{\infty} \frac{(\lambda_k t)^{n-k-1}}{(n-k-1)!}$$

Upon noting that each of the sums in the last expression is equal to $e^{\lambda_k t}$, we can write

$$e^{J_k t} = e^{\lambda_k t}\left(I + \frac{N_k t}{1!} + \frac{N_k^2 t^2}{2!} + \ldots + \frac{N_k^{k-1} t^{k-1}}{(k-1)!}\right)$$

from which (36.7) follows immediately. □.

Recall that a Jordan matrix J is a diagonal block matrix whose diagonal elements consist of one or more matrices of the type shown in (36.6). If P is the $n \times n$ transition matrix whose columns are formed from the generalized eigenvectors of A, then

$$At = P \cdot diag\,(J_1 t, J_2 t, \ldots, J_n t) \cdot P^{-1}$$

so that

$$e^{At} = P \cdot diag\,(e^{J_1 t}, e^{J_2 t}, \ldots, e^{J_n t}) \cdot P^{-1} \tag{36.8}$$

where each of the diagonal blocks $e^{J_k t}$ can be found by (36.7). Thus, (36.7) and (36.8) provide us with a practical recipe for calculating the matrix exponential of any non-diagonalizable matrix. Let's see how this works.

Example-4: Find the solution of the homogeneous system

$$\dot{\mathbf{x}}(t) = A\mathbf{x}(t)$$

$$\mathbf{x}(0) = \mathbf{x}_0$$

where

$$A = \begin{bmatrix} 4 & 2 & -1 \\ 2 & 1 & -2 \\ 3 & 2 & 0 \end{bmatrix}, \quad \mathbf{x}_0 = \begin{bmatrix} 0 \\ 0 \\ 1 \end{bmatrix}$$

Solution: We first find an expression for e^{At}. Here, the matrix A is the same matrix that we put into Jordan canonical form in Example-4 of the previous lesson. We found that

$$P = \begin{bmatrix} 1 & 1/2 & 1 \\ -1 & 1/4 & 0 \\ 1 & 1 & 1 \end{bmatrix}, \quad P^{-1} = \begin{bmatrix} -\frac{1}{2} & -1 & \frac{1}{2} \\ -2 & 0 & 2 \\ \frac{5}{2} & 1 & -\frac{3}{2} \end{bmatrix},$$

$$J = \begin{bmatrix} 1 & 1 & 0 \\ 0 & 1 & 0 \\ 0 & 0 & 3 \end{bmatrix}$$

We also found that the eigenvalues of A were $\lambda_1 = 1$ (of multiplicity 2) and $\lambda_2 = 3$. Consequently, by (36.7) and (36.8), we have

$$
\begin{aligned}
e^{At} &= P \cdot diag(e^{J_1 t}, e^{J_2 t}) \cdot P^{-1} \\
&= P \cdot \begin{bmatrix} e^t & te^t & 0 \\ 0 & e^t & 0 \\ 0 & 0 & e^{3t} \end{bmatrix} \cdot P^{-1} \\
&= \begin{bmatrix} 1 & 1/2 & 1 \\ -1 & 1/4 & 0 \\ 1 & 1 & 1 \end{bmatrix} \begin{bmatrix} e^t & te^t & 0 \\ 0 & e^t & 0 \\ 0 & 0 & e^{3t} \end{bmatrix} \begin{bmatrix} -\frac{1}{2} & -1 & \frac{1}{2} \\ -2 & 0 & 2 \\ \frac{5}{2} & 1 & -\frac{3}{2} \end{bmatrix} \\
&= \begin{bmatrix} \frac{5}{2}e^{3t} - \frac{3}{2}e^t - 2te^t & e^{3t} - e^t & \frac{3}{2}e^t - \frac{3}{2}e^{3t} + 2te^t \\ 2te^t & e^t & -2te^t \\ \frac{5}{2}e^{3t} - \frac{5}{2}e^t - 2te^t & e^{3t} - e^t & \frac{5}{2}e^t - \frac{3}{2}e^{3t} + 2te^t \end{bmatrix}
\end{aligned}
$$

The solution is then $\mathbf{x}(t) = e^{At}\mathbf{x}_0$ so

$$
\begin{aligned}
\mathbf{x}(t) &= \begin{bmatrix} \frac{5}{2}e^{3t} - \frac{3}{2}e^t - 2te^t & e^{3t} - e^t & \frac{3}{2}e^t - \frac{3}{2}e^{3t} + 2te^t \\ 2te^t & e^t & -2te^t \\ \frac{5}{2}e^{3t} - \frac{5}{2}e^t - 2te^t & e^{3t} - e^t & \frac{5}{2}e^t - \frac{3}{2}e^{3t} + 2te^t \end{bmatrix} \begin{bmatrix} 0 \\ 0 \\ 1 \end{bmatrix} \\
&= \begin{bmatrix} \frac{3}{2}e^t - \frac{3}{2}e^{3t} + 2te^t \\ -2te^t \\ \frac{5}{2}e^t - \frac{3}{2}e^{3t} + 2te^t \end{bmatrix}
\end{aligned}
$$

Exercise Set 36

---

1. Use differentiation to show that (36.5) is the solution of (36.4).

2. Use differentiation to show that

$$\mathbf{x}(t) = \int_0^t e^{A(t-u)}\mathbf{f}(u)du$$

is the solution of the non-homogeneous system

$$
\begin{aligned}
\dot{\mathbf{x}}(t) &= A\mathbf{x}(t) + \mathbf{f}(t) \\
\mathbf{x}(0) &= \mathbf{0}
\end{aligned}
$$

3. Compute e^{At} for each of the following matrices:

$$
\textbf{(a)}\quad A = \begin{bmatrix} 1 & 2 \\ 0 & 1 \end{bmatrix} \qquad \textbf{(b)}\quad A = \begin{bmatrix} 1 & 0 & 0 \\ 0 & 2 & 0 \\ 1 & 1 & 2 \end{bmatrix} \qquad \textbf{(c)}\quad \begin{bmatrix} 1 & 0 & 0 \\ 1 & 1 & 1 \\ 2 & 0 & 1 \end{bmatrix}
$$

4. Use the results of Exercise-3 to solve the systems

(a) $\dot{\mathbf{x}} = \begin{bmatrix} 1 & 2 \\ 0 & 1 \end{bmatrix} \mathbf{x} + \begin{bmatrix} 1 \\ 0 \end{bmatrix}$ where $\mathbf{x}_0 = \begin{bmatrix} 0 \\ 1 \end{bmatrix}$.

(b) $\dot{\mathbf{x}} = \begin{bmatrix} 1 & 0 & 0 \\ 0 & 2 & 0 \\ 1 & 1 & 2 \end{bmatrix} \mathbf{x}$ where $\mathbf{x}_0 = \begin{bmatrix} 0 \\ 1 \\ 0 \end{bmatrix}$.

(c) $\dot{\mathbf{x}} = \begin{bmatrix} 1 & 0 & 0 \\ 1 & 1 & 1 \\ 2 & 0 & 1 \end{bmatrix} \mathbf{x}$ where $\mathbf{x}_0 = \begin{bmatrix} 1 \\ 1 \\ 0 \end{bmatrix}$.

5. Consider the second order differential equation

$$
\frac{d^2x}{dt^2} + \omega^2 x = 0, \;\; x(0) = a, \;\; \dot{x}(0) = b
$$

where ω, a, b are all real constants, and $\omega > 0$. (This equation is important in the study of oscillatory motion.)

(a) By making the substitutions $x_1 = x$ and $x_2 = \dot{x}$, show that the original differential equation is equivalent to the system

$$
\begin{bmatrix} \dot{x}_1(t) \\ \dot{x}_2(t) \end{bmatrix} = \begin{bmatrix} 0 & 1 \\ -\omega^2 & 0 \end{bmatrix} \begin{bmatrix} x_1(t) \\ x_2(t) \end{bmatrix}
$$

subject to the initial condition $\mathbf{x}_0 = [a, b]^T$.

(b) By solving the linear system, show that the solution of the original differential equation is given by

$$
x = a\cos\omega t + \frac{b}{\omega}\sin\omega t
$$

(c) Show that the motion is periodic with period $P = 2\pi/\omega$.

37. Stability Analysis of First Order Systems

In this lesson, we develop the tools to discuss the *qualitative behavior* of the solutions to systems of first order differential equations. Such an understanding is extremely useful since these systems have numerous applications to the study of dynamical systems that often appear in engineering and the physical sciences.

Without loss of generality, we shall first confine our analysis to the initial value problem

$$\dot{\mathbf{x}}(t) = A\mathbf{x}(t) \tag{37.1a}$$

$$\mathbf{x}(0) = \mathbf{x}_0 \tag{37.1b}$$

where

$$A = \begin{bmatrix} a_{11} & a_{12} \\ a_{21} & a_{22} \end{bmatrix}, \quad \mathbf{x} = \begin{bmatrix} x_1(t) \\ x_2(t) \end{bmatrix}, \quad \mathbf{x}_0 = \begin{bmatrix} x_1(0) \\ x_2(0) \end{bmatrix} \tag{37.2}$$

and $\det(A) \neq 0$. We can imagine that (37.2) describes the dynamics of of a physical system, and therefore, it is often referred to as a **dynamical system**. The vector $\mathbf{x}(t)$ is often called a **state vector**, and for each value of t, the vector $\mathbf{x}(t)$ is represents a possible **state** of the system.

If we allow the parameter t to assume successive values, then the ordered pairs of solutions $(x_1(t), x_2(t))$ will trace out a curve in the x_1x_2-plane. This curve is called a **trajectory** or **phase portrait** of the system. Clearly, each point on the trajectory represents a possible state of the system.

Example-1: Consider the linear system

$$\begin{bmatrix} \dot{x}_1(t) \\ \dot{x}_2(t) \end{bmatrix} = \begin{bmatrix} 0 & 1 \\ -k^2 & 0 \end{bmatrix} \begin{bmatrix} x_1(t) \\ x_2(t) \end{bmatrix}$$

$$\mathbf{x}_0 = \begin{bmatrix} x_1(0) \\ x_2(0) \end{bmatrix} = \begin{bmatrix} a \\ 0 \end{bmatrix}$$

where the real constants $a, k > 0$. It's easy to show that the solution of the system is

$$\mathbf{x}(t) = \begin{bmatrix} x_1(t) \\ x_2(t) \end{bmatrix} = \begin{bmatrix} a\cos kt \\ -a\sin kt \end{bmatrix}$$

so that for each $a > 0$, the curve

$$[x_1(t)]^2 + [x_2(t)]^2 = a^2$$

is a trajectory of the system. As shown in Figure 37.1, each trajectory is a circle of radius a in the x_1x_2-plane.

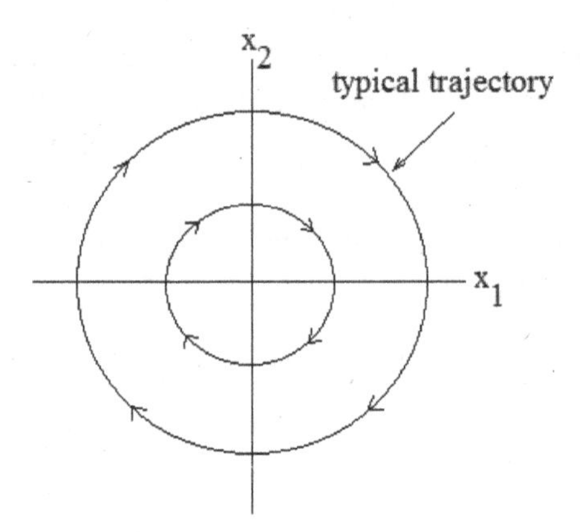

Figure 37.1: Some trajectories of the system in Example-1.

Note that starting from any initial state $(a, 0)$, the successive states $(x_1(t), x_2(t))$ move clockwise around each circle as t increases.

A point $(\bar{x}_1, \bar{x}_2)$ is said to be an **equilibrium point** (or **critical point**) of the system (37.1) if

$$A\begin{bmatrix} \hat{x}_1 \\ \hat{x}_2 \end{bmatrix} = \mathbf{0} \tag{37.3}$$

Since $\det(A) \neq 0$, then the only fixed point of (37.1) is the origin $(0, 0)$. But then, from (37.1), the system has the constant solution $\mathbf{x} = \mathbf{x}_0$ for all time.

In many physical applications, it is useful to know whether any solution of (37.1) that starts close to the constant solution $\mathbf{x} = \mathbf{x}_0$ remains close for all time $t > 0$. If this is the case, then we say that the equilibrium point $(0, 0)$ is **stable**. If, in addition, the successive states of the system *converge* to the constant solution $\mathbf{x} = \mathbf{x}_0$, then we say that the equilibrium point is **asymptotically stable**.

Clearly, the general solution (37.1) depends upon the eigenvalues of A, so we should be able to understand the stability of the system in terms of those eigenvalues. To see this, assume that λ_1, λ_2 are the *distinct* eigenvalues of A, with the corresponding (linearly independent) eigenvectors $\mathbf{u}_1, \mathbf{u}_2$. Now, it's easy to see that

$$\mathbf{x}(t) = c_1\mathbf{u}_1 e^{\lambda_1 t} + c_2\mathbf{u}_2 e^{\lambda_2 t} \tag{37.4}$$

is a solution of (37.1a), since

$$\dot{\mathbf{x}}(t) = c_1\lambda_1\mathbf{u}_1e^{\lambda_1 t} + c_2\lambda_2\mathbf{u}_2e^{\lambda_2 t} = A(c_1\mathbf{u}_1e^{\lambda_1 t} + c_2\mathbf{u}_2e^{\lambda_2 t}) = A\mathbf{x}(t)$$

Furthermore, the constants $c_1$and c_2 are uniquely determined by the initial condition $\mathbf{x}(0) = \mathbf{x}_0$. To see this, observe that since the eigenvectors $\mathbf{u}_1$ and $\mathbf{u}_2$ are linearly independent, there exist unique scalars a, b such that

$$\mathbf{x}(0) = \mathbf{x}_0 = a\mathbf{u}_1 + b\mathbf{u}_2$$

But then, from (37.4), we must have $c_1 = a$, and $c_2 = b$. Thus, we can always satisfy the initial condition $\mathbf{x}(0) = \mathbf{x}_0$ by first expanding $\mathbf{x}_0$ in terms of the eigenvectors of A.

From the above analysis, it is clear that the qualitative behavior of any solution that is initially close to the equilibrium point $(0, 0)$, directly depends on the nature of the eigenvalues of A. So, we summarize the following cases:

1. Two Negative Eigenvalues $(\lambda_1, \lambda_2 < 0)$: Here, both of the terms in (37.4) decay exponentially as t increases. Thus,

$$\lim_{t\to\infty} \mathbf{x}(t) = \mathbf{x}_0$$

and the equilibrium point $(0, 0)$ is asymptotically stable. It is often called a **stable node.**

2. Two Positive Eigenvalues $(\lambda_1, \lambda_2 > 0)$: In this case, each of the terms in (37.4) increase exponentially as $t \to \infty$. The equilibrium point $(0, 0)$ is then unstable, and is called an **unstable node.**

3. Eigenvalues Have Opposite Signs $(\lambda_1\lambda_2 < 0)$: If for example, we have $\lambda_1 > 0$ and $\lambda_2 < 0$, then the term $c_1\mathbf{u}_1e^{\lambda_1 t} \to \infty$ while the term $c_2\mathbf{u}_2e^{\lambda_2 t} \to 0$. The equilibrium point $(0, 0)$ is unstable and is called a **saddle point.**

4. Complex Conjugate Eigenvalues $(\lambda = a \pm bi)$: Here, by Euler's formula, the solution in (37.4) can be written in terms of functions

$$e^{\lambda t} = \exp[(a \pm bi)t] = e^{at}(\cos bt \pm i \sin bt)$$

so there are several cases:
(a) If $a = \operatorname{Re}\lambda < 0$ then $\mathbf{x}(t) \to \mathbf{0}$ as $t \to \infty$, so the equilibrium point $(0, 0)$ is asymptotically stable, and is called a **center**.
(b) If $a = \operatorname{Re}\lambda > 0$ then $\mathbf{x}(t) \to \infty$ as $t \to \infty$, and the equilibrium point is clearly unstable. It is sometimes called a **spiral point**.
(c) If $a = \operatorname{Re}\lambda = 0$ then the solution $\mathbf{x}(t)$ neither grows nor decays as $t \to \infty$, but is purely oscillatory. Hence, the equilibrium point $(0, 0)$ is stable, and is called a **center point**; it is not asymptotically stable.

For easy reference, the results of our analysis have been summarized in

Table 37.1 below.

Eigenvalues of A	Stability
$\lambda_1, \lambda_2 < 0$	Stable Node; Asymptotic Stability
$\lambda_1, \lambda_2 > 0$	Unstable Node
$\lambda_1 \cdot \lambda_2 < 0$	Unstable; Saddle Point
$\lambda = a \pm ib$ where $a < 0$	Stable; Asymptotic Stability
$\lambda = \pm ib$	Stable center point; Oscillatory Solutions
$\lambda = a \pm ib$ where $a > 0$	Unstable; Spiral Point

Table 37.1: Stability Analysis of (37.1).

The above analysis can sometimes help us understand the qualitative behavior of the solutions to *nonlinear* systems of differential equations. Consider the system

$$\begin{aligned} \dot{\mathbf{x}}(t) &= \mathbf{F}(x_1, x_2) \\ \mathbf{x}(0) &= \mathbf{x}_0 \end{aligned} \qquad (37.5$$

where the arguments of $\mathbf{F}(x_1, x_2)$ do not *explicitly* depend upon t, and

$$\mathbf{x}(t) = \begin{bmatrix} x_1(t) \\ x_2(t) \end{bmatrix}, \quad \mathbf{F}(x_1, x_2) = \begin{bmatrix} f_1(x_1, x_2) \\ f_2(x_1, x_2) \end{bmatrix}, \quad \mathbf{x}_0 = \begin{bmatrix} x_1(0) \\ x_2(0) \end{bmatrix} \qquad (37.6$$

We call (37.5) an **autonomous system**. We note that, in general, one or more of the functions $f_1(x_1, x_2)$ and $f_2(x_1, x_2)$ may be nonlinear functions of x_1 and x_2. Although such a system may be impossible to solve, we can, in many cases, gain an understanding of the behavior of its solutions by **linearizing** the system.

To see how this works, we first find any equilibrium points $(\hat{x}_1, \hat{x}_2)$ of the system. Here, we determine the equilibrium points of (37.6) by solving the system of equations

$$f_1(x_1, x_2) = 0, \quad f_2(x_1, x_2) = 0$$

for x_1 and x_2.

Let's assume that this has been done, and $(\hat{x}_1, \hat{x}_2)$ is a known equilibrium point of the system. We now seek solutions to (37.5) in the form

$$\begin{aligned} x_1(t) &= \hat{x}_1 + \varepsilon_1(t) \\ x_2(t) &= \hat{x}_2 + \varepsilon_2(t) \end{aligned} \qquad (37.8$$

where $\varepsilon_1(t)$ and $\varepsilon_2(t)$ are small perturbations. By Taylor series, we can write for $i = 1, 2$:

$$f_i(x_1,x_2) = f_i(\hat{x}_1,\hat{x}_2) + \varepsilon_1 \frac{\partial f_i}{\partial x_1}\bigg|_{(\hat{x}_1,\hat{x}_2)} + \varepsilon_2 \frac{\partial f_i}{\partial x_2}\bigg|_{(\hat{x}_1,\hat{x}_2)} + \text{(higher order terms)}$$

But $f_i(\hat{x}_1,\hat{x}_2) = 0$, so neglecting higher order terms, we have

$$\begin{bmatrix} f_1(x_1,x_2) \\ f_2(x_1,x_2) \end{bmatrix} = \begin{bmatrix} \partial f_1/\partial x_1 & \partial f_1/\partial x_2 \\ \partial f_2/\partial x_1 & \partial f_2/\partial x_2 \end{bmatrix}\Bigg|_{(\hat{x}_1,\hat{x}_2)} \begin{bmatrix} \varepsilon_1(t) \\ \varepsilon_2(t) \end{bmatrix} \tag{37.9}$$

where

$$\mathbf{DF}_{(\hat{x}_1,\hat{x}_2)} \equiv \begin{bmatrix} \partial f_1/\partial x_1 & \partial f_1/\partial x_2 \\ \partial f_2/\partial x_1 & \partial f_2/\partial x_2 \end{bmatrix}\Bigg|_{(\hat{x}_1,\hat{x}_2)} \tag{37.10}$$

is the **Jacobian matrix** of f_1 and f_2, evaluated at the fixed point $(\hat{x}_1,\hat{x}_2)$; note that all of its elements are constant.

Now, we substitute (37.8) and (37.9) into (37.5) to obtain the *linear* system

$$\begin{bmatrix} \dot{\varepsilon}_1(t) \\ \dot{\varepsilon}_2(t) \end{bmatrix} = \begin{bmatrix} \partial f_1/\partial x_1 & \partial f_1/\partial x_2 \\ \partial f_2/\partial x_1 & \partial f_2/\partial x_2 \end{bmatrix}\Bigg|_{(\hat{x}_1,\hat{x}_2)} \begin{bmatrix} \varepsilon_1(t) \\ \varepsilon_2(t) \end{bmatrix} \tag{37.11}$$

Thus, if we stay sufficiently close to the equilibrium point $(\hat{x}_1,\hat{x}_2)$, then the original system (37.5) can be approximated by a linearized system. The main point is that (37.11) is a *linear* first order system whose dynamics can be understood by our previous method of analysis.

Furthermore, it can be shown that the failure to account for the effect of higher order terms doesn't present a problem. *That is, if the analysis of the linearized system implies the existence of a stable node, unstable node, saddle point, spiral point or center point, then the corresponding fixed point of the nonlinear system will have the same characterization.*

Example-2: Find and classify all fixed points of the first order system

$$\begin{bmatrix} \dot{x}_1(t) \\ \dot{x}_2(t) \end{bmatrix} = \begin{bmatrix} x_1^2 - x_2 \\ x_1 - x_2 \end{bmatrix}$$

Solution: Here,

$$\mathbf{F}(x_1,x_2) = \begin{bmatrix} x_1^2 - x_2 \\ x_1 - x_2 \end{bmatrix}$$

so we find the fixed points by solving

$$x_1^2 - x_2 = 0$$
$$x_1 - x_2 = 0$$

We obtain the fixed points $P_1(0,0)$ and $P_2(1,1)$. The Jacobian matrix is

$$D\mathbf{F} = \begin{bmatrix} 2x_1 & -1 \\ 1 & -1 \end{bmatrix}$$

At the fixed point $P_1(0,0)$ we have

$$D\mathbf{F}|_{(0,0)} = \begin{bmatrix} 0 & -1 \\ 1 & -1 \end{bmatrix}$$

so the eigenvalues are the complex numbers $\lambda = -\frac{1}{2} \pm \frac{1}{2}i\sqrt{3}$.
Since $\operatorname{Re}\lambda = -\frac{1}{2} < 0$, then the fixed point $(0,0)$ is stable; in fact, it is asymptotically stable.

At the fixed point $P_1(1,1)$ we have

$$D\mathbf{F}|_{(1,1)} = \begin{bmatrix} 2 & -1 \\ 1 & -1 \end{bmatrix}$$

so the eigenvalues are the real numbers $\lambda = \frac{1}{2} \pm \frac{1}{2}\sqrt{5}$. Since the eigenvalues have opposite signs, then the fixed point $P_2(1,1)$ is unstable; it is a saddle point.

In addition to the method of analysis shown above, it should be pointed out that, in some cases, it may be possible to obtain the trajectories of the system directly from (37.5). To see this, first write out (37.5) as

$$\frac{dx_1}{dt} = f_1(x_1, x_2)$$
$$\frac{dx_2}{dt} = f_2(x_1, x_2)$$

so that if $f_1(x_1, x_2) \neq 0$, then

$$\frac{dx_2}{dx_1} = \frac{f_2(x_1, x_2)}{f_1(x_1, x_2)} \qquad \text{(37.1.}$$

If this equation can be integrated, it will provide the trajectories of the original system (37.5). Similarly, if $f_2(x_1, x_2) \neq 0$ in some region of the phase plane, we can use the equation

$$\frac{dx_1}{dx_2} = \frac{f_1(x_1, x_2)}{f_2(x_1, x_2)} \qquad \text{(37.1.}$$

to compute the trajectories.

Example-3: *(Nonlinear Pendulum)* We examine the frictionless motion of a pendulum bob of mass m, as shown in Figure 37.2. At time $t = 0$, we assume that the pendulum bob is at the position $\theta(0) = 0$ with an angular speed of $\dot{\theta}(0) = \omega_0$.

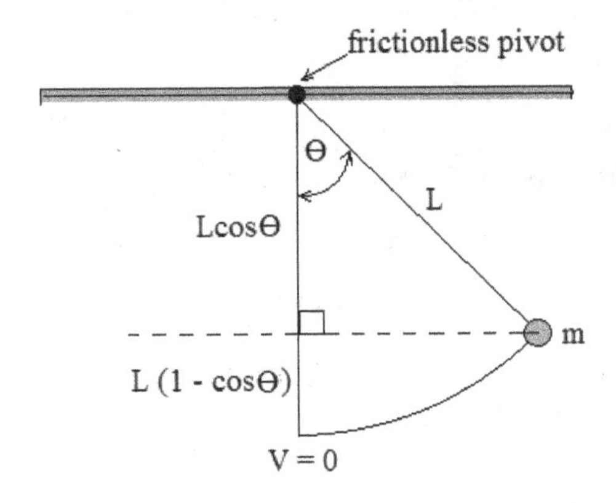

Figure 37.2: A pendulum of arm L.

Solution: From elementary physics, the total energy of the system is $E = T + V$ where T and V are the kinetic and potential energies of the pendulum bob. We assume that $V = 0$ when $\theta = 0$. It's easy to see that

$$T = \frac{1}{2}mv^2 = \frac{1}{2}m\left(L\frac{d\theta}{dt}\right)^2 = \frac{mL^2}{2}\left(\frac{d\theta}{dt}\right)^2$$
$$V = mgL(1 - \cos\theta)$$

where v is the speed of the pendulum bob, and g is its acceleration due to gravity. In the absence of friction, the total energy is conserved, so

$$E = \frac{mL^2}{2}\left(\frac{d\theta}{dt}\right)^2 + mgL(1 - \cos\theta) = \text{Constant} \tag{37.14}$$

and consequently,

$$\frac{dE}{dt} = \frac{d}{dt}\left\{\frac{mL^2}{2}\left(\frac{d\theta}{dt}\right)^2 + mgL(1 - \cos\theta)\right\} = 0$$
$$\Rightarrow \frac{d\theta}{dt}\left\{mL^2\frac{d^2\theta}{dt^2} + mgL\sin\theta\right\} = 0$$

Thus, the equation of motion is seen to be

$$\frac{d^2\theta}{dt^2} + k^2\sin\theta = 0 \tag{37.15}$$

where $k^2 = g/L$. We now write this equation as a nonlinear system. We set $x_1 = \theta$, and $x_2 = \dot{\theta}$ to obtain the nonlinear system

$$\begin{bmatrix} \dot{x}_1 \\ \dot{x}_2 \end{bmatrix} = \begin{bmatrix} x_2 \\ -k^2 \sin x_1 \end{bmatrix}$$

The fixed points are solutions of simultaneous equations:

$$x_2 = 0$$
$$-k^2 \sin x_1 = 0$$

so there are infinitely many fixed points of the form $(0, n\pi)$ where n is an integer, i.e.,

$$\ldots(-2\pi,0),(-\pi,0),(0,0),(0,\pi),(0,2\pi),\ldots$$

Because of the periodic nature of $\sin x_1$, we will confine our study to the fixed points ,$(-\pi,0),(0,0)$, and $(0,\pi)$.

First, let's consider the fixed point $(0,0)$. We find that the Jacobian is

$$DF|_{(0,0)} = \begin{bmatrix} 0 & 1 \\ -k^2 & 0 \end{bmatrix}$$

So, its eigenvalues are $\lambda = \pm ik$. This means that the critical point is a stable point and a center, while the motion is purely oscillatory.

Next, consider the fixed points $(\pm\pi, 0)$. In this case, the Jacobian is

$$DF|_{(\pm\pi,0)} = \begin{bmatrix} 0 & 1 \\ k^2 & 0 \end{bmatrix}$$

and its eigenvalues are $\lambda_1 = -k$, and $\lambda_2 = k$. Thus, the fixed points $(\pm\pi, 0)$ are unstable saddle points.

We find that the trajectories must satisfy the equation

$$\frac{dx_1}{dx_2} = \frac{x_2}{-k^2 x_1}$$

whose solution is

$$x_2^2 - 2k^2 \cos x_1 = C$$

Since $\dot{\theta}(0) = x_2(0) = \omega_0$, and $\theta(0) = x_1(0) = 0$, we find that $C = \omega_0^2 - 2k^2$, so the trajectories may be written as

$$x_2^2 + 2k^2(1 - \cos x_1) = \omega_0^2 \tag{37.1}$$

Now, we want to find the minimum value ω_0 of the initial angular speed which is necessary to cause the pendulum bob to be inverted so that $\theta = \pm\pi$. By conservation of energy, we must have

$$\frac{1}{2}m(L\omega_0)^2 = 2mgL$$

so that $\omega_0^2 = 4k^2$. Thus, we can discuss three separate cases:

(a) $(\omega_0^2 < 4k^2)$: Here, the pendulum bob oscillates back and forth in a simple oscillatory motion. In fact, if we substitute the approximation

$$\cos x_1 = 1 - \frac{x_1^2}{2!} + \ldots$$

into (37.16), we obtain $x_2^2 + k^2x_1^2 = \omega_0^2$. Thus, each trajectory is a closed curve that approximates the ellipse

$$\frac{x_1^2}{(\omega_0/k)^2} + \frac{x_2^2}{\omega_0^2} = 1$$

(b) $(\omega_0^2 = 4k^2)$: Here, the pendulum bob oscillates back and forth between $\theta = -\pi$ and $\theta = \pi$. In fact, if we substitute $\omega_0^2 = 4k^2$ into (37.16) we get

$$x_2^2 = 2k^2(1 + \cos x_1)$$

But $1 + \cos x_1 = 2\cos^2(x_1/2)$, so the trajectories are given by

$$x_2 = \pm 2k\cos^2\left(\frac{x_1}{2}\right)$$

(c) $(\omega_0^2 > 4k^2)$: Here, the pendulum bob makes complete revolutions around the pivot point. The trajectories correspond to the wavy curves that form the outer boundaries of Figure 37.3.

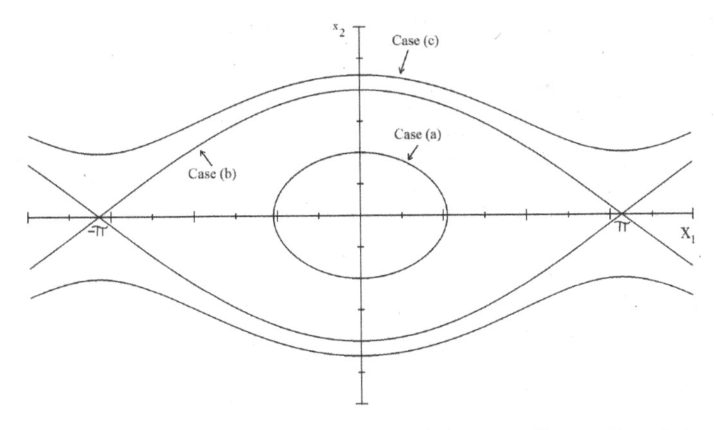

Figure 37.3: Typical Trajectories of the Nonlinear Pendulum.

Exercise Set 37

1. For each dynamical system $\dot{\mathbf{x}}(t) = \mathbf{F}(x_1, x_2)$, find all fixed points and determine their stability.

(a) $\mathbf{F}(x_1,x_2) = \begin{bmatrix} x_1 + x_2^2 \\ x_1 + x_2 \end{bmatrix}$ **(b)** $\mathbf{F}(x_1,x_2) = \begin{bmatrix} x_1^3 - x_2 \\ x_1 - x_2 \end{bmatrix}$

(c) $\mathbf{F}(x_1,x_2) = \begin{bmatrix} x_1^2 + x_2^2 + 2x_1 \\ x_1 + x_2 \end{bmatrix}$ **(d)** $\mathbf{F}(x_1,x_2) = \begin{bmatrix} x_1 + x_2 \\ \sin x_1 \end{bmatrix}$

2. Consider the **van der Pol equation**

$$\ddot{x} + \varepsilon(x^2 - 1)\dot{x} + x = 0 \qquad (\varepsilon > 0) \qquad (37.1$$

It arises in the analysis of the triode vacuum tube used in electronics.

(a) By setting $x_1(t) = x$ and $x_2(t) = \dot{x}$, convert this equation into a nonlinear system.
(b) Show that the only fixed point is $(0,0)$.
(c) Discuss the stability when $\varepsilon < 2$ and when $\varepsilon > 2$.

3. The equation

$$\ddot{x} + b\dot{x} + k^2 x = 0 \qquad (b, k > 0) \qquad (37.1$$

in known to govern **damped oscillations.**

(a) By setting $x_1(t) = x$ and $x_2(t) = \dot{x}$, write this equation as a linear system.
(b) Show that the only fixed point is $(0,0)$.
(c) Discuss the stability when $b^2 < 4k^2$, $b^2 = 4k^2$, and when $b^2 > 4k^2$.
(d) Starting with (37.18), show that

$$\frac{d}{dt}\left(\frac{1}{2}\dot{x}^2 + \frac{1}{2}kx^2\right) = -b\dot{x}^2$$

so the quantity

$$E = \frac{1}{2}\dot{x}^2 + \frac{1}{2}kx^2 \qquad (37.1$$

is *decreasing* for all time t. [**Hint**: Multiply both sides (37.18) by $\dot{x}$.]
(e) What is the physical significance of E? Does the fact that E is strictly decreasing comport with your analysis in **(c)**?

38. Coupled Oscillations

In this lesson, we study systems of second order differential equations that appear in the study of vibrating systems. We begin our study with the system

$$\frac{d^2x_1(t)}{dt^2} + a_{11}x_1(t) + a_{12}x_2(t) = 0$$
$$\frac{d^2x_2(t)}{dt^2} + a_{21}x_1(t) + a_{22}x_2(t) = 0 \qquad (38.1)$$

We can write this system in matrix form as

$$\ddot{\mathbf{x}}(t) + A\mathbf{x} = 0 \qquad (38.2)$$

where

$$\mathbf{x}(t) = \begin{bmatrix} x_1(t) \\ x_2(t) \end{bmatrix}, \quad A = \begin{bmatrix} a_{11} & a_{12} \\ a_{21} & a_{22} \end{bmatrix} \qquad (38.3)$$

We assume the constant matrix A has distinct eigenvalues that are real and positive. This assumption may seem restrictive, but it is often met in practical applications. We say that the system has **two degrees of freedom** since the motion is completely defined by two independent quantities. Also, note that this is a **coupled system**, since from (38.1), we see that the time evolution of each variable depends on another variable, in addition to itself.

In order to "decouple" the system, we first make a change of variables

$$\mathbf{x}(t) = P\mathbf{y}(t) \qquad (38.4)$$

where P is the *transition matrix* corresponding to A. That is, if $\mathbf{v}_1$ and $\mathbf{v}_2$ are the eigenvectors of A corresponding to the eigenvalues λ_1 and λ_2 respectively, then P is the matrix whose column vectors are $\mathbf{v}_1$ and $\mathbf{v}_2$:

$$P = [\mathbf{v}_1 | \mathbf{v}_2]$$

Next, we substitute (38.4) into (38.2) to obtain

$$\ddot{\mathbf{y}}(t) + (P^{-1}AP)\mathbf{y}(t) = \mathbf{0}$$

or,

$$\ddot{\mathbf{y}}(t) + \begin{bmatrix} \lambda_1 & 0 \\ 0 & \lambda_2 \end{bmatrix}\mathbf{y}(t) = \mathbf{0}$$

where $\lambda_1, \lambda_2 > 0$ are the eigenvalues of A. The last equation may be written out as a pair of decoupled equations:

$$\ddot{y}_1(t) + \lambda_1 y_1(t) = 0$$
$$\ddot{y}_2(t) + \lambda_2 y_2(t) = 0$$

Consequently,

$$\mathbf{y}(t) = \begin{bmatrix} y_1(t) \\ y_2(t) \end{bmatrix} = \begin{bmatrix} A_1 \cos\sqrt{\lambda_1}\, t + B_1 \sin\sqrt{\lambda_1}\, t \\ A_2 \cos\sqrt{\lambda_2}\, t + B_2 \sin\sqrt{\lambda_2}\, t \end{bmatrix} \qquad (38.5)$$

where A_1, A_2, B_1, and B_2 are arbitrary constants. Finally, by (38.4) and (38.5), we can write

$$\mathbf{x}(t) = \left(A_1 \cos\sqrt{\lambda_1}\, t + B_1 \sin\sqrt{\lambda_1}\, t\right)\mathbf{v}_1 + \left(A_2 \cos\sqrt{\lambda_2}\, t + B_2 \sin\sqrt{\lambda_2}\, t\right)\mathbf{v}_2$$

or equivalently,

$$\mathbf{x}(t) = \sum_{k=1}^{2}\left(A_k \cos\sqrt{\lambda_k}\, t + B_k \sin\sqrt{\lambda_k}\, t\right)\mathbf{v}_k \qquad (38.6)$$

This equation may be generalized to handle the most general case when A is a square matrix of order n. Furthermore, we see that it describes oscillatory motion where the angular frequencies (radians/sec) of oscillation are $\sqrt{\lambda_1}$ and $\sqrt{\lambda_2}$. The corresponding **characteristic frequencies** (cycles/sec) of oscillation are

$$f_1 = \frac{1}{2\pi}\sqrt{\lambda_1}, \quad f_2 = \frac{1}{2\pi}\sqrt{\lambda_2} \quad \text{(cycles/sec)} \qquad (38.7)$$

The eigenvectors $\mathbf{v}_k$ are called the **normal modes** of oscillation.

Example-1: Find the general solution of the system of differential equations:

$$\ddot{x}_1 + 2x_1 + x_2 = 0$$
$$\ddot{x}_2 + x_1 + 2x_2 = 0$$

Solution: Here, the matrix

$$A = \begin{bmatrix} 2 & 1 \\ 1 & 2 \end{bmatrix}$$

has eigenvalues $\lambda_1 = 1, \lambda_2 = 3$ with corresponding eigenvectors

$$\mathbf{v}_1 = \begin{bmatrix} 1 \\ -1 \end{bmatrix} \quad \text{and} \quad \mathbf{v}_2 = \begin{bmatrix} 1 \\ 1 \end{bmatrix}$$

Thus, by (38.6), the general solution is

$$\mathbf{x}(t) = (A_1 \cos t + B_1 \sin t)\begin{bmatrix} 1 \\ -1 \end{bmatrix} + \left(A_2 \cos \sqrt{3}\, t + B_2 \sin \sqrt{3}\, t\right)\begin{bmatrix} 1 \\ 1 \end{bmatrix}$$

or in component form,

$$\begin{aligned} x_1 &= A_1 \cos t + B_1 \sin t + A_2 \cos \sqrt{3}\, t + B_2 \sin \sqrt{3}\, t \\ x_2 &= -A_1 \cos t - B_1 \sin t + A_2 \cos \sqrt{3}\, t + B_2 \sin \sqrt{3}\, t \end{aligned}$$

The most general vibrating system is one with n degrees of freedom, and it may be described by an equation of the form

$$M\ddot{\mathbf{x}}(t) + K\mathbf{x}(t) = \mathbf{0} \tag{38.8}$$

where M and K are square matrices of order n. If the matrix M is invertible, then we can write

$$\ddot{\mathbf{x}}(t) + (M^{-1}K)\mathbf{x}(t) = \mathbf{0}$$

So, if we set $A = M^{-1}K$, then (38.8) reduces to the form of (38.2). Thus, we can immediately write down its general solution as

$$\mathbf{x}(t) = \sum_{k=1}^{n}\left(A_k \cos \sqrt{\lambda_k}\, t + B_k \sin \sqrt{\lambda_k}\, t\right)\mathbf{v}_k \tag{38.9}$$

where λ_k and $\mathbf{v}_k$ are the eigenvalues and eigenvectors of $A = M^{-1}K$. Once again, the quantities $\sqrt{\lambda_k}$ are the angular frequencies of oscillation, while the eigenvectors $\mathbf{v}_k$ are the normal modes of oscillation. Let's see how this works in practice.

Example-2: *(Two Coupled Masses)* Two equal masses m connected by identical springs with spring constant k lie on a frictionless table. At time $t = 0$, the masses are initially at the positions

$$x_1(0) = 1, \quad x_2(0) = 0$$

and initially at rest, so that $\dot{x}_1(0) = \dot{x}_2(0) = 0$. Describe the motion of the system.

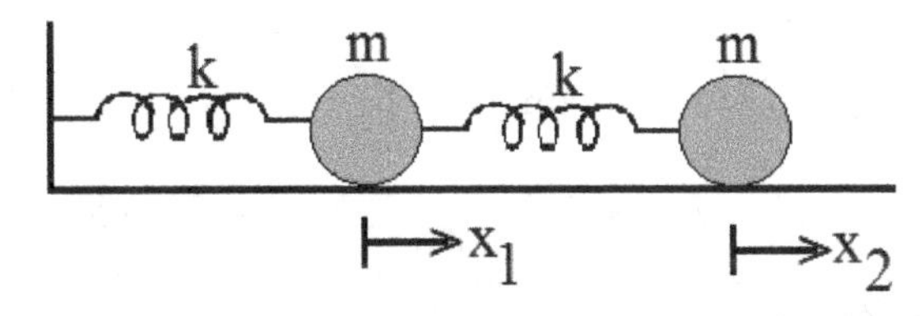

Figure 38.1: Coupled Masses on a Frictionless Table

Solution: According to **Hooke's law**, any mass attached to a spring will experience a *restoring force* given by $-kx$ where x is the displacement of the mass away from the equilibrium position, and k is the *spring constant*; it is a measure of the stiffness of the spring. From Hooke's law, the equations of motion are

$$\begin{aligned} m\ddot{x}_1 &= -kx_1 - k(x_1 - x_2) \\ m\ddot{x}_2 &= -k(x_2 - x_1) \end{aligned}$$

or in matrix form,

$$\begin{bmatrix} m & 0 \\ 0 & m \end{bmatrix}\begin{bmatrix} \ddot{x}_1 \\ \ddot{x}_2 \end{bmatrix} + \begin{bmatrix} 2k & -k \\ -k & k \end{bmatrix}\begin{bmatrix} x_1 \\ x_2 \end{bmatrix} = \mathbf{0}$$

Here,

$$A = M^{-1}K = \begin{bmatrix} 1/m & 0 \\ 0 & 1/m \end{bmatrix}\begin{bmatrix} 2k & -k \\ -k & k \end{bmatrix} = \begin{bmatrix} \frac{2k}{m} & -\frac{k}{m} \\ -\frac{k}{m} & \frac{k}{m} \end{bmatrix}$$

The eigenvalues of A are

$$\lambda_1 = \frac{k}{2m}(3 - \sqrt{5}) \text{ and } \lambda_2 = \frac{k}{2m}(3 + \sqrt{5}),$$

with the corresponding normal modes

$$\mathbf{v}_1 = \begin{bmatrix} \frac{1}{2}(\sqrt{5} - 1) \\ 1 \end{bmatrix} \text{ and } \mathbf{v}_2 = \begin{bmatrix} -\frac{1}{2}(\sqrt{5} + 1) \\ 1 \end{bmatrix}$$

Consequently, the characteristic frequencies are

$$\begin{aligned} f_1 &= \frac{1}{2\pi}\sqrt{\lambda_1} = \frac{1}{2\pi}\sqrt{\frac{k}{2m}(3 - \sqrt{5})} = \frac{(\sqrt{5} - 1)}{4\pi}\sqrt{\frac{k}{m}} \\ f_2 &= \frac{1}{2\pi}\sqrt{\lambda_2} = \frac{1}{2\pi}\sqrt{\frac{k}{2m}(3 + \sqrt{5})} = \frac{(\sqrt{5} + 1)}{4\pi}\sqrt{\frac{k}{m}} \end{aligned}$$

and the general solution of the system is

$$\mathbf{x}(t) = \left(A_1 \cos\sqrt{\lambda_1}\,t + B_1 \sin\sqrt{\lambda_1}\,t\right)\mathbf{v}_1 + \left(A_2 \cos\sqrt{\lambda_2}\,t + B_2 \sin\sqrt{\lambda_2}\,t\right)\mathbf{v}_2$$

so that

$$\begin{aligned} x_1(t) &= \left(A_1 \cos\sqrt{\lambda_1}\,t + B_1 \sin\sqrt{\lambda_1}\,t\right)\left(\frac{1}{2}\sqrt{5} - \frac{1}{2}\right) \\ &\quad - \left(A_2 \cos\sqrt{\lambda_2}\,t + B_2 \sin\sqrt{\lambda_2}\,t\right)\left(\frac{1}{2}\sqrt{5} + \frac{1}{2}\right) \\ x_2(t) &= A_1 \cos\sqrt{\lambda_1}\,t + A_2 \cos\sqrt{\lambda_2}\,t \\ &\quad + B_1 \sin\sqrt{\lambda_1}\,t + B_2 \sin\sqrt{\lambda_2}\,t \end{aligned}$$

The initial conditions $\dot{x}_1(0) = \dot{x}_1(0) = 0$ yield the system of equations

$$\dot{x}_1(0) = \sqrt{\lambda_1}\left(\frac{1}{2}\sqrt{5} - \frac{1}{2}\right)B_1 - \sqrt{\lambda_2}\left(\frac{1}{2}\sqrt{5} + \frac{1}{2}\right)B_2 = 0$$

$$\dot{x}_2(0) = \sqrt{\lambda_1}\,B_1 + \sqrt{\lambda_2}\,B_2 = 0$$

whose solution is $B_1 = B_2 = 0$. The updated solution becomes

$$x_1(t) = \left(A_1 \cos\sqrt{\lambda_1}\,t\right)\left(\frac{1}{2}\sqrt{5} - \frac{1}{2}\right) - \left(A_2 \cos\sqrt{\lambda_2}\,t\right)\left(\frac{1}{2}\sqrt{5} + \frac{1}{2}\right)$$

$$x_2(t) = A_1 \cos\sqrt{\lambda_1}\,t + A_2 \cos\sqrt{\lambda_2}\,t$$

Similarly, the initial conditions $x_1(0) = 1$, $x_2(0) = 0$ yield the system of equations

$$x_1(0) = \left(\frac{1}{2}\sqrt{5} - \frac{1}{2}\right)A_1 - \left(\frac{1}{2}\sqrt{5} + \frac{1}{2}\right)A_2 = 1$$

$$x_2(0) = A_1 + A_2 = 0$$

from which we find $A_1 = \frac{1}{5}\sqrt{5}$, and $A_2 = -\frac{1}{5}\sqrt{5}$. Thus, the final solution to the problem is

$$x_1(t) = \left(\frac{1}{2} - \frac{1}{10}\sqrt{5}\right)\cos\sqrt{\lambda_1}\,t + \left(\frac{1}{2} + \frac{1}{10}\sqrt{5}\right)\cos\sqrt{\lambda_2}\,t$$

$$x_2(t) = \frac{1}{5}\sqrt{5}\cos\sqrt{\lambda_1}\,t - \frac{1}{5}\sqrt{5}\cos\sqrt{\lambda_2}\,t$$

where the angular frequencies are:

$$\sqrt{\lambda_1} = \frac{(\sqrt{5} - 1)}{2}\sqrt{\frac{k}{m}} \quad \text{and} \quad \sqrt{\lambda_2} = \frac{(\sqrt{5} + 1)}{2}\sqrt{\frac{k}{m}}$$

Example-3: *(Coupled Pendulums)* As shown in Figure 28.2, two pendulums of the same length L and mass m are connected by a spring of spring constant k, and undergo *small oscillations*. The system is subject to the initial conditions $\dot{x}$

$$x_1(0) = x_2(0) = 0$$

$$\dot{x}_1(0) = 0,\ \dot{x}_2(0) = 1$$

Determine the motion of the system.

Solution: The equations of motion are

$$m\dot{x}_1 = -k(x_1 - x_2) - \frac{mg}{L}x_1$$

$$m\dot{x}_2 = -k(x_2 - x_1) - \frac{mg}{L}x_2$$

If we let $a^2 = k/m$ and $b^2 = g/L$, then the system can be written in matrix form as

$$\begin{bmatrix} \ddot{x}_1 \\ \ddot{x}_2 \end{bmatrix} + \begin{bmatrix} a^2 + b^2 & -a^2 \\ -a^2 & a^2 + b^2 \end{bmatrix} \begin{bmatrix} x_1 \\ x_2 \end{bmatrix} = 0$$

We find that the eigenvalues are $\lambda_1 = b^2$ and $\lambda_2 = 2a^2 + b^2$ with the corresponding normal modes

$$\mathbf{v}_1 = \begin{bmatrix} 1 \\ 1 \end{bmatrix} \quad \text{and} \quad \mathbf{v}_2 = \begin{bmatrix} 1 \\ -1 \end{bmatrix}$$

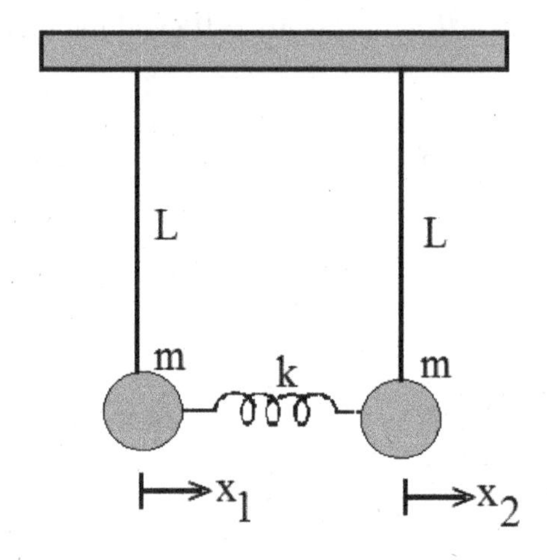

Figure 38.2: Two Coupled Pendulums

Here, the first mode $\mathbf{v}_1$ corresponds to the situation where each pendulum oscillates in phase at the angular frequency a, while the second mode $\mathbf{v}_2$ describes the case where each pendulum bob oscillates π radians out of phase at an angular frequency of $\sqrt{2a^2 + b^2}$. Either normal mode, or a *superposition* of the normal modes can appear in the solution, depending upon the initial conditions.

The general solution of the system is

$$\mathbf{x}(t) = (A_1 \cos at + B_1 \sin at) \begin{bmatrix} 1 \\ 1 \end{bmatrix}$$

$$+ \left\{ A_2 \cos\left(\sqrt{2a^2 + b^2}\, t \right) + B_2 \sin\left(\sqrt{2a^2 + b^2}\, t \right) \right\} \begin{bmatrix} 1 \\ -1 \end{bmatrix}$$

The initial conditions $x_1(0) = x_2(0) = 0$ imply that $A_1 = A_2 = 0$. Thus, the solution simplifies to

$$\mathbf{x}(t) = B_1 \sin at \begin{bmatrix} 1 \\ 1 \end{bmatrix} + B_2 \sin\left(\sqrt{2a^2 + b^2}\, t\right) \begin{bmatrix} 1 \\ -1 \end{bmatrix}$$

Similarly, the initial conditions $\dot{x}_1(0) = 0$, $\dot{x}_2(0) = 1$ imply that

$$B_1 = \frac{1}{2a}, \quad \text{and} \quad B_2 = \frac{1}{2\sqrt{2a^2 + b^2}}$$

Thus, the final form of the solution is

$$x_1(t) = \frac{1}{2\sqrt{\frac{g}{L}}} \sin\left(\sqrt{\frac{g}{L}}\, t\right) + \frac{1}{2\sqrt{\frac{2k}{m} + \frac{g}{L}}} \sin\left(\sqrt{\frac{2k}{m} + \frac{g}{L}}\, t\right)$$

$$x_2(t) = \frac{1}{2\sqrt{\frac{g}{L}}} \sin\left(\sqrt{\frac{g}{L}}\, t\right) + \frac{1}{2\sqrt{\frac{2k}{m} + \frac{g}{L}}} \sin\left(\sqrt{\frac{2k}{m} + \frac{g}{L}}\, t\right)$$

where the characteristic frequencies of oscillation are

$$f_1 = \frac{1}{2\pi}\sqrt{\frac{g}{L}} \quad \text{and} \quad f_2 = \frac{1}{2\pi}\sqrt{\frac{2k}{m} + \frac{g}{L}}$$

Exercise Set 38

---

1. Find the general solution of each 2nd order linear system. Also, find the normal modes and characteristic frequencies.

(a)

$$\ddot{x}_1 = -3x_1 - x_2$$
$$\ddot{x}_2 = -x_1 - 3x_2$$

(b)

$$4\ddot{x}_1 + 9x_1 + 8x_2 = 0$$
$$2\ddot{x}_2 + 3x_1 + 4x_2 = 0$$

2. As shown in Figure 38.3, two equal masses, lying on a frictionless table are connected by a spring of spring constant k.

(a) Write down the equations of motion.
(b) Determine the characteristic frequencies and normal modes.
(c) Find the solution, assuming that the initial conditions are

$$x_1(0) = -x_0, \;\; x_2(0) = x_0$$
$$\dot{x}_1(0) = 0, \;\;\; \dot{x}_2(0) = 0$$

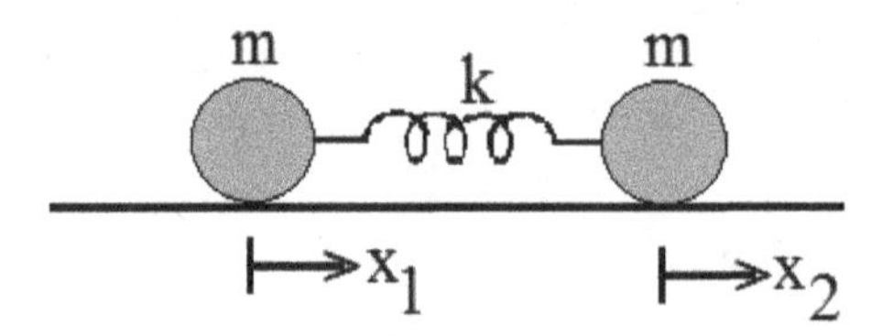

Figure 38.3: Two coupled masses on a frictionless table.

3. As shown in Figure 38.4, two coupled masses are lying on a frictionless table, and are connected by three springs, each of spring constant k.

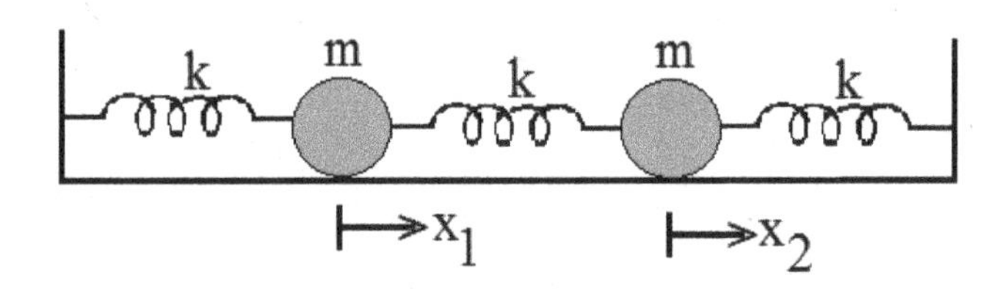

Figure 38.4: Two coupled masses confined between two fixed walls.

(**a**) Write down the equations of motion.
(**b**) Determine the characteristic frequencies and normal modes.
(**c**) Write down the general solution.
(**d**) Find the solution corresponding to the initial conditions

$$x_1(0) = x_2(0) = x_0$$
$$\dot{x}_1(0) = 0, \dot{x}_2(0) = 0$$

Which normal mode manifests itself?
(**e**) Find the solution corresponding to the initial conditions

$$x_1(0) = -x_0, x_2(0) = x_0$$
$$\dot{x}_1(0) = 0, \dot{x}_2(0) = 0$$

Which normal mode corresponds to this solution?

4. Provide an argument which justifies the equations of motion given in Example-3. Remember, we are assuming only small oscillations.

Solutions and Hints to Selected Exercises

Exercise Set 1

1. $\begin{bmatrix} 3 & -1 \\ 2 & 4 \end{bmatrix}$ **3.** $\begin{bmatrix} 3 & 0 \\ 2 & 6 \end{bmatrix}$ **5.** $\begin{bmatrix} 8 & -4 \\ -2 & 10 \end{bmatrix}$ 7. $\begin{bmatrix} 3 & 0 \\ 2 & 6 \end{bmatrix}$

9. $\begin{bmatrix} 0 & 0 \\ 0 & 0 \end{bmatrix}$ **11.** $\begin{bmatrix} 7 \\ 3 \end{bmatrix}$ **13.** $\begin{bmatrix} 4 \\ -4 \end{bmatrix}$ **15.** $\begin{bmatrix} -6 & 12 \\ 18 & 6 \end{bmatrix}$

17. $\begin{bmatrix} 7 & -1 \\ 1 & 13 \end{bmatrix}$ **19.** $\begin{bmatrix} 0 & 0 \\ 0 & 0 \end{bmatrix}$

21. $x = 1, y = 2, z = 3, w = 3$

23. $X = \begin{bmatrix} 1 & 1 \\ 1 & 2 \end{bmatrix}$ **27.** $X = \begin{bmatrix} -1 & -2 \\ -2 & 0 \end{bmatrix}$

Exercise Set 2

1. $\begin{bmatrix} 20 & 6 \\ 30 & 16 \end{bmatrix}$ **3.** $\begin{bmatrix} -11 & 4 \\ 1 & -1 \end{bmatrix}$ **5.** $\begin{bmatrix} 9 & 10 \\ 31 & 15 \end{bmatrix}$

7. $\begin{bmatrix} -118 & 46 \\ -198 & 76 \end{bmatrix}$ **9.** $\begin{bmatrix} 10 & 6 \\ -110 & -52 \end{bmatrix}$ **11.** $\begin{bmatrix} 1 & 4 \\ -76 & 25 \end{bmatrix}$

13. $\begin{bmatrix} -4 & 4 \\ -4 & 0 \end{bmatrix}$ **15.** $\begin{bmatrix} -8 & 0 \\ -8 & -2 \end{bmatrix}$ **17.** $\begin{bmatrix} -17 \\ -10 \end{bmatrix}$

19. $\begin{bmatrix} 1 & 2 \\ -2 & 3 \end{bmatrix}$

Exercise Set 3

1. $\begin{bmatrix} 1 & 1 \\ 1 & 7 \end{bmatrix}$ **3.** $\begin{bmatrix} 6 & 9 \\ 7 & 14 \end{bmatrix}$

5. 20 **7.** 4

9. 2 **11.** 10

13 . $A = \begin{bmatrix} 0 & \frac{3}{2} & \frac{3}{2} \\ \frac{3}{2} & 5 & \frac{3}{2} \\ \frac{3}{2} & \frac{3}{2} & 1 \end{bmatrix} + \begin{bmatrix} 0 & -\frac{1}{2} & -\frac{5}{2} \\ \frac{1}{2} & 0 & \frac{3}{2} \\ \frac{5}{2} & -\frac{3}{2} & 0 \end{bmatrix}$

Exercise Set 4

1. $x = 1, y = 0$ **2.** $x = \frac{1}{3}, y = -1$ **3.** $x = 0, y = 1, z = 2$

4. $x = 0, y = -1, z = 1$ **5.** $x = \frac{1}{3}, y = \frac{1}{3}, z = 0$

6. $x = 1, y = 2, z = 3$ 7. $x = -\frac{3}{2}, y = \frac{1}{3}, z = \frac{3}{4}$

8. $x = -4s, y = 0, z = s$, where s is an arbitrary real number.

9. $x = -\frac{4}{3}, y = \frac{1}{3}, z = \frac{5}{3}, w = \frac{2}{3}$

10. $x = s, y = s, z = s, w = 3s$ where s is an arbitrary real number.

11. $I_1 = \frac{1}{6}$, $I_2 = \frac{7}{9}$, $I_3 = \frac{11}{18}$

Exercise Set 5

1. $\begin{bmatrix} -\frac{1}{2} & \frac{1}{2} \\ \frac{1}{2} & \frac{1}{2} \end{bmatrix}$ **3.** $\begin{bmatrix} -1 & 0 & 1 \\ -1 & -1 & 1 \\ 2 & 0 & -1 \end{bmatrix}$ **5.** $\begin{bmatrix} 1 & 0 & 0 \\ -4 & 1 & 0 \\ 2 & -1 & 1 \end{bmatrix}$

7. Not invertible.

9. $\begin{bmatrix} 1 & 0 & -1 & -\frac{2}{3} \\ 0 & \frac{1}{2} & -\frac{1}{2} & -\frac{1}{3} \\ 0 & 0 & 1 & \frac{1}{3} \\ 0 & 0 & 0 & \frac{1}{3} \end{bmatrix}$ **11.** $A = \begin{bmatrix} -\frac{1}{7} & \frac{3}{7} \\ \frac{2}{7} & \frac{1}{7} \end{bmatrix}$

12.
(a) The solution is $x = -1, y = 1$.
(b) The solution is $x = -1, y = 1, z = 1$.

13. Hint: Let

$$B = \frac{1}{ad - bc}\begin{bmatrix} d & -b \\ -c & a \end{bmatrix}$$

and show by direct multiplication that $AB = BA = I$.

14. Observe that $A^2 - A - I = 0$ implies $A(A - I) = I$ so evidently, $A^{-1} = A - I$.

16. The inverse exists, and is

$$A^{-1} = \begin{bmatrix} 1/a_{11} & 0 & 0 & 0 \\ 0 & 1/a_{22} & 0 & 0 \\ \vdots & \vdots & \vdots & \vdots \\ 0 & 0 & 0 & 1/a_{nn} \end{bmatrix}$$

provided $a_{11}a_{22} \cdots a_{nn} \neq 0$.

18. Hint: Prove the *contrapositive*, i.e., show that if A is non-singular, then the system must have only the trivial solution.

Exercise Set 6

1. 9 **2** −30 **3.** 0 **4.** −9 **5.** 1
6. One finds $\det(A^T) = \det(A) = 2$.
10. 24
11. $\lambda = 0, \frac{3}{2} \pm \frac{1}{2}i\sqrt{3}$

Exercise Set 7

1. We find that $\det(A) = \det(A^T) = 30$.

2. Observe that

$$\det(\lambda A) = \sum_{i_1,i_2,i_3,\ldots,i_n=1}^{n} \varepsilon_{i_1 i_2 i_3 \ldots i_n}(\lambda a_{1i_1})(\lambda a_{2i_2})(\lambda a_{3i_3})\cdots(\lambda a_{ni_n}) = \lambda^n \det(A)$$

3. Follows by repeated application of equation (7.3).

5. We find $\det(AB) = 80$, while $\det(A) = 10$ and $\det(B) = 8$, so $\det(AB) = \det(A)\det(B)$

6. **(a)** -11 **(b)** -64 **(c)** -16

7. If A is an $n \times n$ *skew-symmetric* matrix, then $A^T = -A$, so

$$\det(A) = \det(A^T) = \det[(-1)A] = (-1)^n \det(A).$$

9. If A is a invertible, then $AA^{-1} = I$ so $\det(AA^{-1}) = \det(A)\det(A^{-1}) = 1$, from which the result follows.

Exercise Set 8

2. **(a)** $\begin{bmatrix} \cos\theta & -\sin\theta \\ \sin\theta & \cos\theta \end{bmatrix}$ **(b)** $\begin{bmatrix} -2 & 1 \\ \frac{3}{2} & -\frac{1}{2} \end{bmatrix}$

3. **(a)** $\begin{bmatrix} 1 & 0 & -1 \\ 1 & \frac{1}{2} & -\frac{3}{2} \\ -1 & 0 & 2 \end{bmatrix}$ **(b)** $\begin{bmatrix} -\frac{1}{3} & -\frac{4}{3} & 1 \\ \frac{1}{3} & -\frac{2}{3} & \frac{1}{2} \\ 0 & 1 & -\frac{1}{2} \end{bmatrix}$ **(c)** $\begin{bmatrix} -2 & 2 & 3 \\ 3 & -\frac{5}{2} & -\frac{9}{2} \\ -1 & 1 & 2 \end{bmatrix}$

4. Using Cramer's rule we find:
(a) $x_1 = 1, x_2 = -1$
(b) $x_1 = 3, x_2 = -2, x_3 = 1$
(c) $x_1 = 0, x_2 = 2, x_3 = 1, x_4 = 2$

5. The matrix is not invertible when

$$\begin{vmatrix} 1-\lambda & 0 & 1 \\ 2 & 3-\lambda & -1 \\ 2 & 1 & 2-\lambda \end{vmatrix} = 0 \Rightarrow \lambda^3 - 6\lambda^2 + 10\lambda - 3 = 0$$

So, the matrix is **not invertible** when $\lambda \in \left\{\frac{1}{2}\sqrt{5} + \frac{3}{2}, \frac{3}{2} - \frac{1}{2}\sqrt{5}, 3\right\}$.

6. Follows directly from Cramer's rule.

9.

(a) By direct calculation, we get $\det(R) = \cos^2\theta + \sin^2\theta = 1$.

(b) The inverse is $R^{-1} = \begin{bmatrix} 1 & 0 & 0 \\ 0 & \cos\theta & \sin\theta \\ 0 & -\sin\theta & \cos\theta \end{bmatrix}$.

(c) Observe that $R^{-1} = R^T$.

10. If R is an orthogonal matrix, then $RR^T = I$, so $\det(RR^T) = [\det(R)]^2 = 1$ and consequently, $\det(R) = \pm 1$.

Exercise Set 9

1. $\begin{bmatrix} 10 \\ 4 \\ 16 \end{bmatrix}$ **2.** $\begin{bmatrix} 18 \\ 9 \\ 7 \end{bmatrix}$ **3.** $\begin{bmatrix} -3 \\ -1 \\ 3 \end{bmatrix}$

4. $\begin{bmatrix} 20 \\ 11 \\ 11 \end{bmatrix}$ **5.** $\begin{bmatrix} -6 \\ -1 \\ -1 \end{bmatrix}$

6. The solution is $\mathbf{x} = [2, 2, 4]^T$.

7. We find that $\|a\mathbf{x}\|^2 = a^2\left(x_1^2 + x_2^2 + \ldots + x_n^2\right)$ from which the result follows.

Exercise Set 10

1. Let $\mathbf{x} = [x_1, x_2, 0]^T$ and $\mathbf{y} = [y_1, y_2, 0]^T$ be arbitrary vectors in W. Observe that

$$\mathbf{x} + \mathbf{y} = [x_1, x_2, 0]^T + [y_1, y_2, 0]^T = [x_1 + x_2, y_1 + y_2, 0]^T \in W$$
$$\lambda\mathbf{x} = \lambda[x_1, x_2, 0]^T = [\lambda x_1, \lambda x_2, 0]^T \in W \text{ for any scalar } \lambda$$

So, W is a subspace of V.

3. Let $A = \begin{bmatrix} a & 0 \\ 0 & b \end{bmatrix}$ and $B = \begin{bmatrix} c & 0 \\ 0 & d \end{bmatrix}$ be arbitrary elements in W.

Clearly,

$$A + B = \begin{bmatrix} a & 0 \\ 0 & b \end{bmatrix} + \begin{bmatrix} c & 0 \\ 0 & d \end{bmatrix} = \begin{bmatrix} a+c & 0 \\ 0 & b+d \end{bmatrix} \in W$$

$$\lambda A = \lambda \begin{bmatrix} a & 0 \\ 0 & b \end{bmatrix} = \begin{bmatrix} a\lambda & 0 \\ 0 & b\lambda \end{bmatrix} \in W \text{ for any scalar } \lambda$$

So, W is a subspace of V.

5. No. For example,

$$V = \{[x,y]^T \mid y = 2x\} \text{ and } W = \{[x,y]^T \mid y = x\}$$

are both subspaces of $\mathbb{R}^2$. They both represent lines through the origin $(0,0)$. Observe that $V \cup W$ is not closed under the operation of vector addition. For example, $[1,2]^T$ and $[1,1]^T$ are vectors in $V \cup W$, but their sum is $[2,3]^T$ which is clearly **not** a vector in $V \cup W$.

Exercise Set 11

1. linearly dependent **3**. linearly independent **5**. linearly independent
7. linearly independent **9**. linearly independent
11. Consider the set of n-vectors $S = \{\mathbf{0}, \mathbf{v}_2, \ldots, \mathbf{v}_n\}$. Observe that

$$(1)\mathbf{0} + (0)\mathbf{v}_2 + (0)\mathbf{v}_3 + \ldots + (0)\mathbf{v}_n = \mathbf{0}$$

So, S must be a linearly dependent set.

13. Let

$$\mathbf{v}_1 = \begin{bmatrix} v_{11} \\ v_{21} \\ \vdots \\ v_{n1} \end{bmatrix}, \quad \mathbf{v}_2 = \begin{bmatrix} v_{12} \\ v_{22} \\ \vdots \\ v_{n2} \end{bmatrix}, \quad \ldots, \mathbf{v}_n = \begin{bmatrix} v_{1n} \\ v_{2n} \\ \vdots \\ v_{nn} \end{bmatrix}$$

If we assume that $c_1\mathbf{v}_1 + c_2\mathbf{v}_2 + \ldots + c_n\mathbf{v}_n = \mathbf{0}$, then we are led to consider the homogeneous system, say $A\mathbf{c} = \mathbf{0}$, where

$$A\mathbf{c} \equiv \begin{bmatrix} v_{11} & v_{12} & \cdots & v_{1n} \\ v_{21} & v_{22} & \cdots & v_{2n} \\ \vdots & \vdots & & \vdots \\ v_{n1} & v_{n2} & \cdots & v_{nn} \end{bmatrix} \begin{bmatrix} c_1 \\ c_2 \\ \vdots \\ c_n \end{bmatrix} = \mathbf{0}$$

This system has the *unique* solution $\mathbf{c} = \mathbf{0}$ iff $\det(A) \neq 0$.

Exercise Set 12

1. **(a)** Yes **(b)** Yes **(c)** No, the vectors are not linearly independent.

2. **(a)** Yes **(b)** No, the vectors are linearly dependent. **(c)** Yes.

3. The matrices E_1, E_2, E_3, E_4 span $M(2,2)$ since for any $A \in M(2,2)$, we have

$$A = \begin{bmatrix} a & b \\ c & d \end{bmatrix} = a\begin{bmatrix} 1 & 0 \\ 0 & 0 \end{bmatrix} + b\begin{bmatrix} 0 & 1 \\ 0 & 0 \end{bmatrix} + c\begin{bmatrix} 0 & 0 \\ 1 & 0 \end{bmatrix} + d\begin{bmatrix} 0 & 0 \\ 0 & 1 \end{bmatrix}$$

$$= aE_1 + bE_2 + cE_3 + dE_4$$

Also, E_1, E_2, E_3, E_4 are linearly independent, since if we assume that

$$c_1E_1 + c_2E_2 + c_3E_3 + c_4E_4 = \begin{bmatrix} 0 & 0 \\ 0 & 0 \end{bmatrix}$$

this implies that

$$\begin{bmatrix} c_1 & c_2 \\ c_3 & c_4 \end{bmatrix} = \begin{bmatrix} 0 & 0 \\ 0 & 0 \end{bmatrix}$$

so $c_1 = c_2 = c_3 = c_4 = 0$.

4. **Hint**: Construct an explicit basis analogous to the one given in Exercise-3.

Exercise Set 13

1. The row vectors are $\begin{bmatrix} 1 & 0 & 1 \end{bmatrix}$ and $\begin{bmatrix} 2 & 1 & 3 \end{bmatrix}$. The column vectors are $\begin{bmatrix} 1 \\ 2 \end{bmatrix}, \begin{bmatrix} 0 \\ 1 \end{bmatrix}$, and $\begin{bmatrix} 1 \\ 3 \end{bmatrix}$.

2. A basis for the row space: $\left\{\begin{bmatrix} 1 & 0 & -\frac{1}{3} \end{bmatrix}, \begin{bmatrix} 0 & 1 & \frac{7}{9} \end{bmatrix}\right\}$

A basis for the column space: $\left\{\begin{bmatrix} 1 \\ 0 \end{bmatrix}, \begin{bmatrix} 0 \\ 1 \end{bmatrix}\right\}$

Rank = 2.

3. A basis for the row space: $\left\{\begin{bmatrix} 1 & 0 & \frac{1}{3} \end{bmatrix}, \begin{bmatrix} 0 & 1 & -\frac{5}{3} \end{bmatrix}\right\}$

A basis for the column space: $\left\{ \begin{bmatrix} 1 \\ 0 \\ -1 \end{bmatrix}, \begin{bmatrix} 0 \\ 1 \\ 1 \end{bmatrix} \right\}$

Rank = 2.

4. A basis for the row space is:
$\left\{ \begin{bmatrix} 1 & 0 & 0 & -1 \end{bmatrix}, \begin{bmatrix} 0 & 1 & 0 & -\frac{3}{2} \end{bmatrix}, \begin{bmatrix} 0 & 0 & 1 & \frac{7}{2} \end{bmatrix} \right\}$

A basis for the column space is: $\left\{ \begin{bmatrix} 1 \\ 0 \\ 0 \end{bmatrix}, \begin{bmatrix} 0 \\ 1 \\ 0 \end{bmatrix}, \begin{bmatrix} 0 \\ 0 \\ 1 \end{bmatrix} \right\}$

Rank = 3.

5. A basis for the row space is $\left\{ \begin{bmatrix} 1 & 0 & 0 \end{bmatrix}, \begin{bmatrix} 0 & 1 & 0 \end{bmatrix}, \begin{bmatrix} 0 & 0 & 1 \end{bmatrix} \right\}$
so $\dim(\text{row}(A)) = 3$.

A basis for the column space is $\left\{ \begin{bmatrix} 1 \\ 0 \\ 0 \\ -\frac{1}{7} \end{bmatrix}, \begin{bmatrix} 0 \\ 1 \\ 0 \\ -\frac{8}{7} \end{bmatrix}, \begin{bmatrix} 0 \\ 0 \\ 1 \\ \frac{12}{7} \end{bmatrix} \right\}$

so $\dim(\text{col}(A)) = 3$ as well.

7. Hint: Use Theorem 13.2.

8. Hint: If A is an invertible $n \times n$ matrix, then $\det(A) \neq 0$, so the rows (columns) of A must be linearly independent.

Exercise Set 14

1. Rank = 2 **2.** Rank = 2

3. Rank = 3 **4.** Rank = 4

5. The only solution is $\mathbf{x} = \begin{bmatrix} 0 \\ 0 \end{bmatrix}$, no basis exists.

7. $\mathbf{x} = s\begin{bmatrix} 7 \\ 2 \\ -11 \\ 1 \end{bmatrix}$, so a basis is $\left\{\begin{bmatrix} 7 \\ 2 \\ -11 \\ 1 \end{bmatrix}\right\}$

9. $\mathbf{x} = \mathbf{x}_p + \mathbf{x}_h = \begin{bmatrix} 9 \\ -7 \end{bmatrix} + \begin{bmatrix} 0 \\ 0 \end{bmatrix}$

11. $\mathbf{x} = \mathbf{x}_p + \mathbf{x}_h = \begin{bmatrix} 1 \\ 1 \\ 0 \\ 0 \end{bmatrix} + s\begin{bmatrix} 7 \\ 2 \\ -11 \\ 1 \end{bmatrix}$

13. $\mathbf{x} = \mathbf{x}_p + \mathbf{x}_h = \begin{bmatrix} \frac{4}{3} \\ \frac{1}{3} \\ 0 \\ 0 \end{bmatrix} + s\begin{bmatrix} -\frac{2}{3} \\ \frac{1}{3} \\ 1 \\ 0 \end{bmatrix} + t\begin{bmatrix} -\frac{2}{3} \\ \frac{1}{3} \\ 0 \\ 1 \end{bmatrix}$

Exercise Set 15

1. **(a)** Clearly, $S \cap T = \{\mathbf{0}\}$. Let $\mathbf{x} = [x_1, x_2, x_3]^T \in \mathbb{R}^3$, then

$$\mathbf{x} = \begin{bmatrix} x_1 \\ x_2 \\ 0 \end{bmatrix} + \begin{bmatrix} 0 \\ 0 \\ x_3 \end{bmatrix}$$

where $[x_1, x_2, 0]^T \in S$ and $[0, 0, x_3]^T \in T$. Thus, $\mathbb{R}^3 = S \oplus T$.
(b) S is the set of all vectors in the xy-plane, T is the set of all vectors lying along the z-axis.

2. **(a)** Since a typical element of S must have the form

$$\mathbf{p}(x) = a_4x^4 + a_3x^3 + a_2x^2 + a_1x$$

the complementary subspace $T = \left\{\mathbf{p}(x) \in P_4 : \mathbf{p}(x) = a_0 \text{ where } a_0 \in \mathbb{R}\right\}$.
(b) A basis for T is $\{1\}$.
(c) Observe that $S \cap T = \{\mathbf{0}\}$ and $P_4 = S + T$.

3.
(a) S = all upper triangular 2×2 matrices, T = all lower triangular 2×2 matrices.

(b) $S \cap T$ = all diagonal 2×2 matrices

(c) $V \neq S \oplus T$ since $S \cap T \neq \begin{bmatrix} 0 & 0 \\ 0 & 0 \end{bmatrix}$

(d) $\dim(S+T) = \dim(S) + \dim(T) - \dim(S \cap T) = 3 + 3 - 2.$

4.
(b) S = set of all 2×2 symmetric matrices
T = set of all 2×2 àntisymmetric matrices.
(c) **Hint**: use Theorem 3.2
(d) $\dim S = 3$, $\dim T = 1$.

Exercise Set 16

1.
(a) A typical coset of V/W is a horizontal line in the plane.
(b) A basis for the coset space V/W is $\{[0,1]^T + W\}$.
(c) $\dim(V/W) = 1$
(d) $\dim(V/W) = \dim(V) - \dim(W) = 2 - 1 = 1.$

2.
(a) A typical element of W is.$\mathbf{p}(x) = a_2x^2 + a_4x^4$
(b) A basis for W is $\{x^2, x^4\}$.
(c) A basis for V is $\{1, x, x^2, x^4\}$.
(d) A basis for the quotient space V/W. is $\{1 + W, x + W, x^3 + W\}$
(e) A typical element of the quotient space is $\mathbf{p}(x) = (a_0 + a_1x + a_3x^3) + W$.

3. Let $V = \mathbb{R}^3$ and let W be the subspace of $\mathbb{R}^3$ given by

$$W = \left\{ \begin{bmatrix} a \\ a \\ a \end{bmatrix} \in \mathbb{R}^3 : a \neq 0 \right\}$$

(a) The subspace W is a line through the origin.
(b) A basis for W is $\{[1,1,1]^T\}$.
(c) A basis for V is $\{[1,0,0]^T, [0,1,0]^T, [1,1,1]^T\}$.
(d) A basis for the quotient space V/W is $\{[1,0,0]^T + W, [0,1,0]^T + W\}$.
(e) A typical element of the quotient space is any line parallel to W.
(f) $\dim(V/W) = \dim(V) - \dim(W) = 3 - 1 = 2.$

4.
(a) A basis for W is $\{x, x^2\}$.
(b) **Hint**: Assume that $a(1 + W) + b(x^3 + W) = 0 + W$ and deduce that $a = b = 0$. Then show that the set $\{1 + W, x^3 + W\}$ spans W.

(c) $\dim(V/W) = \dim(V) - \dim(W) = 4 - 2 = 2$.

Exercise Set 17

1. (a) $[\mathbf{x}]_B = \begin{bmatrix} -2 \\ 5 \end{bmatrix}$ (b) $[\mathbf{x}]_B = \begin{bmatrix} 6 \\ -20 \end{bmatrix}$

(c) $[\mathbf{x}]_B = \begin{bmatrix} 1 \\ 0 \end{bmatrix}$

2. (a) $[\mathbf{p}(x)]_B = \begin{bmatrix} 5 \\ 5/2 \\ 1 \end{bmatrix}$ and $[\mathbf{q}(x)]_B = \begin{bmatrix} -2 \\ 0 \\ 1 \end{bmatrix}$.

(b) The transition matrix $P = \begin{bmatrix} 1/3 & 2 & -2/3 \\ 1/3 & 0 & 1/3 \\ -1/3 & 0 & 2/3 \end{bmatrix}$

(c) The coordinate vectors $[\mathbf{p}(x)]_{B'} = \begin{bmatrix} 6 \\ 2 \\ -1 \end{bmatrix}$ and $[\mathbf{q}(x)]_{B'} = \begin{bmatrix} -\frac{4}{3} \\ -\frac{1}{3} \\ \frac{4}{3} \end{bmatrix}$

(d) The inverse transition matrix $P^{-1} = \begin{bmatrix} 0 & 2 & -1 \\ \frac{1}{2} & 0 & \frac{1}{2} \\ 0 & 1 & 1 \end{bmatrix}$

3.

(a) The transition matrix $P = \begin{bmatrix} 3/5 & 2/5 \\ 1/5 & -1/5 \end{bmatrix}$.

(b) The inverse transition matrix $P^{-1} = \begin{bmatrix} 1 & 2 \\ 1 & -3 \end{bmatrix}$.

(c) The coordinate vectors $[\mathbf{x}]_S = \begin{bmatrix} 6 \\ -4 \end{bmatrix}$ and $[\mathbf{y}]_S = \begin{bmatrix} 2 \\ -6 \end{bmatrix}$.

(d) The coordinate vectors $[\mathbf{x}]_T = \begin{bmatrix} 2 \\ 2 \end{bmatrix}$ and $[\mathbf{y}]_T = \begin{bmatrix} -\frac{6}{5} \\ \frac{8}{5} \end{bmatrix}$.

6. Assume that $c_1[\mathbf{x}_1]_B + c_2[\mathbf{x}_2]_B + \ldots + c_r[\mathbf{x}_r]_B = [\mathbf{0}]_B$
Then by Exercise 5,

$$
\begin{aligned}
[c_1\mathbf{x}_1 + c_2\mathbf{x}_2 + \ldots + c_r\mathbf{x}_r]_B &= [\mathbf{0}]_B \\
&\Rightarrow c_1\mathbf{x}_1 + c_2\mathbf{x}_2 + \ldots + c_r\mathbf{x}_r = \mathbf{0} \\
&\Rightarrow c_1 = c_2 = \ldots = c_r
\end{aligned}
$$

Exercise Set 18

1. $\langle \mathbf{x}, \mathbf{y} \rangle = -1$ **3.** $\langle \mathbf{x}, \mathbf{y} + \mathbf{z} \rangle = -1$ **5.** $\|\mathbf{x} + \mathbf{y}\| = \sqrt{6}$

7. $\angle(\mathbf{y}, \mathbf{z}) = \cos^{-1}(1/\sqrt{6}) \approx 1.1503$ **9.** $d(\mathbf{y}, \mathbf{z}) = \sqrt{5}$

11. The set $W^{\perp}$ is not empty, since $\mathbf{0} \in W^{\perp}$. Now, let $\mathbf{v}_1, \mathbf{v}_2 \in W^{\perp}$ so that $\langle \mathbf{v}_1, \mathbf{w} \rangle = 0$ and $\langle \mathbf{v}_2, \mathbf{w} \rangle = 0$ for all $\mathbf{w} \in W$. Observe that we have both

$$
\begin{aligned}
\langle \mathbf{v}_1 + \mathbf{v}_2, \mathbf{w} \rangle &= \langle \mathbf{v}_1, \mathbf{w} \rangle + \langle \mathbf{v}_2, \mathbf{w} \rangle = 0 \\
\langle c\mathbf{v}_1, \mathbf{w} \rangle &= c\langle \mathbf{v}_1, \mathbf{w} \rangle = 0 \text{ for any real scalar } c
\end{aligned}
$$

so $\mathbf{v}_1 + \mathbf{v}_2 \in W^{\perp}$ and $c\mathbf{v}_1 \in W^{\perp}$. The result then follows by Theorem 10.1.

13. Observe that for $k \neq m$, we have

$$
\begin{aligned}
\langle \mathbf{f}_k, \mathbf{f}_m \rangle &= \frac{1}{\pi}\int_{-\pi}^{+\pi} \cos kt \cos mt \, dt = \frac{1}{2\pi}\left(\int_{-\pi}^{+\pi} \cos(k+m)t \, dt + \int_{-\pi}^{+\pi} \cos(k-m)t \, dt\right) \\
&= \frac{1}{2\pi}\left[\frac{\sin(k+m)t}{(k+m)} + \frac{\sin(k-m)t}{(k-m)}\right]_{t=-\pi}^{t=\pi} = 0
\end{aligned}
$$

For $k = m$, we have

$$
\langle \mathbf{f}_k, \mathbf{f}_k \rangle = \frac{1}{\pi}\int_{-\pi}^{+\pi} \cos^2 kt \, dt = \frac{1}{\pi}\int_{-\pi}^{+\pi}\left(\frac{1+\cos 2kt}{2}\right)dt = 1
$$

In other words, $\langle \mathbf{f}_k, \mathbf{f}_m \rangle = \delta_{km}$.

15. By the Cauchy-Schwarz inequality, we have $\langle \mathbf{f}, \mathbf{g} \rangle^2 \leq \langle \mathbf{f}, \mathbf{f} \rangle \langle \mathbf{g}, \mathbf{g} \rangle$ so that

$$
\langle \mathbf{f}, \mathbf{g} \rangle^2 = \left\{\int_a^b f(t)g(t)dt\right\}^2 \leq \int_a^b f^2(t)dt \cdot \int_a^b g^2(t)dt
$$

Exercise Set 19

1.

(a) $\varepsilon_1 = [1/\sqrt{2}, 1/\sqrt{2}]^T$, $\varepsilon_2 = [1/\sqrt{2}, -1/\sqrt{2}]^T$

(b) $\varepsilon_1 = [0, -1]^T$, $\varepsilon_2 = [1, 0]^T$

2.

(a) $\varepsilon_1 = [1, 0, 0]^T$, $\varepsilon_2 = [0, 1, 0]^T$, $\varepsilon_3 = [0, 0, 1]^T$

(b) $\varepsilon_1 = [0,0,1]^T$, $\varepsilon_2 = \left[1/\sqrt{2}, 1/\sqrt{2}, 0\right]^T$, $\varepsilon_3 = \left[1/\sqrt{2}, -1/\sqrt{2}, 0\right]^T$

3. $\varepsilon_1 = 1$, $\varepsilon_2 = \frac{6}{\sqrt{3}}\left(x - \frac{1}{2}\right)$, $\varepsilon_3 = \frac{30}{\sqrt{5}}\left(x^2 - x + \frac{1}{6}\right)$

5. It may be shown that $E^2 = \left\| \mathbf{v} - \sum_{k=1}^{r} a_k \varepsilon_k \right\|^2 = \langle \mathbf{v}, \mathbf{v} \rangle^2 - \sum_{k=1}^{r} \langle \mathbf{v}, \varepsilon_k \rangle^2 + \Delta_k^2$;
so, E^2 is minimized when $\Delta_k^2 = 0$.

7.
(a) A basis is $\{[1,0,-1]^T, [0,1,-2]^T\}$
(b) An orthonormal basis is $\left\{\left[1/\sqrt{2}, 0, -1/\sqrt{2}\right]^T, \left[-1/\sqrt{3}, 1/\sqrt{3}, -1/\sqrt{3}\right]^T\right\}$.
(c) The distance is $\frac{2}{3}\sqrt{6}$

Exercise Set 20

1. linear **3.** linear **5.** linear

7.
(a) $T(-\mathbf{v}) = T(-1 \cdot \mathbf{v}) = -1 \cdot T(\mathbf{v}) = -T(\mathbf{v})$
(b) $T(\mathbf{u} - \mathbf{v}) = T(\mathbf{u} + (-\mathbf{v})) = T(\mathbf{u}) + T(-\mathbf{v}) = T(\mathbf{u}) - T(\mathbf{v})$
(c) $T(\mathbf{0}) = T(\mathbf{v} - \mathbf{v}) = T(\mathbf{v}) - T(\mathbf{v}) = \mathbf{0}$

9. If T is the zero operator, then we have both

$$T(\mathbf{u} + \mathbf{v}) = \mathbf{0} = T(\mathbf{u}) + T(\mathbf{v})$$
$$T(\lambda \mathbf{u}) = \mathbf{0} = \lambda T(\mathbf{u}) \quad \text{(for all scalars } \lambda)$$

since $T(\mathbf{u}) = T(\mathbf{v}) = \mathbf{0}$. So, the zero transformation is a linear operator on V.

10.
(a) Observe that

$$T(\mathbf{p} + \mathbf{q}) = \int_0^x (p(t) + q(t))dt = \int_0^x p(t)dt + \int_0^x q(t)dt$$
$$= T(\mathbf{p}) + T(\mathbf{q})$$

and, for any scalar λ, we have

$$T(\lambda \mathbf{p}) = \int_0^x \lambda p(t)dt = \lambda \int_0^x p(t)dt = \lambda T(\mathbf{p})$$

Thus, T is a linear transformation.

(b) $T(x^2 + 1) = \int_0^x (t^2 + 1)dt = \frac{1}{3}x^3 + x$ **(c)** $T(2x + 3) = \int_0^x (2t + 3)dt = x(x + 3)$

11.
(a) $T(2x + 1) = 2T(x) + T(1) = 2x^2 + 2x + 1$
(b) $T(2x^2 + 3x + 4) = 2T(x^2) + 3T(x) + 4T(1) = 3x^2 + 8x + 6$

12. If $T(\mathbf{v}) = \mathbf{v}$, then

$$\begin{bmatrix} 2x+y \\ x+y \end{bmatrix} = \begin{bmatrix} x \\ y \end{bmatrix}$$

so, $x+y = 0$ and $x = 0$. Thus, the only fixed point of T is $[0,0]^T$.

Exercise Set 21

1. Let $\varphi : P_1 \to \mathbb{R}^2$ be given by $\varphi(a_0 + a_1x) = [a_0, a_1]^T$. We first show that φ is linear. Let

$$\mathbf{p}(x) = a_0 + a_1x \quad \text{and} \quad \mathbf{q}(x) = b_0 + b_1x$$

Then

$$\varphi(\mathbf{p}+\mathbf{q}) = \varphi((a_0+b_0)+(a_1+b_1)x) = \begin{bmatrix} a_0+b_0 \\ a_1+b_1 \end{bmatrix}$$

$$= \begin{bmatrix} a_0 \\ a_1 \end{bmatrix} + \begin{bmatrix} b_0 \\ b_1 \end{bmatrix} = \varphi(\mathbf{p}) + \varphi(\mathbf{q})$$

Also, for any scalar λ,

$$\varphi(\lambda\mathbf{p}) = \varphi(\lambda a_0 + \lambda a_1 x) = \begin{bmatrix} \lambda a_0 \\ \lambda a_1 \end{bmatrix} = \lambda\begin{bmatrix} a_0 \\ a_1 \end{bmatrix} = \lambda\varphi(\mathbf{p})$$

Also, φ is onto, since if $\mathbf{x} = [a_1, a_2]^T \in \mathbb{R}^2$ then $\mathbf{x} = \varphi(a_1 + a_2x)$. Finally, φ is 1-1, since $\varphi(a_0 + a_1x) = \varphi(b_0 + b_1x) \Rightarrow [a_0, a_1]^T = [b_0, b_1]^T$ so $a_0 = b_0, a_1 = b_1$, and consequently, $a_0 + a_1x = b_0 + b_1x$ Thus, $\mathbb{R}^2$ is isomorphic the P_1.

3. We are given φ is the mapping $\varphi : M(2,2) \to M(2,2)$ defined by $\varphi(A) = A^T$. Observe that φ is linear, since if $A, B \in M(2,2)$, then

$$\varphi(A+B) = (A+B)^T = A^T + B^T = \varphi(A) + \varphi(B)$$
$$\varphi(\lambda A) = (\lambda A)^T = \lambda A^T = \lambda\varphi(A) \quad \text{(for any scalar } \lambda)$$

φ is onto, since if $B \in M(2,2)$, then $B = \varphi(B^T)$. Finally, φ is 1-1, since $\varphi(A) = \varphi(B) \Rightarrow A^T = B^T \Rightarrow A = B$.

5. Let $A_0 \in M(n,n)$ be a fixed invertible matrix. We are given the mapping $\varphi : M(n,n) \to M(n,n)$ where $\varphi(X) = A_0X$ for all $X \in M(n,n)$. Clearly, φ is linear, since

$$\varphi(X+Y) = A_0(X+Y) = A_0X + A_0Y = \varphi(X) + \varphi(Y)$$
$$\varphi(\lambda X) = A_0(\lambda X) = \lambda(A_0X) = \lambda\varphi(X) \quad \text{(for any scalar } \lambda\text{)}$$

Also, φ is onto, since if $Y \in M(n,n)$, then $Y = \varphi(A_0^{-1}Y)$. Finally, φ is 1-1, since $\varphi(X) = \varphi(Y) \Rightarrow A_0X = A_0Y \Rightarrow A_0^{-1}(A_0X) = A_0^{-1}(A_0Y) \Rightarrow X = Y$.

6. We are given that φ is an isomorphism from V to V' and S is a subspace of V, where $\varphi(S) = \{\varphi(\mathbf{v}) : \mathbf{v} \in S\}$. Observe that $\varphi(S)$ is not empty, since $\varphi(\mathbf{0}) = \mathbf{0} \in \varphi(S)$. Now, assume that $\mathbf{u}', \mathbf{v}' \in \varphi(S)$, then since φ is onto, there exist vectors $\mathbf{u}, \mathbf{v} \in S$ such that $\varphi(\mathbf{u}) = \mathbf{u}'$ and $\varphi(\mathbf{v}) = \mathbf{v}'$. Thus,

$$\mathbf{u}' + \mathbf{v}' = \varphi(\mathbf{u}) + \varphi(\mathbf{v}) = \varphi(\mathbf{u} + \mathbf{v}) \in \varphi(S)$$
$$\lambda\mathbf{u}' = \lambda\varphi(\mathbf{u}) = \varphi(\lambda\mathbf{u}) \in \varphi(S) \quad \text{(for any scalar } \lambda\text{)}$$

Thus, $\varphi(S)$ is a subspace of V'.

7. **Hint**: let $\varphi : V \to V$ be the identity mapping $\varphi(\mathbf{v}) = \mathbf{v}$, for all $\mathbf{v} \in V$.

8. We are given that $\varphi : V \to V'$ is an isomorphism from V to V'. Let $\mathbf{v}' \in V'$ be arbitrary. Since φ is onto, there exists a vector $\mathbf{v} \in V$ such that $\varphi(\mathbf{v}) = \mathbf{v}'$. Define $\mathbf{v} = \varphi^{-1}(\mathbf{v}')$. Since φ is onto, φ^{-1} is defined for each vector $\mathbf{v}' \in V'$. Moreover, $\mathbf{v}$ is unique, since φ is 1-1. Thus, the inverse mapping φ^{-1} is well defined. Now, assume that $\varphi(\mathbf{u}) = \mathbf{u}'$ and $\varphi(\mathbf{v}) = \mathbf{v}'$. Observe that φ^{-1} is linear, since

$$\varphi(\mathbf{u} + \mathbf{v}) = \mathbf{u}' + \mathbf{v}' \Rightarrow \varphi^{-1}(\mathbf{u}' + \mathbf{v}') = \mathbf{u} + \mathbf{v} = \varphi^{-1}(\mathbf{u}') + \varphi^{-1}(\mathbf{v}')$$
$$\varphi(\lambda\mathbf{u}) = \lambda\mathbf{u}' \Rightarrow \varphi^{-1}(\lambda\mathbf{u}') = \lambda\mathbf{u} = \lambda\varphi^{-1}(\mathbf{u}') \quad \text{(for any scalar } \lambda\text{)}$$

Also, φ^{-1} is onto, since for any $\mathbf{u} \in V$, we have $\mathbf{u} = \varphi^{-1}(\varphi(\mathbf{u}))$. Finally, φ^{-1} is 1-1, since

$$\varphi^{-1}(\mathbf{v}_1') = \varphi^{-1}(\mathbf{v}_2') \Rightarrow \varphi(\varphi^{-1}(\mathbf{v}_1')) = \varphi(\varphi^{-1}(\mathbf{v}_2'))$$
$$\Rightarrow \mathbf{v}_1' = \mathbf{v}_2'$$

9. Consider the isomorphisms $\varphi_1 : U \to V$ and $\varphi_2 : V \to W$. We shall show that the composite mapping $(\varphi_2 \circ \varphi_1) : U \to W$ is an isomorphism. Now, $\varphi_2 \circ \varphi_1$ is linear, since if $\mathbf{u}_1, \mathbf{u}_2 \in U$ then

$$(\varphi_2 \circ \varphi_1)(\mathbf{u}_1 + \mathbf{u}_2) = \varphi_2[\varphi_1(\mathbf{u}_1 + \mathbf{u}_2)] = \varphi_2[\varphi_1(\mathbf{u}_1) + \varphi_1(\mathbf{u}_2)]$$
$$= \varphi_2[\varphi_1\mathbf{u}_1] + \varphi_2[\varphi_1(\mathbf{u}_2)]$$
$$= (\varphi_2 \circ \varphi_1)(\mathbf{u}_1) + (\varphi_2 \circ \varphi_1)(\mathbf{u}_2)$$

and for any scalar λ, we have

$$
\begin{aligned}
(\varphi_2 \circ \varphi_1)(\lambda \mathbf{u}_1) &= \varphi_2[\varphi_1(\lambda \mathbf{u}_1)] = \varphi_2[\lambda \varphi_1 \mathbf{u}_1)] \\
&= \lambda \varphi_2[\varphi_1(\mathbf{u}_1)] \\
&= \lambda(\varphi_2 \circ \varphi_1)(\mathbf{u}_1)
\end{aligned}
$$

To see that $\varphi_2 \circ \varphi_1$ is onto, let $\mathbf{w} \in W$. Since φ_2 is onto, there exists a $\mathbf{v} \in V$ such that $\varphi_2(\mathbf{v}) = \mathbf{w}$. Since φ_1 is onto, there exists a $\mathbf{u} \in U$ such that $\varphi_1(\mathbf{u}) = \mathbf{v}$. But then,

$$
(\varphi_2 \circ \varphi_1)(\mathbf{u}) = \varphi_2[\varphi_1(\mathbf{u})] = \varphi_2(\mathbf{v}) = \mathbf{w}
$$

So, $\mathbf{w}$ is the image of $\mathbf{u}$ under the mapping $\varphi_2 \circ \varphi_1$, and $\varphi_2 \circ \varphi_1$ is onto. To see that $\varphi_2 \circ \varphi_1$ is 1-1, let $\mathbf{w}_1, \mathbf{w}_2 \in W$, $\varphi_2(\mathbf{v}_1) = \mathbf{w}_1$, $\varphi_2(\mathbf{v}_2) = \mathbf{w}_2$, and let $\varphi_1(\mathbf{u}_1) = \mathbf{v}_1$ and $\varphi_1(\mathbf{u}_2) = \mathbf{v}_2$. Observe that

$$
\begin{aligned}
(\varphi_2 \circ \varphi_1)(\mathbf{u}_1) &= (\varphi_2 \circ \varphi_1)(\mathbf{u}_2) \Rightarrow \mathbf{w}_1 = \mathbf{w}_2 \\
&\Rightarrow \varphi_2(\mathbf{v}_1) = \varphi_2(\mathbf{v}_2) \\
&\Rightarrow \mathbf{v}_1 = \mathbf{v}_2 \\
&\Rightarrow \varphi_1(\mathbf{u}_1) = \varphi_1(\mathbf{u}_2) \\
&\Rightarrow \mathbf{u}_1 = \mathbf{u}_2
\end{aligned}
$$

Exercise Set 22

1. Only the vectors in (a) and (c) are in the range of T.

2.

(a) A basis for the null space of A is $\left\{ \begin{bmatrix} 1 \\ -2 \\ 1 \end{bmatrix} \right\}$.

(b) A basis for the range space of A is $\left\{ \begin{bmatrix} 1 \\ 0 \\ 1 \end{bmatrix}, \begin{bmatrix} 0 \\ 1 \\ -1 \end{bmatrix} \right\}$.

(c) The Dimension Theorem is verified since.

$$
n = 3 = 2 + 1 = (\text{rank of } T) + (\text{nullity of } T)
$$

3.

(a) The nullity of $T = 0$ **(b)** The nullity of $T = 1$ **(c)** The nullity of $T = 0$.

4.

(a) A basis for the range of T is $\left\{ \begin{bmatrix} 1 \\ 0 \end{bmatrix}, \begin{bmatrix} 0 \\ 1 \end{bmatrix} \right\}$.

(b) A basis for the kernel of T is $\left\{\begin{bmatrix} 1 \\ -2 \\ 1 \end{bmatrix}\right\}$.

(c) The rank of $T = 2$ and the nullity of $T = 1$.

5. Hint: Since $\det(A) \neq 0$, we are guaranteed that A^{-1} exists.

6. First, assume that $\ker(T) = \{\mathbf{0}\}$ and $A\mathbf{v}_1 = A\mathbf{v}_2$, so $A(\mathbf{v}_1 - \mathbf{v}_2) = \mathbf{0}$. Thus, $\mathbf{v}_1 - \mathbf{v}_2 \in \ker(T)$ which implies $\mathbf{v}_1 - \mathbf{v}_2 = \mathbf{0}$, or in other words, $\mathbf{v}_1 = \mathbf{v}_2$; so T is one-to-one. The proof of the converse is left to the reader.

7. To find the kernel of T, we set $T(A) = A - A^T = 0$ so that $A = A^T$. Thus, the kernel of T consists of all symmetric $n \times n$ matrices.

Exercise Set 23

1.

(a) $A = \begin{bmatrix} a & b \\ c & d \end{bmatrix}$ **(b)** $A = \begin{bmatrix} 2 & 1 \\ 1 & -1 \\ 1 & 3 \end{bmatrix}$

(c) $A = \begin{bmatrix} -1 & 0 \\ 0 & 1 \end{bmatrix}$ (This a reflection through the y-axis.)

2. $A = \begin{bmatrix} \cos\theta & -\sin\theta \\ \sin\theta & \cos\theta \end{bmatrix}$ **3.** $A = \begin{bmatrix} 0 & 1 \\ 1 & 0 \end{bmatrix}$ **4.** $A = \begin{bmatrix} 0 & 1 & 0 \\ 0 & 0 & 2 \end{bmatrix}$

5. $A = \begin{bmatrix} 1 & 0 \\ 0 & 0 \end{bmatrix}$ **6.** $A = \begin{bmatrix} 0 & 0 \\ 1 & 0 \\ 0 & 1 \end{bmatrix}$

7.

(a) $A = \begin{bmatrix} \frac{1}{2} & 1 & -\frac{1}{2} \\ \frac{3}{2} & 0 & \frac{1}{2} \\ -\frac{1}{2} & 0 & \frac{3}{2} \end{bmatrix}$

(b) $T\left(\begin{bmatrix} 1 \\ 1 \\ 1 \end{bmatrix}\right) = \begin{bmatrix} \frac{1}{2} & 1 & -\frac{1}{2} \\ \frac{3}{2} & 0 & \frac{1}{2} \\ -\frac{1}{2} & 0 & \frac{3}{2} \end{bmatrix}\begin{bmatrix} 1 \\ 1 \\ 1 \end{bmatrix} = \begin{bmatrix} 1 \\ 2 \\ 1 \end{bmatrix}_B$

8. Observe that $T(\mathbf{e}_1) = T(\mathbf{e}_2) = T(\mathbf{e}_3) = \begin{bmatrix} 0 \\ 0 \\ 0 \end{bmatrix}$.

9.

(**a**) $A = \begin{bmatrix} 0 & 1 & 0 \\ 0 & 0 & 2 \\ 0 & 0 & 0 \end{bmatrix}$

(**b**) $D(1 + 2x + x^2) = D\left(\begin{bmatrix} 1 \\ 2 \\ 1 \end{bmatrix}\right) = \begin{bmatrix} 0 & 1 & 0 \\ 0 & 0 & 2 \\ 0 & 0 & 0 \end{bmatrix}\begin{bmatrix} 1 \\ 2 \\ 1 \end{bmatrix} =$

$\begin{bmatrix} 2 \\ 2 \\ 0 \end{bmatrix} = 2 + 2x.$

11. **Hint**: Observe that $T(\mathbf{e}_k) = \lambda\mathbf{e}_k$ for $k = 1, 2, 3, \ldots, n$.

Exercise Set 24

1. Here,

$$T(\mathbf{e}_1) = T\left(\begin{bmatrix} 1 \\ 0 \end{bmatrix}\right) = \begin{bmatrix} 1 \\ 2 \end{bmatrix}, \quad T(\mathbf{e}_2) = T\left(\begin{bmatrix} 0 \\ 1 \end{bmatrix}\right) = \begin{bmatrix} 2 \\ 1 \end{bmatrix}$$

so,

$$A = \begin{bmatrix} 1 & 2 \\ 2 & 1 \end{bmatrix}$$

Also,

$$\mathbf{b}_1' = \begin{bmatrix} 1 \\ 1 \end{bmatrix} = (1)\mathbf{e}_1 + (1)\mathbf{e}_2, \quad \mathbf{b}_2' = \begin{bmatrix} -1 \\ 1 \end{bmatrix} = (-1)\mathbf{e}_1 + (1)\mathbf{e}_2$$

So,

$$P = \begin{bmatrix} 1 & -1 \\ 1 & 1 \end{bmatrix}$$

Thus,

$$A' = P^{-1}AP = \begin{bmatrix} 3 & 0 \\ 0 & -1 \end{bmatrix}$$

3. The transition matrix is

$$P = \begin{bmatrix} 1 & 1 & 0 \\ 1 & 0 & 1 \\ 0 & 1 & 1 \end{bmatrix}$$

So,

$$A' = P^{-1}AP = \begin{bmatrix} \frac{1}{2} & \frac{1}{2} & -\frac{1}{2} \\ \frac{1}{2} & -\frac{1}{2} & \frac{1}{2} \\ -\frac{1}{2} & \frac{1}{2} & \frac{1}{2} \end{bmatrix} \begin{bmatrix} 3 & -1 & 1 \\ -2 & 4 & 2 \\ -1 & 1 & 5 \end{bmatrix} \begin{bmatrix} 1 & 1 & 0 \\ 1 & 0 & 1 \\ 0 & 1 & 1 \end{bmatrix}$$

$$= \begin{bmatrix} 2 & 0 & 0 \\ 0 & 4 & 0 \\ 0 & 0 & 6 \end{bmatrix}$$

5. If $A \approx B$ then there exists a non-singular matrix P such that $B = P^{-1}AP$. Observe that

$$\begin{aligned} B^2 &= P^{-1}AP(P^{-1}AP) = P^{-1}A^2P \\ B^3 &= P^{-1}AP(P^{-1}A^2P) = P^{-1}A^3P \\ &\vdots \\ B^k &= P^{-1}A^kP \end{aligned}$$

But $B^k = P^{-1}A^kP \Rightarrow A^k \approx B^k$ for any positive integer k.

6. If $A \approx B$ then observe that

$$\begin{aligned} Tr(A) &= Tr(P^{-1}BP) = Tr((P^{-1}B)P) \\ & Tr(P(P^{-1}B)) \quad \text{(by 3.4c)} \\ &= Tr((PP^{-1})B) \\ &= Tr(IB) = Tr(B) \end{aligned}$$

7. Here,

$$A = \begin{bmatrix} 2 & 0 \\ 0 & 2 \end{bmatrix}$$

Consequently,

$$A' = P^{-1}AP = \begin{bmatrix} \frac{1}{2} & \frac{1}{2} \\ -\frac{1}{2} & \frac{1}{2} \end{bmatrix} \begin{bmatrix} 2 & 0 \\ 0 & 2 \end{bmatrix} \begin{bmatrix} 1 & -1 \\ 1 & 1 \end{bmatrix}$$

$$= \begin{bmatrix} 2 & 0 \\ 0 & 2 \end{bmatrix} = A$$

8. Here,

$$P = \begin{bmatrix} 1 - \sqrt{2} & 1 + \sqrt{2} \\ 1 & 1 \end{bmatrix} \Rightarrow P^{-1} = \begin{bmatrix} -\frac{1}{4}\sqrt{2} & \frac{1}{4}\sqrt{2} + \frac{1}{2} \\ \frac{1}{4}\sqrt{2} & \frac{1}{2} - \frac{1}{4}\sqrt{2} \end{bmatrix}$$

So,

$$A' = P^{-1}AP$$

$$= \begin{bmatrix} -\frac{1}{4}\sqrt{2} & \frac{1}{4}\sqrt{2} + \frac{1}{2} \\ \frac{1}{4}\sqrt{2} & \frac{1}{2} - \frac{1}{4}\sqrt{2} \end{bmatrix} \begin{bmatrix} 1/\sqrt{2} & 1/\sqrt{2} \\ 1/\sqrt{2} & -1/\sqrt{2} \end{bmatrix} \begin{bmatrix} 1 - \sqrt{2} & 1 + \sqrt{2} \\ 1 & 1 \end{bmatrix}$$

$$= \begin{bmatrix} -1 & 0 \\ 0 & 1 \end{bmatrix}$$

Exercise Set 25

1.

(a) $\lambda_1 = -1, \mathbf{x}_1 = \begin{bmatrix} -1 \\ 1 \end{bmatrix}$ and $\lambda_2 = 3, \mathbf{x}_2 = \begin{bmatrix} 1 \\ 1 \end{bmatrix}$

(b) $\lambda_1 = -1, \mathbf{x}_1 = \begin{bmatrix} -2 \\ 1 \end{bmatrix}$ and $\lambda_2 = 1, \mathbf{x}_2 = \begin{bmatrix} 0 \\ 1 \end{bmatrix}$

(c) $\lambda_1 = \left(\frac{5}{2} - \frac{1}{2}\sqrt{5}\right), \mathbf{x}_1 = \begin{bmatrix} \frac{1}{2}\sqrt{5} - \frac{1}{2} \\ 1 \end{bmatrix}$ and

$\lambda_2 = \left(\frac{1}{2}\sqrt{5} + \frac{5}{2}\right), \mathbf{x}_2 = \begin{bmatrix} -\frac{1}{2}\sqrt{5} - \frac{1}{2} \\ 1 \end{bmatrix}$

2.

(a) $\lambda_1 = 1,\ \mathbf{x}_1 = \begin{bmatrix} 0 \\ -1 \\ 1 \end{bmatrix}, \lambda_2 = 2,\ \mathbf{x}_2 = \begin{bmatrix} 1 \\ 0 \\ 1 \end{bmatrix}$, and $\lambda_3 = 3,\ \mathbf{x}_3 = \begin{bmatrix} 2 \\ 0 \\ 1 \end{bmatrix}$

(b) $\lambda_1 = 0,\ \mathbf{x}_1 = \begin{bmatrix} 0 \\ 1 \\ 0 \end{bmatrix}, \lambda_2 = \lambda_3 = 1,\ \mathbf{x}_2 = \begin{bmatrix} 0 \\ 1 \\ 1 \end{bmatrix}$

(c) $\lambda_1 = -1,\ \mathbf{x}_1 = \begin{bmatrix} -1 \\ \frac{3}{2} \\ 1 \end{bmatrix},\ \lambda_2 = \lambda_3 = 1,\ \mathbf{x}_2 = \begin{bmatrix} 0 \\ 1 \\ 0 \end{bmatrix}$, and $\mathbf{x}_2 = \begin{bmatrix} 0 \\ 0 \\ 1 \end{bmatrix}$

3.

(a) $B_1 = \left\{ \begin{bmatrix} 0 \\ -1 \\ 1 \end{bmatrix} \right\}, B_2 = \left\{ \begin{bmatrix} 1 \\ 0 \\ 1 \end{bmatrix} \right\}$, and $B_3 = \left\{ \begin{bmatrix} 2 \\ 0 \\ 1 \end{bmatrix} \right\}$.

(b) $B_0 = \left\{ \begin{bmatrix} 0 \\ 1 \\ 0 \end{bmatrix} \right\}, B_1 = \left\{ \begin{bmatrix} 0 \\ 1 \\ 1 \end{bmatrix} \right\}$

(c) $B_{-1} = \left\{ \begin{bmatrix} -1 \\ \frac{3}{2} \\ 1 \end{bmatrix} \right\},\ B_1 = \left\{ \begin{bmatrix} 0 \\ 1 \\ 0 \end{bmatrix}, \begin{bmatrix} 0 \\ 0 \\ 1 \end{bmatrix} \right\}$

5 Hint: $\det(A) = \lambda_1\lambda_2\cdots\lambda_n$ and A is *invertible* if and only if $\det(A) \neq 0$.

7. Since A is invertible, observe that $A\mathbf{x} = \lambda\mathbf{x}$ implies $(A^{-1}A)\mathbf{x} = \lambda(A^{-1}\mathbf{x})$ so that $A^{-1}\mathbf{x} = (1/\lambda)\mathbf{x}$, and $1/\lambda$ is then an eigenvalue of A^{-1}.

8. Hint: Observe that

$$\det(A - \lambda I) = \det(A - \lambda I)^T = \det(A^T - \lambda I^T) = \det(A^T - \lambda I)$$

Consequently, both A and A^T will have the same characteristic equation.

10. We have

$$\det(A - \lambda I) = \begin{vmatrix} a_{11} - \lambda & a_{12} \\ a_{21} & a_{22} - \lambda \end{vmatrix} = \lambda^2 - (a_{11} + a_{22})\lambda + (a_{11}a_{22} - a_{12}a_{21}) = 0$$

$$\Rightarrow \lambda^2 - Tr(A)\lambda + \det(A) = 0$$

as claimed.

11. We obtain:

$$\lambda_1 = 1,\ \mathbf{x}_1 = \begin{bmatrix} 1 \\ 0 \\ 0 \end{bmatrix},\ \lambda_2 = 2,\ \mathbf{x}_2 = \begin{bmatrix} 0 \\ 1 \\ 0 \end{bmatrix},\ \text{and } \lambda_3 = 4,\ \mathbf{x}_3 = \begin{bmatrix} \frac{2}{3} \\ -\frac{1}{2} \\ 1 \end{bmatrix}$$

Exercise Set 26

1.

(a) $P = \begin{bmatrix} -1 & 0 \\ 1 & 1 \end{bmatrix},\quad P^{-1}AP = \begin{bmatrix} -1 & 0 \\ 0 & 2 \end{bmatrix}$

(b) $P = \begin{bmatrix} 1 & -1/3 \\ 0 & 1 \end{bmatrix},\quad P^{-1}AP = \begin{bmatrix} 1 & 0 \\ 0 & -2 \end{bmatrix}$

(c) $P = \begin{bmatrix} 1 & 0 & 0 \\ 0 & -\frac{1}{2}\sqrt{5} - \frac{1}{2} & -\frac{1}{2} + \frac{1}{2}\sqrt{5} \\ 0 & 1 & 1 \end{bmatrix},$

$$P^{-1}AP = \begin{bmatrix} 1 & 0 & 0 \\ 0 & \frac{5}{2} - \frac{1}{2}\sqrt{5} & 0 \\ 0 & 0 & \frac{5}{2} + \frac{1}{2}\sqrt{5} \end{bmatrix}$$

(d) $P = \begin{bmatrix} 1 & 1 & 0 \\ 1 & 0 & 1 \\ 0 & 1 & 1 \end{bmatrix},\quad P^{-1}AP = \begin{bmatrix} 2 & 0 & 0 \\ 0 & 4 & 0 \\ 0 & 0 & 6 \end{bmatrix}$

3.
(a) The eigenvalues of A are $\lambda_1 = \lambda_2 = 1$ and $\lambda_3 = 2$.
(b) However, there are three linearly independent eigenvectors of A:

$$\mathbf{x}_1^{(1)} = \begin{bmatrix} 0 \\ 0 \\ 1 \end{bmatrix},\ \mathbf{x}_1^{(2)} = \begin{bmatrix} -1 \\ 1 \\ 0 \end{bmatrix},\ \text{and } \mathbf{x}_3 = \left\{\begin{bmatrix} 0 \\ 1 \\ 0 \end{bmatrix}\right\}.$$

(c) $P = \begin{bmatrix} 0 & -1 & 0 \\ 0 & 1 & 1 \\ 1 & 0 & 0 \end{bmatrix}$ and $P^{-1}AP = \begin{bmatrix} 1 & 0 & 0 \\ 0 & 1 & 0 \\ 0 & 0 & 2 \end{bmatrix}.$

(d) No, because Theorem 26.2 only provides a sufficient condition for diagonalizability, not a necessary condition for the same.

5. Hint: Observe that $A = PDP^{-1}$ so that

$$\begin{aligned} A^2 &= (PDP^{-1})(PDP^{-1}) = PD^2P^{-1} \\ A^3 &= (PDP^{-1})(PD^2P^{-1}) = PD^3P^{-1} \\ &\vdots \\ A^k &= PD^kP^{-1} \end{aligned}$$

6.

$$\begin{bmatrix} -3 & -4 \\ 1 & 2 \end{bmatrix}^4 = \begin{bmatrix} -1 & -4 \\ 1 & 1 \end{bmatrix}\begin{bmatrix} 1^4 & 0 \\ 0 & (-2)^4 \end{bmatrix}\begin{bmatrix} \frac{1}{3} & \frac{4}{3} \\ -\frac{1}{3} & -\frac{1}{3} \end{bmatrix} = \begin{bmatrix} 21 & 20 \\ -5 & -4 \end{bmatrix}$$

7. If A and B are square matrices of order n, and $A \approx B$, then $A = P^{-1}BP$ where P is an invertible square matrix of order n. Observe that

$$\begin{aligned} \det(A - \lambda I) &= \det(P^{-1}BP - \lambda I) = \det(P^{-1}(BP - \lambda P)) \\ &= \frac{1}{\det(P)}\det((B - \lambda I)P) = \frac{1}{\det(P)}\det(B - \lambda I) \cdot \det(P) \\ &= \det(B - \lambda I) \end{aligned}$$

So, $\det(A - \lambda I) = \det(B - \lambda I)$; thus, A and B have the same characteristic equation, and consequently, they have the same eigenvalues.

Exercise Set 27

1. **(a)** $\lambda = -1, 3$ **(b)** $\lambda = -1, 7$ **(c)** $\lambda = 0, 6$ **(d)** $\lambda = 0, 3, 6$

2.

(a) $P = \begin{bmatrix} -1 & 1 \\ 1 & 1 \end{bmatrix}$, $P^{-1}AP = \begin{bmatrix} -1 & 0 \\ 0 & 3 \end{bmatrix}$

(b) $P = \begin{bmatrix} -1 & 1 \\ 1 & 1 \end{bmatrix}$, $P^{-1}AP = \begin{bmatrix} -1 & 0 \\ 0 & 7 \end{bmatrix}$

(c) $P = \begin{bmatrix} -1 & -1 & 1 \\ 1 & 0 & 1 \\ 0 & 1 & 1 \end{bmatrix}$, $P^{-1}AP = \begin{bmatrix} 0 & 0 & 0 \\ 0 & 0 & 0 \\ 0 & 0 & 6 \end{bmatrix}$

(d) $P = \begin{bmatrix} -2/3 & -1/3 & 2/3 \\ 2/3 & -2/3 & 1/3 \\ 1/3 & 2/3 & 2/3 \end{bmatrix}$, $P^{-1}AP = \begin{bmatrix} 0 & 0 & 0 \\ 0 & 3 & 0 \\ 0 & 0 & 6 \end{bmatrix}$

3. If A and B are symmetric matrices that commute, then

$$(AB)^T = B^T A^T = BA = AB$$
$$(BA)^T = A^T B^T = AB = BA$$

Thus, AB and BA are also symmetric matrices.

5. If P and Q are $n \times n$ orthogonal matrices, then

$$PQ \cdot (PQ)^T = PQ \cdot (Q^T P^T) = PQ \cdot Q^{-1} P^{-1} = I$$

Thus, $(PQ)^{-1} = (PQ)^T$, and so, PQ is orthogonal as well. The fact that QP is orthogonal follows in a similar fashion.

6. Observe that since P is orthogonal, we have

$$\begin{aligned}\langle P\mathbf{x}, P\mathbf{y}\rangle &= (P\mathbf{x})^T P\mathbf{y} = \mathbf{x}^T (P^T P)\mathbf{y} \\ &= \mathbf{x}^T \mathbf{y} = \langle \mathbf{x}, \mathbf{y}\rangle\end{aligned}$$

Thus, the inner product of any two vectors is **invariant** under an orthogonal transformation.

7. We have

$$AA^T = \begin{bmatrix} \cos\theta & \sin\theta & 0 \\ -\sin\theta & \cos\theta & 0 \\ 0 & 0 & 1 \end{bmatrix} \begin{bmatrix} \cos\theta & -\sin\theta & 0 \\ \sin\theta & \cos\theta & 0 \\ 0 & 0 & 1 \end{bmatrix} = \begin{bmatrix} 1 & 0 & 0 \\ 0 & 1 & 0 \\ 0 & 0 & 1 \end{bmatrix} = I$$

so A is orthogonal for any real number θ.

Exercise Set 28

1. **(a)** $\langle \mathbf{u}, \mathbf{v}\rangle = 2$ **(b)** $\langle \mathbf{v}, \mathbf{u}\rangle = 2$ **(c)** $\langle \mathbf{v}, \mathbf{w}\rangle = 1 - 2i$
(d) $\|\mathbf{u} + \mathbf{v}\| = \sqrt{5}$ **(e)** $\|\mathbf{u}\| + \|\mathbf{v}\| = \sqrt{2} + \sqrt{3}$
(f) $\langle \mathbf{u} + \mathbf{v}, \mathbf{u} - \mathbf{v}\rangle = -1 - 2i$

2. **(a)** $A^\dagger = \begin{bmatrix} 1 - i & -5i \\ 3 - 2i & 2i \end{bmatrix}$ **(b)** $A^\dagger = \begin{bmatrix} -2i & 2 - i \\ 3 - 4i & 0 \\ 1 & -4i \end{bmatrix}$

4. We can write:

$$\begin{aligned}\langle \mathbf{u}, c_1\mathbf{v}+c_2\mathbf{w}\rangle &= \overline{\langle c_1\mathbf{v}+c_2\mathbf{w}, \mathbf{u}\rangle} = \overline{(\overline{c_1}\langle \mathbf{v},\mathbf{u}\rangle + \overline{c_2}\langle \mathbf{w},\mathbf{u}\rangle)}\\ &= c_1\overline{\langle \mathbf{v},\mathbf{u}\rangle} + c_2\overline{\langle \mathbf{w},\mathbf{u}\rangle}\\ &= c_1\langle \mathbf{u},\mathbf{v}\rangle + c_2\langle \mathbf{u},\mathbf{w}\rangle\end{aligned}$$

7.

(a) $\langle \mathbf{f}(x), \mathbf{g}(x)\rangle = 3/4$ **(b)** $\langle \mathbf{g}(x), \mathbf{h}(x)\rangle = 2i - (1+2i)e^i$
(c) $\|\mathbf{g}(x)\| = 2/\sqrt{3}$ **(d)** $\|\mathbf{h}(x)\| = 1$

9. Since the vectors $\mathbf{u}, \mathbf{v}$ are orthogonal, then

$$\begin{aligned}\|\mathbf{u}+\mathbf{v}\|^2 &= \langle \mathbf{u}+\mathbf{v}, \mathbf{u}+\mathbf{v}\rangle\\ &= \langle \mathbf{u},\mathbf{u}\rangle + \langle \mathbf{u},\mathbf{v}\rangle + \langle \mathbf{v},\mathbf{u}\rangle + \langle \mathbf{v},\mathbf{v}\rangle\\ &= \langle \mathbf{u},\mathbf{u}\rangle + \langle \mathbf{v},\mathbf{v}\rangle \quad [\text{ Since } \langle \mathbf{u},\mathbf{v}\rangle = \langle \mathbf{v},\mathbf{u}\rangle = 0\,]\\ &= \|\mathbf{u}\|^2 + \|\mathbf{v}\|^2\end{aligned}$$

Exercise Set 29

1.

(a) $A^\dagger = \begin{bmatrix} 1 & -2i \\ 1-i & 3 \end{bmatrix}$

(b) $A^\dagger = \begin{bmatrix} 0 & -i & -2i \\ 1-i & 0 & 0 \\ -i & i & -4i \end{bmatrix}$

4.

(a) $\lambda_1 = 0, \lambda_2 = 2$; the eigenvectors are $\mathbf{x}_1 = \begin{bmatrix} -i \\ 1 \end{bmatrix}$ and $\mathbf{x}_2 = \begin{bmatrix} i \\ 1 \end{bmatrix}$

and $U = \begin{bmatrix} -i/\sqrt{2} & i/\sqrt{2} \\ 1/\sqrt{2} & 1/\sqrt{2} \end{bmatrix}$

(b) $\lambda_1 = -\sqrt{3}, \lambda_2 = 0, \lambda_3 = \sqrt{3}$; the eigenvectors are

$\mathbf{x}_1 = \begin{bmatrix} -\frac{1}{2}i\sqrt{3} - \frac{1}{2} \\ \frac{1}{2}i\sqrt{3} - \frac{1}{2} \\ 1 \end{bmatrix}$, $\mathbf{x}_2 = \begin{bmatrix} 1 \\ 1 \\ 1 \end{bmatrix}$, and $\mathbf{x}_3 = \begin{bmatrix} \frac{1}{2}i\sqrt{3} - \frac{1}{2} \\ -\frac{1}{2}i\sqrt{3} - \frac{1}{2} \\ 1 \end{bmatrix}$

and $U = \frac{1}{\sqrt{3}}\begin{bmatrix} -\frac{1}{2}i\sqrt{3} - \frac{1}{2} & 1 & \frac{1}{2}i\sqrt{3} - \frac{1}{2} \\ \frac{1}{2}i\sqrt{3} - \frac{1}{2} & 1 & -\frac{1}{2}i\sqrt{3} - \frac{1}{2} \\ 1 & 1 & 1 \end{bmatrix}$

(c) $\lambda_1 = -\sqrt{2}, \lambda_2 = 1, \lambda_3 = \sqrt{2}$; the eigenvectors are

$$\mathbf{x}_1 = \begin{bmatrix} -\left(\frac{1}{2} - \frac{1}{2}i\right)\sqrt{2} \\ 1 \\ 0 \end{bmatrix}, \mathbf{x}_2 = \begin{bmatrix} 0 \\ 0 \\ 1 \end{bmatrix}, \text{ and } \mathbf{x}_3 = \begin{bmatrix} \left(\frac{1}{2} - \frac{1}{2}i\right)\sqrt{2} \\ 1 \\ 0 \end{bmatrix}$$

and

$$U = \begin{bmatrix} -\frac{1}{2} + \frac{1}{2}i & 0 & \frac{1}{2} - \frac{1}{2}i \\ 1/\sqrt{2} & 0 & 1/\sqrt{2} \\ 0 & 1 & 0 \end{bmatrix}$$

(d) $\lambda_1 = -1, \lambda_2 = 3$; the eigenvectors are $\mathbf{x}_1 = \begin{bmatrix} i \\ 1 \end{bmatrix}$, $\mathbf{x}_2 = \begin{bmatrix} -i \\ 1 \end{bmatrix}$

and $U = \frac{1}{\sqrt{2}}\begin{bmatrix} i & -i \\ 1 & 1 \end{bmatrix}$

5. We can write

$$\begin{aligned} \langle T\mathbf{x}, T\mathbf{y}\rangle &= \langle U\mathbf{x}, U\mathbf{y}\rangle = (U\mathbf{x})^\dagger U\mathbf{y} = \mathbf{x}^\dagger(U^\dagger U)\mathbf{y} \\ &= \mathbf{x}^\dagger I\mathbf{y} = \mathbf{x}^\dagger I\mathbf{y} = \langle \mathbf{x}, \mathbf{y}\rangle \end{aligned}$$

6. Since A and B are Hermitian, and $AB = BA$, then

$$(AB)^\dagger = B^\dagger A^\dagger = BA = AB$$

8. Let U be a unitary matrix and assume that $U\mathbf{x} = \lambda\mathbf{x}$. Then we have both

$$\begin{aligned} \langle U\mathbf{x}, U\mathbf{x}\rangle &= (U\mathbf{x})^\dagger(U\mathbf{x}) = \mathbf{x}^\dagger(U^\dagger U)\mathbf{x} = \|\mathbf{x}\|^2 \\ \langle U\mathbf{x}, U\mathbf{x}\rangle &= \langle \lambda\mathbf{x}, \lambda\mathbf{x}\rangle = |\lambda|^2 \cdot \|\mathbf{x}\|^2 \end{aligned}$$

so $\|\mathbf{x}\|^2 = |\lambda|^2 \cdot \|\mathbf{x}\|^2$, and consequently, $|\lambda| = 1$.

9. If U is a unitary matrix, and H is a Hermitian matrix, then the matrix $U^{-1}HU$ is Hermitian since

$$(U^{-1}HU)^\dagger = U^\dagger H^\dagger(U^{-1})^\dagger = U^{-1}HU$$

10. If H is Hermitian, then $(H - iI)(H + iI)^{-1}$ is unitary since

$$\begin{aligned} [(H - iI)(H + iI)^{-1}]^\dagger &= \left[(H + iI)^{-1}\right]^\dagger (H - iI)^\dagger = \left[(H + iI)^\dagger\right]^{-1}(H - iI) \\ &= (H + iI)^{-1}(H - iI) = [(H - iI)(H + iI)^{-1}]^{-1} \end{aligned}$$

11. If U_1 and U_2 are unitary matrices then U_1U_2 is also a unitary since

$$(U_1U_2)^\dagger = U_2^\dagger U_1^\dagger = U_2^{-1}U_1^{-1} = (U_1U_2)^{-1}$$

15. If U is unitary, then $UU^\dagger = I = U^\dagger U$, so U is normal as well.

Exercise Set 30

1. (a) $A^4 = \begin{bmatrix} 16 & 15 \\ 0 & 1 \end{bmatrix}$ **(b)** $A^4 = \begin{bmatrix} 209 & 208 & 208 \\ 208 & 209 & 208 \\ 208 & 208 & 209 \end{bmatrix}$

2. For the matrix in (1a), the characteristic equation of $A = \begin{bmatrix} 2 & 1 \\ 0 & 1 \end{bmatrix}$, is $\lambda^2 - 3\lambda + 2 = 0$. We find that

$$A^2 - 3A + 2I = \begin{bmatrix} 2 & 1 \\ 0 & 1 \end{bmatrix}^2 - 3\begin{bmatrix} 2 & 1 \\ 0 & 1 \end{bmatrix} + 2\begin{bmatrix} 1 & 0 \\ 0 & 1 \end{bmatrix}$$
$$= \begin{bmatrix} 4 & 3 \\ 0 & 1 \end{bmatrix} - \begin{bmatrix} 6 & 3 \\ 0 & 3 \end{bmatrix} + \begin{bmatrix} 2 & 0 \\ 0 & 2 \end{bmatrix}$$
$$= \begin{bmatrix} 0 & 0 \\ 0 & 0 \end{bmatrix}$$

as it should, according to the Cayley-Hamilton theorem.

3. (a) $\begin{bmatrix} \frac{2}{3} & -\frac{1}{3} \\ -\frac{1}{3} & \frac{2}{3} \end{bmatrix}$ **(b)** $\begin{bmatrix} \frac{1}{2} & -\frac{1}{2} \\ -\frac{1}{4} & \frac{3}{4} \end{bmatrix}$ **(c)** $\begin{bmatrix} 0 & -1 & 1 \\ 0 & 1 & 0 \\ -1 & -1 & 1 \end{bmatrix}$

Exercise Set 31

1. $p(A) = A^2 - 6A + 9I = \begin{bmatrix} 4 & 0 \\ 0 & 4 \end{bmatrix}$. **2**. $\sqrt{B} = \frac{1}{2}\begin{bmatrix} \sqrt{3}+\sqrt{5} & \sqrt{5}-\sqrt{3} \\ \sqrt{5}-\sqrt{3} & \sqrt{3}+\sqrt{5} \end{bmatrix}$

3. $2^A = \frac{1}{16}\begin{bmatrix} 19 & 45 \\ 15 & 49 \end{bmatrix}$

5.

(a) $\cos B = \frac{1}{2}\begin{bmatrix} \cos 3 + \cos 5 & \cos 5 - \cos 3 \\ \cos 5 - \cos 3 & \cos 3 + \cos 5 \end{bmatrix}$

(b) $\sin B = \frac{1}{2}\begin{bmatrix} \sin 3 + \sin 5 & \sin 5 - \sin 3 \\ \sin 5 - \sin 3 & \sin 3 + \sin 5 \end{bmatrix}$

7. $B^k = \begin{bmatrix} 1 & 0 & 0 \\ 2^k - 1 & 2^k & 0 \\ \frac{3}{2}3^k - 2^k - \frac{1}{2} & 3^k - 2^k & 3^k \end{bmatrix}$ where k is any positive integer.

Exercise Set 32

1. (a) $\|A\| = 3$ (b) $\|A\| = \sqrt{2\sqrt{17} + 18} = 1 + \sqrt{17}$ (c) $\| A \| = 3$

2.
(a) H is Hermitian, since $H^\dagger = H$
(b) $\|H\| = 2$.

3. Observe that $\|A^k\| \leq \|A\|^k$ so $\|A^k\| \to 0$ as $k \to \infty$, since $\|A\| < 1$ by hypothesis.

4. If U is unitary matrix, then $H = U^\dagger U = I$. By Theorem 32.2, we have $\|U\| = \sqrt{1} = 1$ since *every* eigenvalue of H is equal to one.

5. Let A be a square matrix, with eigenvalues $\lambda_1, \lambda_2, \ldots, \lambda_n$. By Theorem 32.5, the eigenvalues of e^A are $e^{\lambda_1}, e^{\lambda_2}, \ldots, e^{\lambda_n}$. Hence, $\det(e^A) = e^{\lambda_1}e^{\lambda_2}\cdots e^{\lambda_n} \neq 0$, so e^A is invertible. Equation (32.28) follows by multiplying the power series expansions for e^A and e^{-A}, where we find, by brute force, $e^A e^{-A} = I$.

7. Hint: Use the power series expansions for e^{A+B}, e^A and e^B; then simplify your work by using the fact that $AB = BA$.

8.
(a) $\sin A = A - \frac{1}{3!}A^3 + \frac{1}{5!}A^5 - \ldots$
(b) The matrix series converges for any square matrix A.
(c) $\sin A = \begin{bmatrix} -\frac{1}{2}\sin 2 & \frac{1}{2}\sin 2 \\ \frac{3}{2}\sin 2 & \frac{1}{2}\sin 2 \end{bmatrix}$

9. We have, by Theorem 32.5,

$$\frac{d}{dt}(e^{At}) = \sum_{k=0}^{\infty} \frac{1}{k!} A^k \frac{d}{dt}(t^k) = A \cdot \sum_{k=1}^{\infty} \frac{1}{(k-1)!} A^{k-1} t^{k-1}$$
$$= Ae^{At}$$

Exercise Set 33

1.

(a) $\Delta(x) = x^2 - 3x + 5 = \mu(x)$

(b) $\Delta(x) = x^3 - 3x^2 + 3x - 1, \ \ \mu(x) = x^2 - 2x + 1$

(c) $\Delta(x) = x^3, \ \ \mu(x) = x^2$

(d) $\Delta(x) = \mu(x) = x^3 - x^2 - 4x - 4$

3. $A = \lambda I \Rightarrow A - \lambda I = 0 \Rightarrow \mu(x) = (x - \lambda)$ must be the minimal polynomial.

5. If A has n distinct eigenvalues, then each of its eigenvalues must be *simple*. Hence, the result then follows immediately from Theorem 33.2, and the *definition* of minimal polynomial.

6. Let $\mu(x) = x^s + a_{s-1}x^{s-1} + a_{s-2}x^{s-2} + \ldots + a_0$ be the minimal polynomial of a square matrix A where $a_0 \neq 0$. Then we must have

$$A^s + a_{s-1}A^{s-1} + a_{s-2}A^{s-2} + \ldots + a_0 I = 0$$

then, since $a_0 \neq 0$, we have

$$I = -\frac{1}{a_0}(A^s + a_{s-1}A^{s-1} + a_{s-2}A^{s-2} + \ldots + a_1 A)$$

Multiply the last equation on the left by A^{-1} to obtain

$$A^{-1} = -\frac{1}{a_0}(A^{s-1} + a_{s-1}A^{s-2} + a_{s-2}A^{s-3} + \ldots + a_1 I)$$

So A^{-1} exists, and A^{-1} is a polynomial in A of degree $s - 1$.

7. We can write:

$$\begin{aligned} T[p(T)]\mathbf{v} &= T\left(a_n T^n\mathbf{v} + a_{n-1}T^{n-1}\mathbf{v} + \ldots + a_1 T\mathbf{v} + a_0\mathbf{v}\right) \\ &= a_n T^{n+1}\mathbf{v} + a_{n-1}T^n\mathbf{v} + \ldots + a_1 T^2\mathbf{v} + a_0 T\mathbf{v} \\ &= \left(a_n T^n + a_{n-1}T^{n-1} + \ldots + a_1 T + a_0 I\right) T\mathbf{v} \\ &= p(T)[T(\mathbf{v})] \end{aligned}$$

Thus, the linear operators $P(T)$ and T commute.

Exercise Set 34

2. Clearly, if $\mathbf{u} \in S$ then there exists a c such that $\mathbf{u} = c\mathbf{v}$. But then $A\mathbf{u} = (\lambda c)\mathbf{u} \in S$, so S is invariant under T.

3. The invariant subspaces of T are $\{\mathbf{0}\}, \mathbb{R}^2$ and the eigenspaces

$$E_1 = span\left\{\begin{bmatrix} -1 \\ 1 \end{bmatrix}\right\} \text{ and } E_2 = span\left\{\left\{\begin{bmatrix} 1 \\ 1 \end{bmatrix}\right\}\right\}.$$

4. Let $\mathbf{v} \in S_1 \cap S_2$ so that $\mathbf{v} \in S_1$ and $\mathbf{v} \in S_2$. Clearly, $T(\mathbf{v}) = 0$ by hypothesis, so $S_1 \cap S_2$ is invariant under T.

5.
(**b**) Bases are $B_1 = \{[-1/2, 0, 1]^T, [0, 1, 0]^T\}$ and $B_2 = [0, 1, 1]^T$.
(**c**) Clearly, $B = B_1 \cup B_2$ is a basis for $\mathbb{R}^3$ since the vectors in B are linearly independent.
(**d**) The transition matrix is

$$P = \begin{bmatrix} -1/2 & 0 & 0 \\ 0 & 1 & 1 \\ 1 & 0 & 1 \end{bmatrix}$$

so that the matrix representation of T in terms of the basis B is

$$P^{-1}AP = \begin{bmatrix} -2 & 0 & 0 \\ -2 & 1 & -1 \\ 2 & 0 & 1 \end{bmatrix}\begin{bmatrix} 1 & 0 & 0 \\ 0 & 1 & 1 \\ 2 & 0 & 2 \end{bmatrix}\begin{bmatrix} -1/2 & 0 & 0 \\ 0 & 1 & 1 \\ 1 & 0 & 1 \end{bmatrix} = \begin{bmatrix} 1 & 0 & 0 \\ 1 & 1 & 0 \\ 0 & 0 & 2 \end{bmatrix}$$

Note the block diagonal form of $P^{-1}AP$.

Exercise Set 35

2. Bt direct computation, we find

$$N^2 = \begin{bmatrix} 0 & 0 & 1 & 1 \\ 0 & 0 & 0 & 1 \\ 0 & 0 & 0 & 0 \\ 0 & 0 & 0 & 0 \end{bmatrix}, \quad N^3 = \begin{bmatrix} 0 & 0 & 0 & 1 \\ 0 & 0 & 0 & 0 \\ 0 & 0 & 0 & 0 \\ 0 & 0 & 0 & 0 \end{bmatrix}, \quad N^4 = 0$$

So, N is a nilpotent matrix of index 4.

3. Jordan basis: $\left\{\begin{bmatrix} 1 \\ 0 \end{bmatrix}, \begin{bmatrix} 1 \\ 1/3 \end{bmatrix}\right\}$, $J = \begin{bmatrix} 2 & 1 \\ 0 & 2 \end{bmatrix}$

4. Jordan basis: $\left\{ \begin{bmatrix} 0 \\ 0 \\ 1 \end{bmatrix}, \begin{bmatrix} 0 \\ 1 \\ 1 \end{bmatrix}, \begin{bmatrix} 1 \\ 1 \\ 1 \end{bmatrix} \right\}$, $J = \begin{bmatrix} 1 & 1 & 0 \\ 0 & 1 & 1 \\ 0 & 0 & 1 \end{bmatrix}$

5. Jordan basis: $\left\{ \begin{bmatrix} -1 \\ 0 \\ 1 \end{bmatrix}, \begin{bmatrix} 0 \\ 0 \\ 1 \end{bmatrix}, \begin{bmatrix} 0 \\ 1 \\ 1 \end{bmatrix} \right\}$, $J = \begin{bmatrix} 1 & 0 & 0 \\ 0 & 2 & 1 \\ 0 & 0 & 2 \end{bmatrix}$

Exercise Set 36

1. Taking the time derivative of (36.4), we obtain, by the Liebnitz rule for the differentiation of integrals,

$$\begin{aligned} \dot{\mathbf{x}}(t) &= \frac{d}{dt}\mathbf{e}^{At}\mathbf{x}_0 + \frac{d}{dt}\int_0^t e^{A(t-u)}\mathbf{f}(u)du \\ &= Ae^{At}\mathbf{x}_0 + \mathbf{f}(t) + A\int_0^t e^{A(t-u)}\mathbf{f}(u)du \\ &= A\left(e^{At}\mathbf{x}_0 + \int_0^t e^{A(t-u)}\mathbf{f}(u)du\right) + \mathbf{f}(t) \\ &= A\mathbf{x}(t) + \mathbf{f}(t) \end{aligned}$$

so (36.5) is a solution of the differential equation. Also, observe that

$$\begin{aligned} \mathbf{x}(0) &= \mathbf{e}^{A(0)}\mathbf{x}_0 + \frac{d}{dt}\int_0^0 e^{A(t-u)}\mathbf{f}(u)du \\ &= I\mathbf{x}_0 + \mathbf{0} = \mathbf{x}_0 \end{aligned}$$

so the initial condition $\mathbf{x}(0) = \mathbf{x}_0$ is satisfied as well.

3.

(a) $e^{At} = \begin{bmatrix} e^t & 2te^t \\ 0 & e^t \end{bmatrix}$

(b) $e^{At} = \begin{bmatrix} e^t & 0 & 0 \\ 0 & e^{2t} & 0 \\ e^t(e^t - 1) & te^{2t} & e^{2t} \end{bmatrix}$

(c) $e^{At} = \begin{bmatrix} e^t & 0 & 0 \\ te^t(t+1) & e^t & te^t \\ 2te^t & 0 & e^t \end{bmatrix}$

4.
(a) $x_2(t) = 0, x_1(t) = 2e^t - 1,\ x_2(t) = 0.$
(b) $x_1(t) = 0,\ x_2(t) = e^{2t},\ x_3(t) = te^{2t}.$

(c) $x_1(t) = e^t,\ x_2(t) = e^t + t^2e^t + te^t,\ x_3(t) = 2te^t.$

5.
(b) We find that

$$e^{At} = \exp\left(\begin{bmatrix} 0 & 1 \\ -\omega^2 & 0 \end{bmatrix} t\right) = \begin{bmatrix} \cos\omega t & \frac{1}{\omega}\sin\omega t \\ -\omega\sin\omega t & \cos\omega t \end{bmatrix}$$

so the solution is

$$\mathbf{x}(t) = e^{At}\mathbf{x}_0 = \begin{bmatrix} \cos\omega t & \frac{1}{\omega}\sin\omega t \\ -\omega\sin\omega t & \cos\omega t \end{bmatrix}\begin{bmatrix} a \\ b \end{bmatrix} = \begin{bmatrix} a\cos\omega t + \frac{b}{\omega}\sin\omega t \\ b\cos\omega t - a\omega\sin\omega t \end{bmatrix}$$

Thus, the solution of the original differential equation is

$$x_1(t) = x(t) = a\cos\omega t + \frac{b}{\omega}\sin\omega t$$

(c) The motion is periodic with period $P = 2\pi/\omega$, since

$$\begin{aligned} x\left(t + \tfrac{2\pi}{\omega}\right) &= a\cos\omega\left(t + \tfrac{2\pi}{\omega}\right) + \tfrac{b}{\omega}\sin\omega\left(t + \tfrac{2\pi}{\omega}\right) \\ &= a\cos(\omega t + 2\pi) + \frac{b}{\omega}\sin(\omega t + 2\pi) \\ &= a\cos\omega t + \frac{b}{\omega}\sin\omega t = x(t) \end{aligned}$$

Exercise Set 37

1.
(a) The fixed points are $P_1(0,0)$, and $P_2(-1,1)$.
$P_1(0,0)$ is an unstable node, $P_2(-1,1)$ is an unstable saddle point.
(b) The fixed points are $P_1(-1,-1), P_2(0,0)$, and $P_3(1,1)$.
$P_1(-1,-1)$ is an unstable saddle point,
$P_2(0,0)$ is a stable point (asymptotically stable)
$P_1(1,1)$ is an unstable saddle point.
(c) $P_1(-1,1)$ and $P_2(0,0)$ are the fixed points.
$P_1(-1,1)$ is an unstable saddle point.
$P_2(0,0)$ is an unstable node.
(d) $P_1(0,0)$ is the only fixed point. It is an unstable node.

2.
(a) The nonlinear system we obtain is

$$\frac{d}{dt}\begin{bmatrix} x_1 \\ x_2 \end{bmatrix} = \begin{bmatrix} x_2 \\ -\varepsilon x_1^2 x_2 - x_1 + \varepsilon x_2 \end{bmatrix} = \mathbf{F}(x_1,x_2)$$

(b) The only solution of $\mathbf{F}(x_1,x_2) = \mathbf{0}$, is $x_1 = x_2 = 0$. So, the only fixed

point is $(0,0)$.
(c) The Jacobian matrix, evaluated at $(0,0)$ is

$$DF|_{(0,0)} = \begin{bmatrix} 0 & 1 \\ -1 & \varepsilon \end{bmatrix}$$

Its eigenvalues are $\lambda_1 = \frac{1}{2}\varepsilon + \frac{1}{2}\sqrt{\varepsilon^2 - 4}$, and $\lambda_2 = \frac{1}{2}\varepsilon - \frac{1}{2}\sqrt{\varepsilon^2 - 4}$.
If $\varepsilon < 2$, then $\lambda_1 = \frac{1}{2}\varepsilon + \frac{i}{2}\sqrt{4 - \varepsilon^2}$, and $\lambda_2 = \frac{1}{2}\varepsilon - \frac{i}{2}\sqrt{4 - \varepsilon^2}$, so the fixed point $(0,0)$ is an unstable spiral point.
If $\varepsilon > 2$, then the eigenvalues are both real and positive, so the fixed point $(0,0)$ is an unstable node.

3.
(a) We obtain the linear system

$$\frac{d}{dt}\begin{bmatrix} x_1 \\ x_2 \end{bmatrix} = \begin{bmatrix} 0 & 1 \\ -k^2 & -b \end{bmatrix}\begin{bmatrix} x_1 \\ x_2 \end{bmatrix} \equiv A\mathbf{x}$$

(b) The only fixed point is $(0,0)$, since

$$\begin{bmatrix} 0 & 1 \\ -k^2 & -b \end{bmatrix}\begin{bmatrix} x_1 \\ x_2 \end{bmatrix} = \begin{bmatrix} 0 \\ 0 \end{bmatrix} \Rightarrow \begin{bmatrix} x_1 \\ x_2 \end{bmatrix} = \begin{bmatrix} 0 \\ 0 \end{bmatrix}$$

The eigenvalues of A are $\lambda = -\frac{1}{2}b \pm \frac{1}{2}\sqrt{b^2 - 4k^2}$
If $b^2 < 4k^2$, then both eigenvalues are complex, so $(0,0)$ is an asymptotically stable point.
If $b^2 = 4k^2$, then both eigenvalues are negative, so $(0,0)$ is an asymptotically stable point.
If $b^2 > 4k^2$, then both eigenvalues are negative, so $(0,0)$ is an asymptotically stable point.
(e) Here, E is the total energy of the system and it is strictly decreasing as $t \to \infty$. Since from part (c), the solution $x(t) \to 0$ as $t \to \infty$, this comports with our analysis in **(c)**.

Exercise Set 38

1.
(a) $\mathbf{x}(t) = \left(A_1 \cos\sqrt{2}\,t + B_1 \sin\sqrt{2}\,t\right)\mathbf{v}_1 + (A_2 \cos 2t + B_2 \sin 2t)\mathbf{v}_2$

The normal modes are $\mathbf{v}_1 = \begin{bmatrix} -1 \\ 1 \end{bmatrix}$, $\mathbf{v}_2 = \begin{bmatrix} 1 \\ 1 \end{bmatrix}$,

The characteristic frequencies are $f_1 = \frac{1}{2\pi}\sqrt{2}$, $f_2 = \frac{1}{\pi}$
(b) $\mathbf{x}(t) = \left(A_1 \cos\sqrt{3}\,t + B_1 \sin\sqrt{3}\,t\right)\mathbf{v}_1 + \left(A_2 \cos\frac{1}{2}t + B_2 \sin\frac{1}{2}t\right)\mathbf{v}_2$

The normal modes are $\mathbf{v}_1 = \begin{bmatrix} \frac{8}{3} \\ 1 \end{bmatrix}$, $\mathbf{v}_2 = \begin{bmatrix} -1 \\ 1 \end{bmatrix}$,

The characteristic frequencies are $f_1 = \frac{1}{2\pi}\sqrt{3}$, $f_2 = \frac{1}{4\pi}$

2.

(a) The equations of motion are $\ddot{x}_1 + \frac{k}{m}(x_1 - x_2) = 0, \ \ddot{x}_2 + \frac{k}{m}(x_2 - x_1) = 0$

(b) The normal modes are $\mathbf{v}_1 = \begin{bmatrix} 1 \\ 1 \end{bmatrix}$, $\mathbf{v}_2 = \begin{bmatrix} -1 \\ 1 \end{bmatrix}$,

The characteristic frequencies are $f_1 = 0$ (no oscillation), and $f_2 = \frac{1}{2\pi}\sqrt{\frac{2k}{m}}$

(c) $x_1 = -x_0 \cos\sqrt{\frac{2k}{m}}t$ and $x_2 = x_0 \cos\sqrt{\frac{2k}{m}}t$.

3.

(a) The equations of motion are $m\ddot{x}_1 = -kx_1 - k(x_1 - x_2), \ \ddot{x}_2 = -kx_2 - k(x_2 - x_1)$

(b) The normal modes are $\mathbf{v}_1 = \begin{bmatrix} 1 \\ 1 \end{bmatrix}$, $\mathbf{v}_2 = \begin{bmatrix} -1 \\ 1 \end{bmatrix}$,

The characteristic frequencies are $f_1 = \frac{1}{2\pi}\sqrt{\frac{k}{m}}$, and $f_2 = \frac{1}{2\pi}\sqrt{\frac{3k}{m}}$

(c) $\mathbf{x}(t) = \left(A_1 \cos\sqrt{\frac{k}{m}}\,t + B_1 \sin\sqrt{\frac{k}{m}}\,t\right)\mathbf{v}_1 + \left(A_2 \cos\sqrt{\frac{3k}{m}}\,t + B_2 \sin\sqrt{\frac{3k}{m}}\,t\right)\mathbf{v}_2$

(d) $x_1 = x_2 = x_0 \cos\sqrt{\frac{3k}{m}}\,t$. Here, the first mode is manifested.

(e) $x_1 = -x_0 \cos\sqrt{\frac{3k}{m}}\,t$, $x_2 = x_0 \cos\sqrt{\frac{3k}{m}}\,t$. This time the second mode appears.

References

[**1**] Anton, Howard, *Elementary Linear Algebra*, John Wiley & Sons, 1977.

[**2**] Brinkmann, Heinrich W., and Klotz, Eugene A., *Linear Algebra and Analytic Geometry*, Addison-Wesley Publishing Co., 1971.

[**3**] Chen, Chi-Tsong, *Introduction to Linear System Theory*, Holt, Reinhart and Winston, Inc., New York, 1970.

[**4**] Curtiss, Charles, W., *Linear Algebra - An Introductory Approach*, Allyn and Bacon, Inc., 1968.

[**5**] Jackson, John David, *Mathematics for Quantum Mechanics*, Dover Publications, Inc., Mineola, New York, 1990.

[**6**] Kolman, Bernard, *Introductory Linear Algebra with Applications*, Macmillan Publishing Co., New York, 1993.

[**7**] Kline, Morris, *Mathematical Thought from Ancient to Modern Times*, Oxford University Press, 1972.

[**8**] Larson, Roland, E. and Edwards, Bruce, H., *Elementary Linear Algebra*, D.C. Heath and Co., 1991.

[**9**] May, W. Graham, *Linear Algebra*, Scott, Foresman and Co., Glenview, Illinois, 1970.

[**10**] Sharma, Vinod, A., *Matrix Methods and Vector Spaces in Physics*, PHI Learning Private Limited, New Delhi, 2009.

[**11**] Shilov, Georgi, E., *Introduction to the Theory of Linear Spaces*, Prentice-Hall Inc., Englewood Cliffs, New Jersey, 1961.

[**12**] Usmani, Riaz, A., *Applied Linear Algebra*, Marcel Dekker, Inc., New York and Basel, 1987.

Index

E

N

O